CARDIOVASCULAR GENOMICS: NEW PATHOPHYSIOLOGICAL CONCEPTS

Proceedings of the 2001 European Science Foundation Workshop in Maastricht

Developments in Cardiovascular Medicine

232. A. Bayés de Luna, F. Furlanello, B.J. Maron and D.P. Zipes (eds.): *Arrhythmias and Sudden Death in Athletes*. 2000 ISBN: 0-7923-6337-X
233. J-C. Tardif and M.G. Bourassa (eds): *Antioxidants and Cardiovascular Disease*. 2000. ISBN: 0-7923-7829-6
234. J. Candell-Riera, J. Castell-Conesa, S. Aguadé Bruiz (eds): *Myocardium at Risk and Viable Myocardium Evaluation by SPET*. 2000. ISBN: 0-7923-6724-3
235. M.H. Ellestad and E. Amsterdam (eds): *Exercise Testing: New Concepts for the* New Century. 2001. ISBN: 0-7923-7378-2
236. Douglas L. Mann (ed.): *The Role of Inflammatory Mediators in the Failing Heart*. 2001 ISBN: 0-7923-7381-2
237. Donald M. Bers (ed.): *Excitation-Contraction Coupling and Cardiac Contractile Force, Second Edition*. 2001 ISBN: 0-7923-7157-7
238. Brian D. Hoit, Richard A. Walsh (eds.): *Cardiovascular Physiology in the Genetically Engineered Mouse, Second Edition*. 2001 ISBN 0-7923-7536-X
239. Pieter A. Doevendans, A.A.M. Wilde (eds.): *Cardiovascular Genetics for Clinicians* 2001 ISBN 1-4020-0097-9
240. Stephen M. Factor, Maria A.Lamberti-Abadi, Jacobo Abadi (eds.): *Handbook of Pathology and Pathophysiology of Cardiovascular Disease*. 2001 ISBN 0-7923-7542-4
241. Liong Bing Liem, Eugene Downar (eds): *Progress in Catheter Ablation*. 2001 ISBN 1-4020-0147-9
242. Pieter A. Doevendans, Stefan Kääb (eds): *Cardiovascular Genomics: New Pathophysiological Concepts*. 2002 ISBN 1-4020-7022-5

Previous volumes are still available

CARDIOVASCULAR GENOMICS: NEW PATHOPHYSIOLOGICAL CONCEPTS

Proceedings of the 2001 European Science Foundation Workshop in Maastricht

edited by

Pieter A. Doevendans, MD
Academic Hospital Maastricht

and

Stefan Kääb
Klinikum Grosshadern

SPRINGER SCIENCE+BUSINESS MEDIA, LLC

Library of Congress Cataloging-in-Publication Data

European Science Foundation. Workshop (2001 : Maastricht, Netherlands)
Cardiovascular genomics : new pathophysiological concepts : proceedings of the 2001 European Science Foundation Workshop in Maastricht / edited by Pieter A. Doevendans and Stefan Kääb.
p. ; cm. -- (Developments in cardiovascular medicine ; 242)
Includes bibliographical references and index.
ISBN 978-1-4613-5347-8 ISBN 978-1-4615-1005-5 (eBook)
DOI 10.1007/978-1-4615-1005-5
1. Cardiovascular system--Diseases--Genetic aspects--Congresses. 2. Cardiovascular system--Diseases--Molecular aspects--Congresses. 3. Gene expression--Congresses. I. Doevendans, Pieter A. II. Kääb, Stefan, 1963- III. Title. IV. Developments in cardiovascular medicine ; v. 242.
[DNLM: 1. Cardiovascular Diseases--genetics--Congresses. 2. Genomics--Congresses. WG 120 E887c 2002]
RC669.E77 2001
616.1'042--dc21

2002066877

Printed on acid-free paper.

Table of Contents

List of Contributors ix

Preface xix

1. Genomics and clinical cardiology: Hype or hope? 1-4
H.J. Wellens

Session 1: Gene expression analysis 5

2. Salt-sensitive hypertension: Identification of downstream renal targets in the aldosterone-signaling pathway by Serial Analysis of Gene Expression (SAGE) 7-19
B.C. Rossier, D. Firsov

3. Cardiovascular proteomics: Effect of hypoxic conditions 21-28
C. Michiels, D. Mottet, E. Delaive, M. Dieu, J-F. Dierick, J. Remacle

4. Using comparative genome analysis to find interaction partners for frataxin 29-40
M.A. Huynen

Session 2: Genomics and pathophysiology of atherosclerosis 41

5. In vitro-in vivo gene expression analysis in atherosclerosis 43-52
A.J.G. Horrevoets, R.J. Dekker, R.D. Fontijn, S. van Soest, H. Pannekoek

Session 3: Genomics in hypertension 53

6. Human essential hypertension: Role of the genes of the renin-aldosterone system 55-63
X. Jeunemaitre, S. Disse-Nicodème, A. Gimenez-Roqueplo, P. Corvol

7. Gene therapy for hypertension: Future or fiction? 65-75
J.P. Fennell, M.J. Brosnan, A.J. Frater, A.H. Baker, A.F. Dominiczak

8. Hormones and signalling pathways 77-81
G. Lembo

9. Endothelial changes in hypertension 83-94
C. Zaragoza, S. Lamas

Session 4: Genomics in cardiac hypertrophy and failure 95

10. Towards elucidation of genetic pathways in cardiac hypertrophy: Technique to develop microarrays 97-105
B.J.C. van den Bosch, P.A. Doevendans, D.J. Lips, J.M.W. Geurts, H.J.M. Smeets

11. Gene expression in cardiac hypertrophy and failure: Role of G protein-coupled receptors 107-114
G. Esposito, A. Rapacciuolo, S.V. Naga Prasad, H.A. Rockman

12. Little mice with big hearts: Finding the molecular basis for dilated cardiomyopathy 115-129
L.J. de Windt, M.A. Sussman

13. Cardiac hypertrophic signaling the good, the bad and the ugly 131-155
O.F. Bueno, E. van Rooij, D.J. Lips, P.A. Doevendans, L.J. de Windt

Session 5: Molecular remodelling in arrhythmias 157

14. Ion channel regulation: From arrhythmias to genes to channels (to cures?) 159-165
J. Donahue

15. Differential expression and functional regulation of delayed rectifier channels 167-185
M. Stengl, P.G.A. Volders, M.A. Vos

16. Genetic polymorphisms and their role in ventricular arrhythmias 187-198
S. Kääb, M. Näbauer, A. Pfeufer

17. Molecular mechanisms of remodeling in human atrial fibrillation 199-212
B.J.J.M. Brundel, R.H. Henning, H.H. Kampinga, I.C. van Gelder, H.J.G.M. Crijns

18. G-protein β3-subunit polymorphism and atrial fibrillation 213-222
U. Ravens, E. Wettwer, T. Christ, D. Dobrev

Session 6: Towards cellular transplantion 223

19. Human stem cell gene therapy 225-230
A.A.F. de Vries

20. Towards human embryonic stem cell derived cardiomyocytes 231-243
C. Mummery, D. Ward, C.E. van den Brink, S.D. Bird, P.A. Doevendans, D.J.Lips, T. Opthof, A. Brutel de la Riviere, L. Tertoolen, M. van der Heyden, M. Pera

21. Use of mesenchymal stem cells for regeneration of cardiomyocytes and its application to the treatment of congestive heart failure 245-256
K. Fukuda

22.	Cellular cardiac reinforcement P. Menasché	257-264
23.	Appendix to session 3 hypertension Adducin paradigm: An approach to the complexity of hypertension genetics G. Bianchi	265-271
	Index	273

List of Contributors

H. Baker
Gardiner Institute, Department of Medicine and Therapeutics,
Western Infirmary, GLASGOW, G- 11 6 NT, United Kingdom
Co-authors: M. Fennel, J. Brosnan, A. Frater, A. Dominiczak

G. Bianchi
Chair and School of Nephrology, University Vita Salute San Raffaele,
Division of Nephrology, Dialysis and hypertension,
San Raffaele Hospital, MILAN, Italy

S. Bird
Hubrecht Laboratorium, Interuniversity Cardiology Institute of the Netherlands,
P.O. Box 19008,
3501 DA UTRECHT, The Netherlands

B. van den Bosch
Department of Molecular Genetics, University Maasticht,
Universiteitssingel 50,
P.O. Box 616, 6200 MD MAASTRICHT, The Netherlands
Co-authors: P. Doevendans, H. Smeets, D. Lips, J. Geurts

C. van den Brink
Hubrecht Laboratorium, Interuniversity Cardiology Institute of the Netherlands,
P.O. Box 19008,
3501 DA UTRECHT, The Netherlands

M. Brosnan
Gardiner Institute, Department of Medicine and Therapeutics,
Western Infirmary, GLASGOW, G- 11 6 NT, United Kingdom
Co-authors: M. Fennel, A. Dominiczak, A. Frater, A. Baker

B. Brundel
AZG,
P.O. Box 30001,
9700 RB GRONINGEN, The Netherlands
Co-authors: R. Henning, H. Kampinga, I. van Gelder, H. Crijns

A. Brutel de la Riviere
Department of Cardiothoracacic Surgery, University Medical Centre
P.O. Box 18008,
3501 DA UTRECHT, The Netherlands

O. Bueno
The Childrens Hospital and Research Foundation,
Division of Molecular Cardiovascular Biology,
Department of Pediatics,
3333 Burnet Avenue,
CINCINNATI OH 45229-3039, United States
Co-authors: L. de Windt, P. Doevendans, E. van Rooy, D. Lips

T. Christ
Institut für Pharmakologie und Toxikologie, Medizinische Fakultät der TU Dresden,
Fletscherstrasse 74,
D-01307 DRESDEN, Germany
Co-authors: E. Wettwer, U. Ravens, D. Dobrev

P. Corvol
INSERM U 36, College de France,
11 place Marcelin Berthelot,
75231 PARIS Cedex 05, France
Co-authors: X. Jeunemaitre, S. Disse-Nicodème, A. Gimenez-Roqueplo

H. Crijns
Department of Cardiology, Academic Hospital Maastricht, P.O. Box 5800,
6202 AZ MAASTRICHT, The Netherlands
Co-authors: R. Henning, H. Kampinga, I. van Gelder, B. Brundel

E. Delaive
Laboratoire de Biochimie et Biologie Cellulaire, Facultes Universitaires Notre Dame de la Paix 61, Rue de Bruxelles,
5000 NAMUR, Belgium
Co-authors: D. Mottet, C. Michiels, M. Dieu, J-F. Dierick, J. Remacle

R. Dekker
AMC, Academic Medical Center,
Department of Biochemistry, Rm. KL-161,
Meibergdreef 15,
1105 AZ AMSTERDAM, The Netherlands
Co-authors: A. Horrevoets, R. Fontijn, S. van Soest, H. Pannekoek

M. Dieu
Laboratoire de Biochimie et Biologie Cellulaire, Facultes Universitaires Notre Dame de la Paix 61, Rue de Bruxelles,
5000 NAMUR, Belgium
Co-authors: D. Mottet, E. Delaive, C. Michiels, J-F. Dierick, J. Remacle

J-F. Dierick
Laboratoire de Biochimie et Biologie Cellulaire, Facultes Universitaires Notre Dame de la Paix 61, Rue de Bruxelles,
5000 NAMUR, Belgium
Co-authors: D. Mottet, E. Delaive, M. Dieu, C. Michiels, J. Remacle

S. Disse-Nicodème
INSERM U 36, College de France,
11 place Marcelin Berthelot,
75231 PARIS Cedex 05, France
Co-authors: P. Corvol, X. Jeunemaitre, A. Gimenez-Roqueplo

P. Doevendans
Department of Cardiology, Academic Hospital Maastricht, P.O. Box 5800,
6200 AZ MAASTRICHT, and ICIN, P.O. Box 19258, 3501 DG UTRECHT,
The Netherlands
Co-authors: O. Bueno, L. de Windt

A. Dominiczak
Gardiner Institute, Department of Medicine and Therapeutics,
Western Infirmary, GLASGOW, G-11 6 NT, United Kingdom
Co-authors: M. Fennel, J. Brosnan, A. Frater, A. Baker

K. Donahue
Department of Medicine, Johns Hopkins University School of Medicine,
N. Wolfe St. BALTIMORE, MD 21287, United States

D. Dobrev
Institut für Pharmakologie und Toxikologie, Medizinische Fakultät der TU Dresden,
Fletscherstrasse 74,
D-01307 DRESDEN, Germany
Co-authors: E. Wettwer, T. Christ, U. Ravens

G. Esposito
Duke University Medical Center, DUMC 3104, CARL Building, Rm. 226,
DURHAM, NC 27710, United States
Co-authors: H. Rockman, A. Rapacciuolo, S. Naga Prasad

J. Frater
Gardiner Institute, Department of Medicine and Therapeutics,
Western Infirmary, GLASGOW, G- 11 6 NT, United Kingdom
Co-authors: M. Fennel, J. Brosnan, A. Dominiczak, A. Baker

J. Fennel
Gardiner Institute, Department of Medicine and Therapeutics,
Western Infirmary, GLASGOW, G- 11 6 NT, United Kingdom
Co-authors: A. Dominiczak, J. Brosnan, A. Frater, A. Baker

R. Fontijn
AMC, Academic Medical Center,
Department of Biochemistry, Rm. KL-161,
Meibergdreef 15,
1105 AZ AMSTERDAM, The Netherlands
Co-authors: A. Horrevoets, R. Dekker, S. van Soest, H. Pannekoek

D. Firsov
Institut de Pharmacologie et de Toxicologie,
Université de Lausanne, Bugnon 27,
CH-1005- LAUSANNE, Switserland
Co-authors: B. Rossier

K. Fukuda
Keio University, School of Medicine, 35 Shinanomachi,
Shinjuku-ku TOKYO 160-8582, Japan

A. Gimenez-Roqueplo
INSERM U 36, College de France,
11 place Marcelin Berthelot,
75231 PARIS Cedex 05, France
Co-authors: P. Corvol, S. Disse-Nicodème, X. Jeunemaitre

I. van Gelder
AZG, P.O. Box 30001,
9700 RB GRONINGEN, The Netherlands
Co-authors: R. Henning, H. Kampinga, B. Brundel, H. Crijns

J. Geurts
Department of Molecular Genetics, University Maastricht,
Universiteitssingel 50, P.O. Box 616, 6200 MD MAASTRICHT,
The Netherlands
Co-authors: P. Doevendans, H. Smeets, D. Lips, B. van den Bosch

R. Henning
AZG, P.O. Box 30001,
9700 RB GRONINGEN, The Netherlands
Co-authors: B. Brundel, H. Kampinga, I. van Gelder, H. Crijns

M. van der Heyden
Department medical Physiology, University Medical Centre
P.O. Box 19008,
3501 DA UTRECHT, The Netherlands

A. Horrevoets
AMC, Academic Medical Center,
Department of Biochemistry, Rm. KL-161,
Meibergdreef 15,
1105 AZ AMSTERDAM, The Netherlands
Co-authors: R. Dekker, R. Fontijn, S. van Soest, H. Pannekoek

A. Howard
Duke University Medical Center, DUMC 3104, CARL Building, Rm. 226,
DURHAM, NC 27710, United States
Co-authors: G. Esposito, A. Rapacciuolo, S. Naga Prasad, H. Rockman

H. Huynen
Biocomputing Group, EMBL,
P.O. Box 10.2209,
69012 HEIDELBERG, Germany

X. Jeunemaitre
INSERM U 36, College de France,
11 place Marcelin Berthelot,
75231 PARIS Cedex 05, France
Co-authors: P. Corvol, S. Disse-Nicodème, A. Gimenez-Roqueplo

S. Kääb
Klinikum Grosshadern, Medizinische Klinik,
81366 MUNCHEN, Germany
Co-author: A. Pfeufer, M. Näbauer

H. Kampinga
AZG, P.O. Box 30001,
9700 RB GRONINGEN, The Netherlands
Co-authors: R. Henning, B. Brundel, van Gelder, H. Crijns

S. Lamas
CIB/CSIC, Velazquez 144,
28006 MADRID, Spain
Co-authors: C. Zaragoza

G. Lembo
Departement of Experimental Medicine & Pathology,
"La Sapienza" University of Rome, c/o irccs neuromed,
Loc. Camerelle, 86077 POZZILLI (IS), Italy

D. Lips
Department of Cardiology, Academic Hospital Maastricht,
P.O. Box 5800, 6202 AZ MAASTRICHT, The Netherlands
Co-authors: L. de Windt, P. Doevendans, E. van Rooy, O. Bueno

P. Menasché
Department of Cardiovascular Surgery
Hopital Bichat Claude Bernard,
46, rue Henri Huchard
75018 PARIS, France

C. Michiels
Laboratoire de Biochimie et Biologie Cellulaire, Facultes Universitaires Notre Dame de la Paix 61, Rue de Bruxelles,
5000 NAMUR, Belgium
Co-authors: D. Mottet, E. Delaive, M. Dieu, J-F. Dierick, J. Remacle

D. Mottet
Laboratoire de Biochimie et Biologie Cellulaire, Facultes Universitaires Notre Dame de la Paix 61, Rue de Bruxelles,
5000 NAMUR, Belgium
Co-authors: C. Michiels, E. Delaive, M. Dieu, J-F. Dierick, J. Remacle

C. Mummery
Hubrecht Laboratorium, Interuniversity Cardiology Institute of the Netherlands,
P.O. Box 19008,
3501 DA UTRECHT, The Netherlands
Co-authors: D. Ward, C. van den Brink, S. Bird, P. Doevendans, D. Lips, T. Opthof, A. Brutel de la Riviere, L. Tertoolen, M. van der Heyden, M. Pera

M. Näbauer
Klinikum Grosshadern, Medizinische Klinik,
81366 MUNCHEN, Germany
Co-author: A. Pfeufer, S. Kääb

S. Naga Prasad
Duke University Medical Center, DUMC 3104, CARL Building, Rm. 226,
DURHAM, NC 27710, United States
Co-authors: G. Esposito, A. Rapacciuolo, H. Rockman

T. Opthof
Department medical Physiology, University Medical Centre
P.O. Box 19008,
3501 DA UTRECHT, The Netherlands

H. Pannekoek
AMC, Academic Medical Center,
Department of Biochemistry, Rm. KL-161,
Meibergdreef 15,
1105 AZ AMSTERDAM, The Netherlands
Co-authors: R. Dekker, R. Fontijn, S. van Soest, A. Horrevoets

M. Pera
Monash Institute of Reproduction and Development,
Monash University, 246 Clayton Road, Clayton, VICTORIA, Australia

A. Pfeufer
Klinikum Grosshadern, Medizinische Klinik,
81366 MUNCHEN, Germany
Co-author: S. Kääb, M. Näbauer

A. Rappacciuolo
Duke University Medical Center, DUMC 3104, CARL Building, Rm. 226,
DURHAM, NC 27710, United States
Co-authors: G. Esposito, H. Rockman, S. Naga Prasad

U. Ravens
Institut für Pharmakologie und Toxikologie, Medizinische Fakultät der TU
Dresden,
Fletscherstrasse 74,
D-01307 DRESDEN, Germany
Co-authors: E. Wettwer, T. Christ, D. Dobrev

J. Remacle
Laboratoire de Biochimie et Biologie Cellulaire, Facultes Universitaires Notre
Dame de la Paix 61, Rue de Bruxelles,
5000 NAMUR, Belgium
Co-authors: D. Mottet, E. Delaive, M. Dieu, J-F. Dierick, C. Michiels

H. Rockman
Duke University Medical Center, DUMC 3104, CARL Building, Rm. 226,
DURHAM, NC 27710, United States
Co-authors: G. Esposito, A. Rapacciuolo, S. Naga Prasad

E. van Rooy
Department of Cardiology, Cardiovascular Research Institute Maastricht
Academic Hospital Maastricht,
P.O. Box 5800,
6200 AZ MAASTRICHT, The Netherlands
Co-authors: L. de Windt, P. Doevendans, O. Bueno, D. Lips

B. Rossier
Institut de Pharmacologie et de Toxicologie,
Université de Lausanne, Bugnon 27,
CH-1005- LAUSANNE, Switserland
Co-author: D. Firsov

H. Smeets
Department of Molecular Genetics, University Maastricht, Universiteitssingel 50, P.O. Box 616, 6200 MD MAASTRICHT, The Netherlands

S. van Soest
AMC, Academic Medical Center,
Department of Biochemistry, Rm. KL-161,
Meibergdreef 15,
1105 AZ AMSTERDAM, The Netherlands
Co-authors: A. Horrevoets, R. Fontijn, R. Dekker, H. Pannekoek

M. Stengl
Department of Cardiology, Academic Hospital Maastricht,
P.O. Box 5800, 6202 AZ MAASTRICHT, The Netherlands
Co-authors: M. Vos, P. Volders

Prof. H. Struijker Boudier
Director of CARIM
Department of Farmacology and Toxicology, University Hospital Maastricht,
Universiteitssingel 50, 6229 ER MAASTRICHT, The Netherlands

M. Sussman
The Childrens Hospital and Research Foundation,
Division of Molecular Cardiovascular Biology,
Department of Pediatics,
3333 Burnet Avenue,
CINCINNATI 45229-3039, United States
Co-author: L. de Windt

L. Tertoolen
Department of Cardiothoracic Surgery, University Medical Centre
P.O. Box 19008,
3501 DA UTRECHT, The Netherlands

P. Volders
Department of Cardiology, Academic Hospital Maastricht,
P.O. Box 5800, 6202 AZ MAASTRICHT, The Netherlands
Co-authors: M. Stengl, M. Vos

M. Vos
Department of Cardiology, Academic Hospital Maastricht, P.O. Box 5800,
6202 AZ MAASTRICHT, The Netherlands
Co-authors: M. Stengl, P. Volders

T. de Vries
Department of Molecular Cell Biology, Gene Therapy Section,
Leiden University medical Center, P.O. Box 9503,
2300 RA LEIDEN, The Netherlands

D. Ward
Hubrecht Laboratorium, Interuniversity Cardiology Institute of the Netherlands,
P.O. Box 19008,
3501 DA UTRECHT, The Netherlands

H. Wellens
ICIN, P.O. Box 19258,
3501 DG UTRECHT, The Netherlands

E. Wettwer
Institut für Pharmakologie und Toxikologie, Medizinische Fakultät der TU Dresden,
Fletscherstrasse 74,
D-01307 DRESDEN, Germany
Co-authors: U. Ravens, T. Christ, D. Dobrev

L. de Windt
Department of Cardiology, Cardiovascular Research Institute Maastricht
Academic Hospital Maastricht,
P.O. Box 5800,
6202 AZ MAASTRICHT, The Netherlands
Co-authors: O. Bueno, P. Doevendans, E. van Rooy, D. Lips

C. Zaragoza
CIB/CSIC, Velazquez 144,
28006 MADRID, Spain
Co: S. Lamas

PREFACE

Four years ago-in December 1997-the first European Science Foundation Workshop on Cardiovascular Specific Gene Expression was held in Maastricht. It was hardly possible to imagine the progress in the field in those four years. In 1997, gene expression was still an art focused on individual genes; in 2001, many labs have access to micro-array facilities to determine the expression of thousands of genes simultaneously. In 1997, gene expression was an area of fundamental research in basis molecular biology laboratories; in 2001, clinical cardiovascular research has incorporated gene expression approaches. In 1997, the interpretation of a gene expression experiment was usually straightforward; in 2001, advanced bioinformatics tools are needed to approach the extreme complexities of genetic control of cell and tissue function.

The second symposium in this series is focused on Cardiovascular Genomics. New Pathophysiological Concepts. The organizing committee chose to invite a group of renown scientists and young investigators around four topics of eminent importance in cardiovascular research. These topics reflect the major present-day clinical cardiovascular problems: atherosclerosis, hypertension, arrhythmias and heart failure. In addition to these four disease-driven topics, the workshop has sessions on gene expression methodologies and cellular transplant approaches to cardiovascular disease.

The workshop and the papers contained in these proceedings give an in-depth review on how large-scale gene-expression approaches affect basis concepts of cardiovascular disease. The days on one gene – one function – one disease seem far behind. A more modern view is that of integrated pathways with redundancies to allow stability so characteristic for the cardiovascular system. These are exciting days to witness the physiome emerge from the genome!

Although the whole organizing committee takes full responsibility for the program of the workshop and these proceedings, Pieter Doevendans deserved a special compliment for his unflagging energy to organize the workshop and edit these proceedings.

Prof H.A Struijker- Boudier

Acknowledgements: we would like to thank Ireen van der Velden, Nicole van Geleen and Xander Wehrens for preparing this publication.

1. GENOMICS AND CLINICAL CARDIOLOGY: HYPE OR HOPE?

H. J. Wellens

In my 35 years as a cardiologist I have seen tremendous changes in our diagnostic and therapeutic approaches to the patient with heart disease. Echocardiography, coronary angiography and programmed electrical stimulation of the heart, to name a few techniques, have enormously advanced our ability to correctly diagnose what is wrong with the heart. Also our non-invasive and invasive therapeutic possibilities have improved to such an extent that death from heart disease has been reduced considerably in recent years.

Still, most of our therapeutic interventions are palliative. They result in the patient living longer with less complaints but they do not cure the disease. Also our abilities to prevent cardiac disease are very limited. Therefore the challenge in the coming years will be to move from palliative to curative and ultimately to preventive therapy (figure. 1).

In recent years the unraveling of the human genome opened new ways to reach that goal. Information that can be obtained through molecular biologic, cell biologic and molecular genetic techniques will bring us closer to understanding the genetic control of normal and abnormal function of our different organs, how they interact and how the are influenced by external factors. At this point in time it is impossible to predict how long will it take to reach the stage of cure and prevention. How will we be able to speed up that process? I would like to address that question looking at this through the eyes of a clinical cardiologist. So far monogenetic diseases are being unraveled. The genetic background of patients with the long QT syndrome and hypertrophic cardiomyopathy have been identified. This clearly has been an important step forward. Carriers of the gene can be recognized, a risk profile can be constructed and pharmacologic or device management can be selected likely to protect the carrier against possible complications like sudden death.

However we have learned that spontaneous mutations may occur , that in diseases considered monogenetic more than one gene may be abnormal and that we are unable at the present time to definitely remove the disease gene and replace it by the correct one.

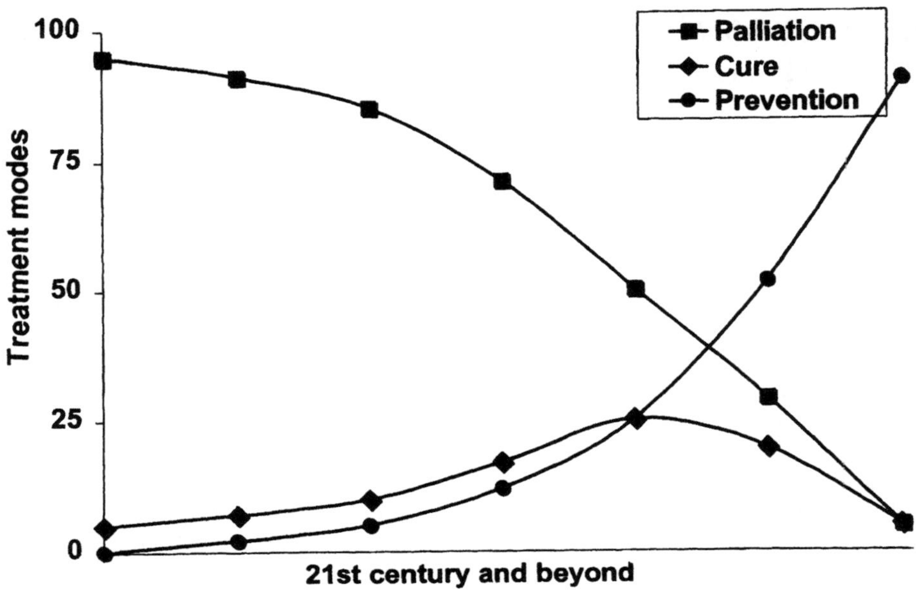

Figure 1 Graph showing the future of medicine. Palliation will decrease and cure and prevention will increase. However, nobody knows the exact time required to reach the end of the graph! (1)

Since we cannot do this in single gene disease at this point in time it might be much more rewarding to direct our research to processes that on one hand play a role in the development of a cell in a certain direction and to processes that gradually impair cardiac function,for example during ageing. Preferably, these processes should be influenced without the use of alien material to prevent immunological repercussions. An example of the first approach is to promote the development of cardiomyocytes from stemcells. When possible these stemcells should be harvested from the patient himself, multiplied, directed towards becoming cardiomyocytes and then deposited in the heart at the required location.
Ageing results in important changes in the heart both as far as cardiomyocyte content and matrix composition is concerned. When one has become 50 years of age each year, each year 1% of heart muscle will be replaced by fibrotic tissue. This is the basis for rhythm and conduction disturbances such as atrial fibrillation and heart block but also for diastolic dysfunction. It is extremely important to know how to delay or prevent these changes with ageing. Realistically one should not expect that gene therapy will be realized soon of we consider replacement of the "sick" gene by the "correct" one as the definition of gene therapy. Of great importance however, is to develop the knowledge which protein and where is produced by which gene (genomics) and the role of these proteins for functioning of the organ (proteomics). This approach is within our one reach using DNA chips and microarrays. They will

allow us to unravel the underlying gene regulation network and help us to control disease processes and ageing by up or down regulation of the underlying network. A major problem to solve will be to understand how a mutation in a gene known to be important for a certain organ all of a sudden may interfere with metabolic processes in a different organ.

These surprises will be with us for a long time. Growing knowledge of genomics and proteomics should not let us astray from the importance of external factors. Genetic epidemioloy is important but because of all the difficulties mentioned above it is clear that recognition, correction and prevention of external factors will remain a key factor in disease management.

In the clinical setting this requires the training of a new breed of clinicians having better knowledge and understanding of basis processes than our current generation. It means that the physician who wants to become a cardiologist should receive structured education to understand the significance and possible implications of molecular genetic information for future clinical management of the patient. In a small country like ours this should be done centrally by modern telecommunication techniques using the best teachers. Also as indicated in table 1, to implement this new knowledge into clinical medicine, the clinical scientist should have protected research time.

Table 1

Necessary changes in the cardiac training programme and clinical research activities:

1. Better education in cel biology and molecular genetics, preferably by central teaching using modern telecommunications techniques and the best teachers.
2. Protected research time for clinical scientists.

At this moment activities applying genomics and proteomics into clinical medicine are being performed in several Dutch centers. To achieve the maximum with the least amount of money we should approach these activities on a national basis. Both the Netherlands Heart Foundation and the Interuniversity Cardiology Institute are firmly supporting such an attitude.

In summary genomics and proteomics offer new possibilities for the management of cardiac disease. They give hope for the future. It is important to realize though, that it will take quite a while before they can be fully implemented in clinical cardiology.

Reference

1. Wellens HJ, Cardiology: Where to go from here? The Lancet 1999;354:Sup IV8.

ESF workshop Maastricht 2001: Session 1

Gene expression analysis

2. SALT-SENSITIVE HYPERTENSION:

Identification of downstream renal targets in the aldosterone-signaling pathway by Serial Analysis of Gene Expression (SAGE)

B.C. Rossier and D. Firsov

Introduction

Hypertension is the most frequent disease in human populations. Genetic and non-genetic factors are involved and high salt intake has been proposed as a major risk factor [1]. Large epidemiological and clinical studies, however, have led to controversial conclusions [2,3]. We have proposed that these contradictory results may be explained by two kinds of confounding factors [4]. First, the study of low salt or high salt intake as a single experimental variable in human populations is obviously difficult if not impossible. Second, epidemiological studies have not taken into account the possibility that *susceptibility* genes could confer salt-sensitivity or salt-resistance in a proportion of the population. Regarding the first factor, Denton et al. [5] were recently able to demonstrate in chimpanzees (the species genetically the closest to us) that the addition of salt within the human dietetic range causes a highly significant rise in their blood pressure. Interestingly, the effect of salt differed between individuals and only 60% of the cohort developed high blood pressure, defining two sub-populations of salt-sensitive or salt-resistant animals. Regarding the second factor, the identification of mutations in the epithelial sodium channel (ENaC) β subunit as a cause of a monogenic form of salt-sensitive hypertension in humans (Liddle's syndrome) has recently highlighted the importance of a single mutated gene as sufficient to induce large changes in blood pressure [6].

According to the hypothesis put forward by Guyton, over 20 years ago [7], control of blood pressure at steady state and on a long-term basis is critically dependent on renal mechanisms. During the last 6 years, a number of genes expressed in various parts of the nephron have been shown to be directly involved in the control of blood pressure. The identification of mutations in monogenic diseases such as the Bartter's [8] or the Gitelman's [9] syndromes clearly indicate that defects in ion transporters expressed in the thick ascending limb (TAL) or in the distal convoluted tubule (DCT) may lead to a severe salt-loosing syndrome with a hypotensive phenotype. Recently, two genes causing pseudohypoaldosteronism type II, a Mendelian trait featuring hypertension, increased renal salt reabsorption, and impaired K+ and H+ excretion have been identified [10]. Both genes encode members of the WNK family of serine-threonine kinases and are expressed in DCT/CNT and /or CCD. In the connecting tubule (CNT), in the cortical collecting duct (CCD) and, to some extent, in the outer medullary collecting duct (OMCD) and inner medullary collecting duct (IMCD), the final control of sodium reabsorption is achieved through an amiloride-sensitive electrogenic sodium reabsorption which is under tight hormonal control, aldosterone playing the key role [11]. The main limiting factor in sodium reabsorption in this part of the nephron also termed Aldosterone Sensitive Distal Nephron (ASDN) [12] is the apically located amiloride-sensitive epithelial sodium channel (ENaC) [13-15]. Two monogenic diseases have been linked to ENaC subunit genes; first, pseudohypoaldosteronism Type 1, a severe autosomal recessive form of a salt-loosing syndrome is due to loss (or partial loss) of function mutations in the α, β or γ subunit genes of ENaC [16,17]. Gain of function mutations in the β or γ subunit of ENaC lead to a hypertensive phenotype (Liddle syndrome), a paradigm for salt-sensitive hypertension [6,18-20].

The most important conserved functions of aldosterone are to promote sodium reabsorption and potassium and hydrogen secretion across the so called tight epithelia, that is epithelia that displays a high transepithelial electrical resistance and an amiloride-sensitive electrogenic sodium transport in ASDN [11,21]. According to the well accepted model of epithelial sodium transport [21], the 2 major steps in sodium reabsorption by renal epithelia are: a) facilitated transport driven by an electrochemical potential difference across the apical membrane from urine to cell and, b) active transport driven by metabolic energy across the basolateral membrane, from cell to interstitium. The apical membrane entry step is mediated by the sodium selective amiloride-sensitive ion channel ENaC and the exit step is catalyzed by Na,K-ATPase, the ouabain-sensitive sodium pump [21]. These two in series mechanisms are the rate-determining steps for sodium transport and they are thus the most likely final effectors of action for aldosterone or for any other hormone that regulates the overall reabsorption process. The classic model of the mechanism of aldosterone action in tight epithelia proposes the following steps (figure 1): aldosterone crosses the plasma membrane and binds to its cytosolic receptor, either the mineralocorticoid (MR) or glucocorticoid (GR) receptor. MR and GR are protected from illicit occupation by high levels of plasma glucocorticoids (cortisol or corticosterone), thanks to the metabolizing action of 11-β-HSD2 that transforms the active cortisol into cortisone, an inactive metabolite, unable to bind either to MR or GR. The receptor-hormone complex is translocated to the nucleus where it interacts with the promoter region of target genes activating or repressing their transcriptional activity. Aldosterone-induced or -repressed proteins (AIPs or ARPs, respectively) mediate an increase in transepithelial sodium transport. Early effects are produced by the activation of pre-existing transport proteins (ENaC, Na,K-ATPase) via yet uncharacterized mediators. The late effect is characterized by an accumulation of additional transport proteins and other elements of the sodium transport machinery. Three monogenic forms of salt-sensitive hypertension map to genes expressed in principal cells: I) ENaC (Liddle's syndrome) [6], II) SAME (Syndrome of Apparent Mineralocorticoid Excess) due to gene inactivation of 11-β-HSD2 [22], III) gain of function mutations in the mineralocorticoid binding site of MR [23]. It is therefore remarkable to observe that mutations of genes involved in the two main "entry" steps in the aldosterone signaling pathway (MR and 11-β-HSD2) and in the main effector (ENaC) are indeed able to induce very severe salt-sensitive hypertensive phenotypes. These genes are therefore also candidate genes as *susceptibility* genes for salt-sensitivity or salt-resistance in the general population. A number of groups are presently actively looking for single nucleotide polymorphisms (SNPs) in these candidate genes. Although there is not yet a consensus among scientists, because linkage and association studies are notoriously difficult to perform in human populations, there is already evidence that some alleles of ENaC are associated with salt-sensitive hypertension in some populations [24,25]. The same has also been proposed for a specific allele of the 11-β-HSD2 [26] and it will be interesting to see whether this is also true for the mineralocorticoid receptor. Besides ENaC and 11-β-HSD2, other genes have been associated with essential hypertension: angiotensinogen [27] and genes on the renin-angiotensin-aldosterone axis [28], α and β subunits of adducin [29,30], glucagon receptor [31] and $G_s\alpha$. [32]. Interestingly, most of these genes are expressed in the kidney or in principal cells and may participate in a salt-sensitive phenotype. Recently, Pierre Corvol and colleagues [33] summarized in a well thought review their current opinion about the candidate gene approach to the understanding of the genetics of human essential hypertension. They draw our attention on some difficulties and pitfalls of this approach.

1. Many linkage or association studies have a limited statistical power.
2. The genetic findings may vary greatly according to the populations studied.
3. There is a need for better phenotyping of the hypertensive population.
4. The causal relationship between molecular variance and hypertension is and will be difficult to establish firmly.
5. The contribution of the genetic studies in rodents to the molecular genetics of human hypertension must be re-examined.
6. Most molecular variance leads to a low attributable risk in the population or a low individual effect at the individual levels.
7. It is too early to propose dietary recommendation and specific drug treatment, according to patients' phenotype.

We can agree with most of these cautious conclusions. As experimentalist trying to understand one form of essential hypertension (salt-sensitive), we believe that the weakest spots are points 3 and 4. We believe that point 5 is well taken but recent findings from our laboratory [34] and others [22] demonstrate the pathophysiological relevance of mouse models in recapitulating human diseases and in understanding the complex interaction between a gene and the environment. We would like to stress that what we need most is a better understanding of cellular signaling pathways involved in salt sensitive hypertension. A complete understanding of the signaling pathways for aldosterone and their cross-talk with interacting cascades (i.e. insulin and vasopressin) is one of the key points in better understanding the pathophysiology of salt-sensitive hypertension.

Working hypothesis

Our *primary* working hypothesis is that the genes involved in the aldosterone signaling pathway are *potentially candidate genes* for salt-sensitivity or salt-resistance, depending on their relative importance and their limiting character in the cascade (figure 2). Thus, gain of function mutations of activators or loss of function mutations of repressors of sodium transport may lead to abnormal and constitutive sodium reabsorption in CCD cells. Our *secondary* working hypothesis is that other signaling pathways, leading to increased sodium reabsorption in principal cells, in an additive or synergistic fashion with aldosterone, must also be considered. In this respect, vasopressin and insulin have been well characterized as promoting sodium reabsorption in these cells. The effect of insulin on sodium transport has been characterized in amphibian kidney cells [35-37]. The effect of vasopressin on sodium transport is mediated by a V2 receptor coupled to a G protein (GαS), leading to increased cyclic AMP with transcriptional effects [38] and PKA mediated effects [39,40]. The signaling pathway for aldosterone has just begun to be understood, thanks to the seminal work of François Verrey [41,42] in Zurich and David Pearce

[43] in San Francisco. Using differential display RT-PCR techniques, they have been able to identify a few candidate genes in the aldosterone-signaling pathway. One of the most promising candidate is a serum- glucose- induced kinase (sgk) which appears to be induced early on in the aldosterone action [43,45,46] and can mediate an increased activity of ENaC in a heterologous expression system such as the *Xenopus* oocyte [43,46,47]. Interestingly, the insulin and the aldosterone pathways could impact on the same target sgk enzyme (figure 2). Recently, Olivier Staub and colleagues presented evidence (Gordon Conference, June 2001) that sgk phophorylates Nedd4-2, an ubiquitin ligase protein that binds to ENaC and promotes its internalization and degradation. Phosphorylation of Nedd4-2 decreases the affinity of the protein to ENaC preventing its internalization and thus leading to an increased cell surface expression of ENaC[48]. These data for the first time provide experimental evidence for an important missing link in the aldosterone-signaling cascade. As discussed by Loffing and colleagues [46], however, it is likely that the induction of sgk is critical for ENaC localization in the apical membrane in the early segment of ASDN but is not sufficient to explain the establishment of amiloride-sensitive electrogenic sodium transport in the most distal parts of ASDN.

Analysis of the aldosterone-dependent signaling cascade by SAGE

In principal cells, aldosterone increases the amiloride-sensitive electrogenic sodium transport by transcriptional modulation of a gene network, which may comprise up to a few hundreds of different genes [21,42]. Aldosterone increases the transcription and mRNA abundance for the α and β subunits of Na,K-ATPase and for ENaC subunits. However, these transcriptional effects are translated into a physiological sodium transport response relatively late (6 to 24 hours after hormone addition). During the early phase of aldosterone action, 60 min. to 4 hours, we postulate the existence of regulatory proteins controlling either channel or Na,K-ATPase activities.

Using Serial Analysis of Gene Expression (SAGE), we have identified the *transcriptome* of principal cells [44] in a highly differentiated aldosterone-responsive cell line derived from mouse kidney [49]. The rationale for using the SAGE method was described by Kinzler and colleagues in 1995 [50] and is based on two principles: a) a short sequence tag (9 bp), contains sufficient information to identify a transcript in a unique way in the genome. b) concatenation of short sequence tags permits efficient analysis of transcripts by sequencing multiple tags in a single clone. Statistically, a 9 bp tag permits to distinguish over 250.000 different transcripts. The SAGE method has been validated experimentally in the yeast system [51]: the technique is efficient to quantitatively and qualitatively describe the transcriptome conveying the identity of each expressed gene and its level of

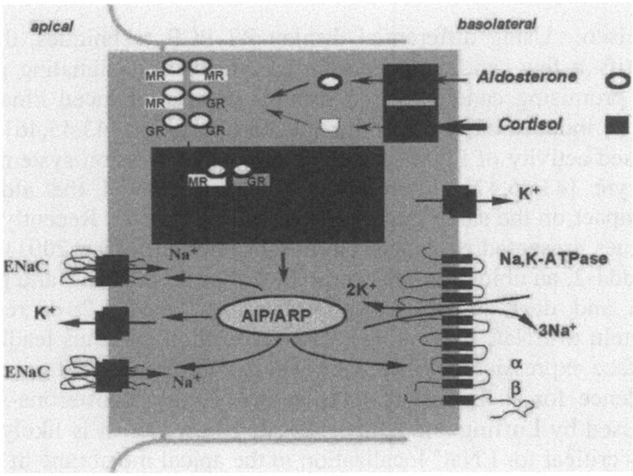

***Figure 1** Aldosterone regulated Na reabsorption in the principal cell of collecting ducts*

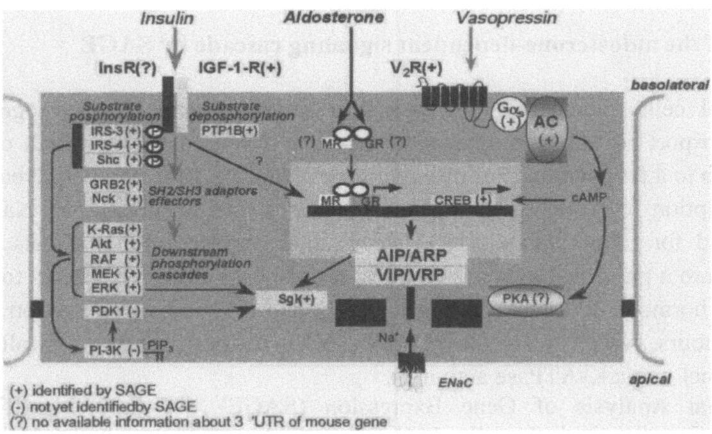

***Figure 2** Downstream renal target in the aldosterone, insulin- and vasopressin regulating pathways*

expression for a defined population of cells. The advantages of SAGE are the following:

1. Identification of transcripts by tag has no bias. There is no preferential tag redundancy due to the DNA length or nucleotide composition.
2. The technique allows the measurement of the number of copies of mRNA originally expressed in the cell.
3. SAGE allows not only to define the transcriptome of a given cell but also transcripts which were not considered as potentially expressed, as shown

for instance in yeast [51]. SAGE is independent of sample manipulation and pre-judgement of the scientist. It is therefore a convenient method to assess the effect of any factors or hormones modulating part of the genes expressed in a given cell [52].

The major drawbacks of the SAGE methods are the following.

1. High cost for sequencing (5000 sequence reactions are required per SAGE library).
2. Difficulty to detect induction or repression of low copy number mRNA in a given cell.
3. Requirement for a relatively large amount of poly A+ RNA as starting material and necessity to consider a homogenous cell population. A micro SAGE assay has recently been designed by Elalouf and colleagues [53], allowing to study the transcriptome of manually dissected nephron segments but the detection of low copy number transcripts is probably a limiting factor in this approach. Only cell culture of highly differentiated epithelial cells may overcome the problem at the expense, however, of possible significant changes (i.e. dedifferentiation) in gene profiling due to the culture as already shown in colonic epithelial cells [52]. In our cell line [49] the physiological endpoint (i.e. aldosterone-dependent sodium transport) is fully expressed and recapitulates the normal response in vivo.

In our recent study[44], we analyzed the *transcriptome* of a highly differentiated mouse clonal CCD principal cell line ($mpkCCD_{cl4}$) and the changes in the transcriptome induced by aldosterone and vasopressin. SAGE was performed on untreated cells and on cells treated with either aldosterone or vasopressin for 4 h. The transcriptomes in these three experimental conditions were determined by sequencing 166,927 transcript tags from the corresponding SAGE libraries. Limiting the analysis to that occurred twice or more in the data set, 14,654 different transcripts were identified, 3,642 of which do not match known mouse sequences [44] corresponding to as many genes or their spliced forms. Statistical comparison (at $p<0.05$ level) of the three SAGE libraries revealed 34 AITs (Aldosterone-Induced Transcripts), 29 ARTs (Aldosterone-Repressed Transcripts), 48 VITs (Vasopressin-Induced Transcripts) and 11 VRTs (Vasopressin-Repressed Transcripts).

Candidate AIPs and ARPs have been validated by the following approaches:

1. Northern blot analysis of control and induced cells or, alternatively, semi-quantitative RT-PCR on the same tissues if the level of expression is too low to be reliably detected by Northern blot.
2. Induction or repression of the candidate gene *in vivo*, using poly A+ RNA extracted from animal kidneys under different physiological

conditions (adrenalectomy, aldosterone-injected or fed under low salt diet for one or 2 weeks).

3. Dose-response curves for aldosterone and corticosterone in the CCD cell line, in order to determine whether the effect is mediated through MR, GR or through both [54].

At present, 5 genes distinct from previously identified AIPs (ENaC, NaK-ATPase, ras, sgk) have been validated, according to the criteria mentioned above. One of them will be discussed here in more details

GILZ : a novel early aldosterone induced protein

Analysis of our SAGE data has revealed that the tag "CCTTTGGGGT" is present in a significantly higher copy number in the aldosterone treated cells (Control (1 copy), Aldosterone 30 min (1 copy) and Aldosterone 4 h (12 copies)). Blast search of the NCBI database using this tag has identified a corresponding full-length mRNA encoding for the Glucocorticoid-Induced Leucine Zipper protein (GILZ) [55]. Northern blot analysis using GILZ probe has demonstrated that GILZ mRNA is strongly up-regulated by aldosterone both in the mpkCCD cells and in the kidney of adrenalectomized mice. Importantly, the increase in the GILZ mRNA abundance is already detected after 30 min of aldosterone treatment in the mpkCCD cells, making this protein as a novel example of an early AIP. The GILZ protein has recently been cloned from the dexamethasone-treated murine thymocytes and the few available publications on GILZ function identifies GILZ as an anti-apoptotic protein in T lymphocytes [55,56]. As of today, our knowledge of GILZ function is limited by the immunological studies. No data are available on the role of the protein in the aldosterone-induced transepithelial sodium reabsorption. Sequence similarity analysis indicates that GILZ protein belongs to the TSC22/DIP/BUN leucine-zipper protein subfamily sharing 60-90% homology in their amino acid sequences. Several common protein domains may play important functional roles. Some features are worthwhile noting:
This subfamily of leucine zippers differs from other leucine zippers mainly by their low molecular weight and by the absence of a typical basic DNA-binding domain. In the bZIP family of transcription factor, the leucine-zipper acts as a dimerization domain and the upstream basic region as a DNA-binding domain [57]. Based on the presence of the leucine zipper structure it has been hypothesized that GILZ is a transcription factor [55], however, the leucine zipper structure is implicated in protein-protein interaction and does not predict *per se* protein-DNA interaction. Cellular localization of these proteins remains unclear since TSC-22 (TGFβ Stimulated Clone-22) has been reported to have a

nuclear localization, whereas SHS, another member of this subfamily, demonstrates cytoplasmic localization [58,59].
In the C-terminal sequences of GILZ, we have identified a typical signature for a PDZ-binding domain (X-S/T-X-V), characteristic for PDZ-binding domains observed in a large number of ion channels and G protein-coupled receptors [60]. To our knowledge, the presence of this type of PDZ-binding domain has never been described in any members of transcription factors. The cytosolic (or periplasmic) expression of GILZ and its potential interaction with ENaC activity must now be tested experimentally in heterologous or homologous (principal cell of CCD) expression systems.

Conclusions and perspectives

A full understanding of the aldosterone-signaling pathway will require not only the analysis of the transcriptome by SAGE or RT-PCR differential display but a combination of approaches that include protein-protein interaction methodology (i.e. yeast-two-hybrid technique), proteomics and finally functional expression cloning. No less important will be the understanding of the interacting signaling pathways among vasopressin and insulin. This will ultimately allow to define the most significant limiting factors in these pathways and in all likelihood the most important genes conferring salt-sensitivity or salt-resistance in the general population.

References

1. Lifton RP. Molecular genetics of human blood pressure variation. Science 1996;272:676-80.
2. Midgley JP, Matthew AG, Greenwood CM, Logan AG. Effect of reduced dietary sodium on blood pressure: a meta-analysis of randomized controlled trials. JAMA 1996;275:1590-7.
3. Dyer AR, Elliott P, Shipley M. Urinary electrolyte excretion in 24 hours and blood pressure in the Intersalt study. Am J Epidemiol 1994;139:940-51.
4. Rossier BC: Cum grano salis: the epithelial sodium channel and the control of blood pressure. J Am Soc Nephrol 1997;9:980-92.
5. Denton D, Weisinger R, Mundy NI, Wickings EJ, Dixson A, Moisson P, Pingard AM, Shade R, Carey D, Ardaillou R, et al. The effect of increased salt intake on blood pressure of chimpanzees. Nature Med 1995;1:1009-1016.
6. Shimkets RA, Warnock DG, Bositis CM, Nelson-Williams C, Hansson JH, Schambelan M, Gill JR, Jr., Ulick S, Milora RV, Findling JW, Canessa CM, Rossier BC, Lifton RP. Liddle's syndrome: heritable human hypertension caused by mutations in the beta subunit of the epithelial sodium channel. Cell 1994;79:407-14.
7. Guyton AC. Blood pressure control - Special role of the kidneys and body fluid. Science 1991;252:1813-6.
8. Simon DB, Karet FE, Hamdan JM, DiPietro A, Sanjad SA, Lifton RP. Barrter's syndrome, hypokalaemic alkalosis with hypercalciuria, is caused by mutations in the Na-K-2Cl cotransporter NKCC2. Nature Genet 1996;13:183-8.
9. Simon DB, Nelson-Williams C, Bia MJ, Ellison D, Karet FE, Molina AM, Vaara I, Iwata F, Cushner HM, Koolen M, Gainza FJ, Gitleman HJ, Lifton RP. Gitelman's variant of Bartter's syndrome, inherited hypokalaemic alkalosis, is caused by mutations in the thiazide-sensitive Na-Cl cotransporter. Nature Genet 1996;12:24-30.
10. Wilson FH, Disse-Nicodeme S, Choate KA, Ishikawa K, Nelson-Williams C, Desitter I, Gunel M, Milford DV, Lipkin GW, Achard JM, Feely MP, Dussol B, Berland Y, Unwin RJ, Mayan H, Simon DB, Farfel Z, Jeunemaitre X, Lifton RP: Human hypertension caused by mutations in WNK kinases. Science 2001; 293:1107-12.
11. Rossier BC, Palmer LG. Mechanisms of Aldosterone Action on Sodium and Potassium Transport, in The Kidney, Physiology and Pathophysiology, edited by Seldin DW, Giebiesch G, New York, Raven Press, 1992, pp 1373-409.
12. Loffing J, Pietri L, Aregger F, Bloch-Faure M, Ziegler U, Meneton P, Rossier BC, Kaissling B. Differential subcellular localization of ENaC subunits in mouse kidney in response to high- and low-Na diets. Am J Physiol 2000;279:F252-8.
13. Garty H, Palmer L. Epithelial sodium channels: function, structure, and regulation. Physiol Rev 1997;77:359-96.
14. Barbry P, Lazdunski M. Structure and regulation of the amiloride-sensitive epithelial sodium channel, in Ion Channels (vol 4), edited by Narahashi T, New York, Plenum Press, 1996, pp 115-67.
15. Rossier BC. 1996 Homer Smith Award Lecture. Cum grano salis: the epithelial sodium channel and the control of blood pressure. J Am Soc Nephrol 1997;8:980-92.
16. Chang SS, Grunder S, Hanukoglu A, Rosler A, Mathew PM, Hanukoglu I, Schild L, Lu Y, Shimkets RA, Nelson-Williams C, Rossier BC, Lifton RP. Mutations in subunits of the epithelial sodium channel cause salt wasting with hyperkalaemic acidosis, pseudohypoaldosteronism type 1. Nature Genet 1996;12:248-53.
17. Gründer S, Firsov D, Chang SS, Fowler Jaeger N, Gautschi I, Schild L, Lifton RP, Rossier BC. A mutation causing pseudohypoaldosteronism type 1 identifies a conserved glycine that is involved in the gating of the epithelial sodium channel. EMBO J 1997;16:899-907.
18. Schild L, Canessa CM, Shimkets RA, Gautschi I, Lifton RP, Rossier BC. A mutation in the epithelial sodium channel causing Liddle disease increases channel activity in the Xenopus laevis oocyte expression system. Proc Natl Acad Sci USA 1995;92:5699-703.

19. Hansson JH, Nelsonwilliams C, Suzuki H, Schild L, Shimkets R, Lu Y, Canessa C, Iwasaki T, Rossier B, Lifton RP. Hypertension caused by a truncated epithelial sodium channel gamma subunit: Genetic heterogeneity of Liddle syndrome. Nature Genet 1995;11:76-82.
20. Hansson JH, Schild L, Lu Y, Wilson TA, Gautschi I, Shimkets R, Nelson-Williams C, Rossier BC, Lifton RP. A de novo missense mutation of the beta subunit of the epithelial sodium channel causes hypertension and Liddle syndrome, identifying a proline-rich segment critical for regulation of channel activity. Proc Natl Acad Sci USA 1995;92:11495-9.
21. Verrey F, Hummler E, Schild L, Rossier BC. Control of Na+ transport by aldosterone. In: The Kidney: Physiology and pathophysiology. Seldin, D.W., Giebisch, G. eds. Lippincott Williams & Wilkins, Philadelphia, Volume one, 3rd Edition, 2000, 53: 1441-71.
22. Kotelevtsev Y, Brown RW, Fleming S, Kenyon C, Edwards CRW, Seckl JR, Mullins AJ. Hypertension in mice lacking 11 beta-hydroxysteroid dehydrogenase type 2. J Clin Invest 1999;103:683-9.
23. Geller DS, Farhi A, Pinkerton N, Fradley M, Moritz M, Spitzer A, Meinke G, Tsai FTF, Sigler PB, Lifton RP. Activating mineralocorticoid receptor mutation in hypertension exacerbated by pregnancy. Science 2000;289:119-23.
24. Baker EH, Dong YB, Sagnella GA, Rothwell M, Onipinla AK, Markandu ND, Cappuccio FP, Cook DG, Persu A, Corvol P, Jeunemaitre X, Carter ND, MacGregor GA. Association of hypertension with T594M mutation in beta subunit of epithelial sodium channels in black people resident in London. Lancet 1998;351:1388-92.
25. Ambrosius WT, Bloem LJ, Zhou LF, Rebhun JF, Snyder PM, Wagner MA, Guo CL, Pratt JH. Genetic variants in the epithelial sodium channel in relation to aldosterone and potassium excretion and risk for hypertension. Hypertension 1999;34:631-7.
26. Lovati E, Ferrari P, Dick B, Jostarndt K, Frey BM, Frey FJ, Schorr U, Sharma AM. Molecular basis of human salt sensitivity: the role of 11 beta-hydroxysteroid dehydrogenase Type 2. J Clin Endocr Metab 1999;84:3745-9.
27. Jeunemaitre X, Soubrier F, Kotelevtsev Y, Lifton R, Williams CS, Charru A, Hunt SC, Hopkins PN, Williams RR, Lalouel JM, Corvol P. Molecular basis of human hypertension: role of angiotensinogen. Cell 1992;71:169-80.
28. Paillard F, Chansel D, Brand E, Benetos A, Thomas F, Czekalski S, Ardaillou R, Soubrier F. Genotype-phenotype relationships for the renin-angiotensin-aldosterone system in a normal population. Hypertension 1999;34:423-9.
29. Bianchi G, Tripodi G, Casari G, Salardi S, Barber BR, Garcia R, Leoni P, Torielli L, Cusi D, Ferrandi M, Pinna LA, Baralle FE, Ferrari P. Two Point Mutations Within the Adducin Genes Are Involved in Blood Pressure Variation. Proc Natl Acad Sci USA 1994;91:3999-4003.
30. Manunta P, Cusi D, Barlassina C, Righetti M, Lanzani C, D'Amico M, Buzzi L, Citterio L, Stella P, Rivera R, Bianchi G. alpha-Adducin polymorphisms and renal sodium handling in essential hypertensive patients. Kidney Int 1998;53:1471-8.
31. Brand E, Bankir L, Plouin P-F, Soubrier F. Glucagon receptor gene mutation (Gly40Ser) in human essential hypertension. Hypertension 1999;34:15-7.
32. Jia H, Hingorani AD, Sharma P, Hopper R, Dickerson C, Trutwein D, Lloyd DD, Brown MJ. Association of the G(s)alpha gene with essential hypertension and response to beta-blockade. Hypertension 1999;34:8-14.
33. Corvol P, Persu A, Gimenez-Roqueplo AP, Jeunemaitre X. Seven lessons from two candidate genes in human essential hypertension - Angiotensinogen and epithelial sodium channel. Hypertension 1999;33:1324-31.
34. Pradervand S, Wang Q, Burnier M, Beermann F, Horisberger J-D, Hummler E, Rossier BC. A mouse model for Liddle's syndrome. J Am Soc Nephrol 1999;10:2527-33.

35. Blazer-Yost BL, Liu X, Helman SI. Hormonal regulation of ENaCs: insulin and aldosterone. Am J Physiol 1998;274:C1373-79.
36. Record RD, Froelich LL, Vlahos CJ, Blazer-Yost BL. Phosphatidylinositol 3-kinase activation is required for insulin-stimulated sodium transport in A6 cells. Am J Physiol 1998;274:E611-7.
37. Blazer-Yost BL, Paunescu TG, Helman SI, Lee KD, Vlahos CJ. Phosphoinositide 3-kinase is required for aldosterone-regulated sodium reabsorption. Am J Physiol 1999;46:C531-6.
38. Djelidi S, Fay M, Cluzeaud F, Escoubet B, Eugene E, Capurro C, Bonvalet JP, Farman N, Blot-Chabaud M. Transcriptional regulation of sodium transport by vasopressin in renal cells. J Biol Chem 1997;272:32919-24.
39. Schafer JA, Troutman SL. cAMP mediates the increase in apical membrane Na+ conductance produced in rat CCD by vasopressin. Am J Physiol 1990;259:F823-31.
40. Schafer JA, Troutman SL, Schlatter E. Vasopressin and mineralocorticoid increase apical membrane driving force for K+ secretion in rat CCD. Am J Physiol 1990;258:F199-210.
41. Spindler B, Mastroberardino L, Custer M, Verrey F. Characterization of early-induced RNAs identified in A6 kidney epithelia. Pflügers Arch 1997;434:323-31.
42. Verrey F: Early aldosterone action: toward filling the gap between transcription and transport. Am J Physiol 1999;46:F319-27.
43. Chen SY, Bhargava A, Mastroberardino L, Meijer OC, Wang J, Buse P, Firestone GL, Verrey F, Pearce D. Epithelial sodium channel regulated by aldosterone-induced protein sgk. Proc Natl Acad Sci USA 1999;96:2514-9.
44. Robert-Nicoud M, Flahaut M, Elalouf JM, Nicod M, Salinas M, Bens M, Doucet A, Wincker P, Artiguenave F, Horisberger JD, Vandewalle A, Rossier BC, Firsov D. Transcriptome of a mouse kidney cortical collecting duct cell line: Effects of aldosterone and vasopressin. Proc Natl Acad Sci USA 2001;98:2712-6.
45. Naray-Fejes-Toth A, Canessa C, Cleaveland ES, Aldrich G, Fejes-Toth G. sgk is an aldosterone-induced kinase in the renal collecting duct - Effects on epithelial Na+ channels. J Biol Chem 1999;274:16973-8.
46. Loffing J, Zecevic M, Feraille E, Kaissling B, Asher C, Rossier BC, Firestone GL, Pearce D, Verrey F. Aldosterone induces rapid apical translocation of ENaC in early portion of renal collecting system: possible role of SGK. Am J Physiol 2001;280:F675-82.
47. Alvarez de la Rosa D, Zhang P, Naray-Fejes-Toth A, Fejes-Toth G, Canessa CM. The serum and glucocorticoid kinase sgk increases the abundance of epithelial sodium channels in the plasma membrane of Xenopus oocytes. J Biol Chem 1999;274:37834-9.
48. Debonneville C, Flores SY, Kamynina E, Plant PJ, Tauxe C, Thomas MA, Münster C, Chraïbi A, Pratt JH, Horisberger J-D, Pearce D, Loffing J, Staub O. Phosphorylation of Nedd4-2 by Sgk1 regulates epithelial Na+ channel cell surface expression. EMBO J (in press) 2001.
49. Bens M, Vallet V, Cluzeaud F, Pascual-Letallec L, Kahn A, Rafestin-Oblin ME, Rossier BC, Vandewalle A. Corticosteroid-dependent sodium transport in a novel immortalized mouse collecting duct principal cell line. J Am Soc Nephrol 1999;10:923-34.
50. Velculescu V, Zhang L, Vogelstein B, Kinzler K. Serial analysis of gene expression. Science 1995;270:484-7.
51. Velculescu V, Zhang L, Zhou W, Vogelstein J, Basrai M, Bassett Jr D, Hieter P, Vogelstein B, Kinzler K. Characterization of the Yeast Transcriptome. Cell 1997;88:243-51.
52. Zhang L, Zhou W, Velculescu V, Kern S, Hruban R, Hamilton S, Vogelstein B, Kinzler K. Gene expression profiles in normal and cancer cells. Science 1997;276:1268-72.
53. Virlon B, Cheval L, Buhler JM, Billon E, Doucet A, Elalouf JM. Serial microanalysis of renal transcriptomes. Proc Natl Acad Sci USA 1999;96:15286-91.
54. Geering K, Claire M, Gaeggeler HP, Rossier BC. Receptor occupancy vs induction of Na-K-ATPase and Na+ transport by aldosterone. Am J Physiol 1985;248:C102-8.

55. D'Adamio F, Zollo O, Moraca R, Ayroldi E, Bruscoli S, Bartoli A, Cannarile L, Migliorati G, Riccardi C. A new dexamethasone-induced gene of the leucine zipper family protects T lymphocytes from TCR/CD3-activated cell death. Immunity 1997;7:803-12.
56. Ayroldi E, Migliorati G, Bruscoli S, Marchetti C, Zollo O, Cannarile L, D'Adamio F, Riccardi C. Modulation of T-cell activation by the glucocorticoid-induced leucine zipper factor via inhibition of nuclear factor kappaB. Blood 2001;98:743-53.
57. Vogel P, Magert HJ, Cieslak A, Adermann K, Forssmann WG. hDIP--a potential transcriptional regulator related to murine TSC-22 and Drosophila shortsighted (shs)--is expressed in a large number of human tissues. Biochim Biophys Acta 1996;1309:200-4.
58. Shibanuma M, Kuroki T, Nose K. Isolation of a gene encoding a putative leucine zipper structure that is induced by transforming growth factor beta 1 and other growth factors. J Biol Chem 1992;267:10219-24.
59. Treisman JE, Lai ZC, Rubin GM. Shortsighted acts in the decapentaplegic pathway in Drosophila eye development and has homology to a mouse TGF-beta-responsive gene. Development 1995;121:2835-45.
60. Kornau HC, Schenker LT, Kennedy MB, Seeburg PH. Domain interaction between NMDA receptor subunits and the postsynaptic density protein PSD-95. Science 1995;269:1737-40.

3. CARDIOVASCULAR PROTEOMICS: EFFECT OF HYPOXIC CONDITIONS

C.Michiels , D.Mottet , E.Delaive , M.Dieu , J-F.Dierick , J.Remacle .

Introduction

Diseases of the cardiovascular system result from a complex mixture of genetic and environmental factors. Much has been learned in the remodeling processes in heart and/or arteries during these diseases. Moreover, large epidemiological studies have provided insights in environmental as well genetic risk factors and will continue to unravel new interactions. Progress in defining the cellular and molecular interactions involved, however has been hindered by the disease's etiological complexity. In addition, although pathological disorders are believed to be due to alterations in gene expression, relatively few modifications are yet reported. This is partly due to the classical, labor-intensive approaches used until now which require a reiteration of the detection procedure for each gene. However, the new techniques recently developed allow the detection of hundreds to thousands genes (DNA microarrays, differential display, SAGE) or proteins (2D gel electrophoresis, protein chips) at one time. It leads to the discovery of unexpected genes whose expression is modified but also to the development of new hypotheses concerning the etiology of the disease. It also provides a useful tool to study the effect of drugs or treatments in patients.

Proteomics, the large scale analysis of proteins, will remain the complementary approach to genomics. It arised from the completion of the human genome sequencing. Indeed, genomics offers limited insight regarding the nature and the expression level of the actual effectors, i.e. the proteins. In addition, proteomics brings information about the various modifications of these proteins that occur posttranslationally.

Proteomics traditionally refers to a display of proteins from a given cell type or tissue on two-dimensional polyacrylamide gels (2D gels) [1,2,3]. This results in maps of spots from 2D gels to build databases of all expressed proteins. Identification of these spots is now possible through mass spectrometry (see below). More recently, proteomics evolved into functional proteomics including the determination of the subcellular localization of proteins and of protein-protein interactions. Most proteins are thought to exist in the cell as part of cellular complexes which perform cellular functions cooperatively. Systematic identification of these complexes is of great value for the assignment of protein function, the principal problem of the post-genomic area, but also to understand the fine tuning of these functions according to the metabolic and/or activated states of the cells [4].

Experimental approaches

The first step of proteomic analysis is obtaining protein samples and resolve them on 2D gels. During the first dimension, or iso-electric focusing, proteins are separated via an ampholyte pH gradient and allow to migrate to their respective iso-electric points on the gel on the basis of their ionic property. The first-dimensional gel is then subjected to a second dimensional SDS-PAGE and the proteins are separated on the basis of mass [5,6]. Modern large-format gels are highly reproducible and Coomassie-blue or silver staining allows protein quantitation. Fluorescent dyes as well as radioactive labeling can be used to quantitatively visualize thousands of proteins on a wider dynamic range. A typical gel can reliably separate about 2,000 protein spots in this way while using high-sensitivity staining methods, refinements of ampholyte technology to accommodate extended pI ranges; i.e. IPG (immobilized pH gradient) and larger gel formats can increase the resolution up to 10,000 protein spots. However, despite these major improvements, some classes of proteins remain very difficult to analyze by this approach. This is the case for high molecular weight proteins, low-abundance proteins as well as for hydrophobic proteins such as membrane spanning receptors, that are very difficult to dissolve.

Once proteins have been separated, visualized and quantified, they must be identified. Classical Edman amino acid sequencing, which requires large amount of proteins, is now replaced by high sensitive mass spectrometry techniques. Practically, spots are excised from the gel and proteins are digested in-gel, into fragments by specific proteases such as trypsin. The fragments are analyzed by mass spectrometry, generating a peptide mass fingerprint or "peptide map". The protein is then identified by comparing this experimental peptide map to the theoretical peptide maps obtained by in silico digestion of all the proteins present in sequence databases. MALDI-TOF (matrix-assisted laser-desorption-ionization time of flight) mass spectrometry is usually used for peptide mass fingerprinting [7,8]. Using the mass accuracy now available (better than 10 ppm), this technique alone is sufficient to identify proteins from completely sequenced genomes, as the masses of all the tryptic peptides from the open reading frames can be predicted. A more time consuming method is needed for the identification of novel protein via decoding the primary amino acid sequence of some of the different peptidic fragments. These peptides are further fragmented and then detected through tandem mass spectrometry (electrospray-MS-MS).

One of the unique features of proteomics analysis is the ability to analyze the post-translational modifications of proteins. Protein phosphorylation represents one of the most prevalent covalent modifications, but glycosylation, acetylation or sulphation are also of importance. Detection of phosphorylation on a given peptide fragment from one protein is possible using mass spectrometry: phosphopeptide are 80 Da heavier than their unmodified counterparts but the task is much more difficult than mere determination of protein identity: more material is required and of higher purity. Enrichment of the samples in phosphoproteins may provide a better alternative, for example using anti-phosphotyrosine antibodies [9], I-MAC columns or selective modifications of phosphopeptides allowing more stringent purification [10]. In addition, electrospray-MS-MS analysis can determine which amino acid in the sequence was modified.

Proteomic analysis generates large sets of data, and consequently relies on bioinformatics technologies. After staining, the gels are recorded and the resulting spots quantified through image analysis. This allows detailed comparison and determination of spots whose intensity increases or decreases in one condition in comparison to another. Secondly, bioinformatics generates theoretical tryptic profiles to which data from MS can be compared. It is also required for the use of existing proteomic or genomic databases.

Effect of hypoxia on protein expression

Biomedical applications of the comparative 2D gel approach are numerous but the objective is to identify proteins that are up- or down-regulated in a disease-specific manner to get insight in the etiology of the disease, for use as diagnostic markers or for identifying potential therapeutic targets. Vascular genomics has already yielded contribution to the understanding of the mechanisms that are central to cardiovascular diseases (for a review, see [11, 12]. The proteomics, both classical 2D gel comparison and functional genomics, that follows will certainly be as promising [11]. With regards to functional genomics, the work of Ping and colleagues is one such remarkable example that led to the identification of binding partners of one major component of the signal transduction pathway, PKCε, that mediates cardioprotection in ischemic preconditioning (for a review, see [13])

Our current work is devoted to understanding the effect of hypoxia on gene transcription regulation. The origin of several vascular pathologies involves sudden or recurrent oxygen deficiency. This is certainly the case for myocardial infarction, cerebral ischemia or pulmonary hypertension [14]. Neo-angiogenesis in tumors like vasculogenesis during embryonic development are both triggered by hypoxia [15].

Hypoxic conditions are associated with differential expression of genes allowing cells and tissues to adapt to low oxygen concentration. Specific increased expression of glycolytic enzymes, glucose transporter GLUT-1, VEGF (vascular endothelial growth factor), tyrosine hydroxylase, transferrin and erythropoietin is observed, all proteins involved in the adaptative response of cells to hypoxia. The modification of gene expression is dependent on the activation of specific transcription factors: HIF-1 being the most important (for a review, see [16]). HIF-1 is a heterodimeric transcription factor consisting of HIF-1α and HIF-1ß/ARNT (aryl hydrocarbon receptor nuclear translocator). Both subunits belong to the Per-ARN/Ahr-Sim family of the bHLH (basic helix-loop-helix) transcription factors. HIF-1α as well as ARNT are constitutively expressed. However, HIF-1α appears to be the HIF-1 subunit regulated by hypoxia. HIF−1α protein level is specifically upregulated in hypoxic conditions, through inhibition of HIF−1α rapid degradation by the proteasome pathway. HIF−1α then translocates into the nucleus where it dimerizes with ARNT and becomes phosphorylated, to form the transcriptionally active complex HIF-1.

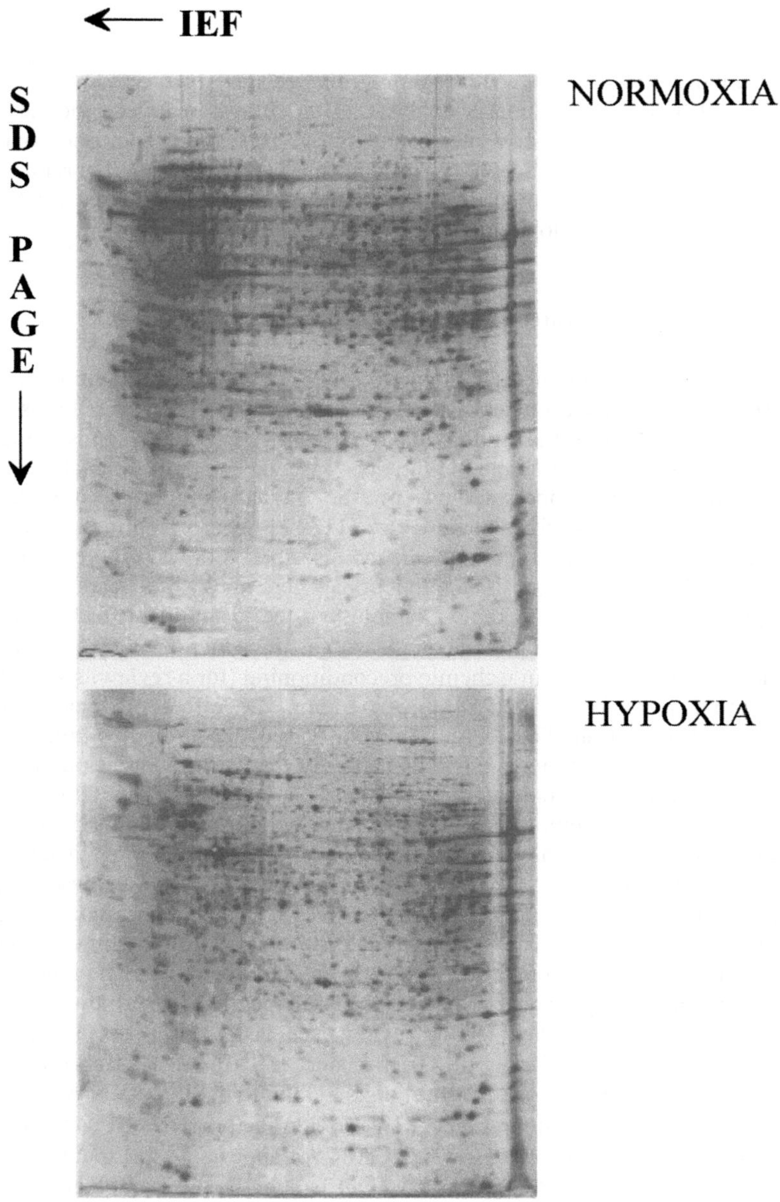

Figure 1 *2D patterns of silver-stained proteins from COS-7 cells incubated 6h in normoxia and hypoxia*

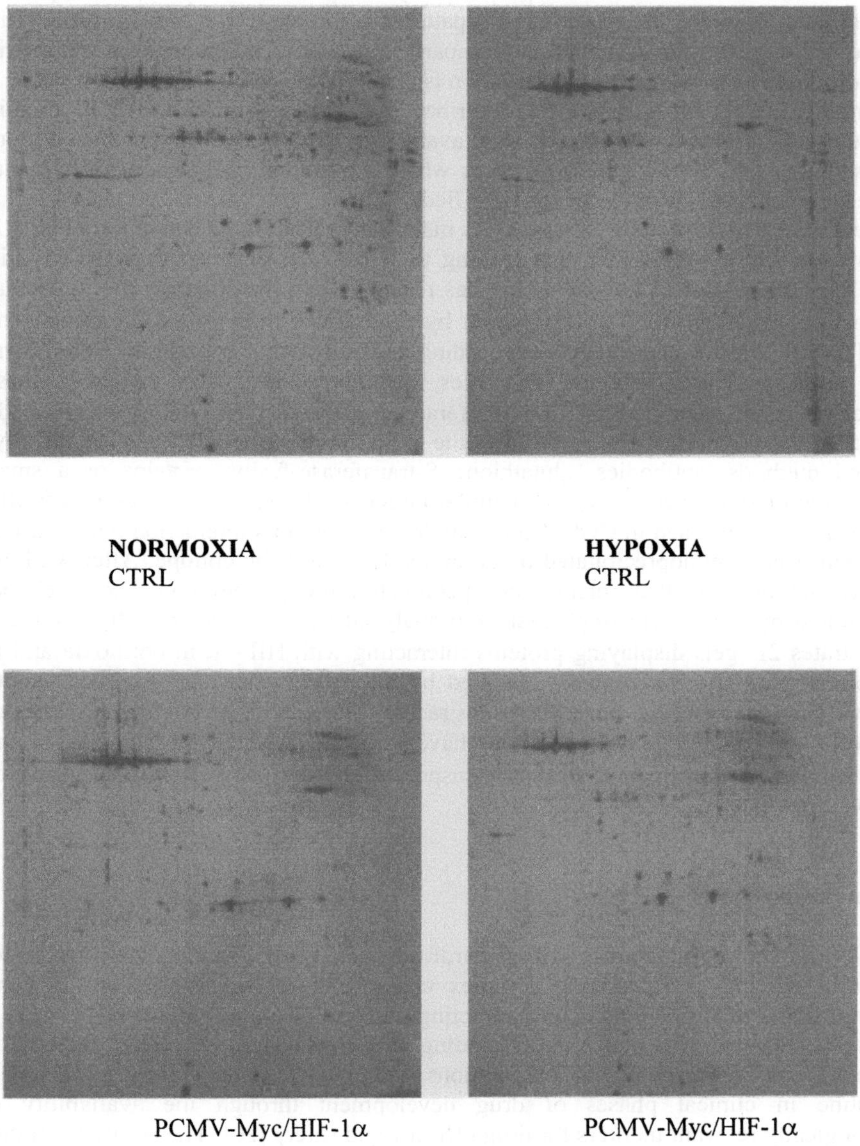

Figure2 *2D patterns of silver-painted proteins co-immunoprecipitated with HIF-1α. Extracts from normoxic or hypoxic COS-7 cells that were transfected with a HIF-1α expressing vector (pCMV-Myc/HIF-1α) or with the control vector (CTRL) were incubated with an anti-HIF-1a antibody and proteins interacting with HIF-1a were then precipitated and separated on 2D gels.*

Up to now very few genes upregulated by hypoxia are described. An overall approach to visualize the proteome of hypoxic cells in comparison to normoxic cells has thus been initiated in an attempt to identify new proteins upregulated in hypoxic conditions. Figure 1 illustrates the 2D patterns of proteins from cells incubated in vitro in hypoxia for 6 hours in comparison to cells maintained in normoxic conditions. The most obvious observation is that hypoxia decreases the expression of numerous proteins. It is well described that hypoxia inhibits overall protein synthesis [17], due to the limited ATP availability. However, when analyzed more closely, the gels reveal several proteins whose expression is increased in hypoxia. These proteins are currently being identified.

Overexpression of proteins in hypoxia is mainly the result of the activation of HIF-1. However, the exact mechanisms leading to HIF-1 activation in hypoxia are still unclear. While proline hydroxylation has recently been identified as the molecular signature targeting HIF-1α degradation by proteasome through the protooncogene pVHL [18,19], the subsequent steps leading to a fully active protein are not known. We decided to use functional genomics in order to bring new insight in these mechanisms through identification of interacting partners of HIF-1α. A way to study protein-protein interactions is to purify the entire multi-protein complex by affinity-based methods, antibodies, glutathione S-transferase-fusion proteins or a small molecule binding specifically to a cellular target [4]. In this case, Myc-tagged HIF-1α was overexpressed in COS-7 cells and together with its interaction partners, this protein was immunoprecipitated by an antibody against the epitope. After washing away the proteins that interact nonspecifically, the protein complex is eluted, separated by 2D gel electrophoresis and analyzed by mass spectrometry. Figure 2 illustrates 2D gels displaying proteins interacting with HIF-1α in normoxia and in hypoxia. Non specific proteins obtained by the mock procedure using the empty expressing vector pMyc have been substracted. The approach provide new lines of investigation because no assumptions have to be made about the complex before performing the purification so that unsuspected HIF-1α partners will be unraveled soon.

Conclusion

Incorporation of proteomics within cardiovascular research offers new means of identifying and characterizing complex changes in protein expression, in post-translational modifications or in interacting partners of a particular protein. This is possible not only for better understanding cardiovascular dysfunction but also to characterize pharmacological interventions. Moreover, it is anticipated to become a routine in clinical phases of drug development through the availability of biologically relevant markers for drug efficacy and safety [20]. The ability to stratify patients, through genomics and proteomics, into clinically useful risk and treatments groups together with the development of new and customized treatments should greatly improve the outcome of these patients.

Acknowledgements

C. Michiels Research Associate of the FNRS (Fonds National de la Recherche Scientifique, Belgium). D. Mottet and J-F Dierick are fellows of FRIA (Fonds pour la Recherche dans l'Industrie et l'Agriculture, Belgium). This text presents results of the Belgian Programme on Interuniversity Poles of Attraction initiated by the Belgian State, Prime Minister's Office, Science Policy Programming. The scientific responsability is assumed by the authors.

References

1. Anderson NG and Anderson NL Twenty years of two-dimensional electrophoresis: past, present and future. Electrophoresis 1996;17:443-53.
2. Celis JE, Gromov P, Ostergaard M, Madsen P, Honore B, Dejgaard K, Olsen E, Vorum H, Kristensen DB, Gromova I, Haunso A, Van Damme J, Puype M, Vandekerckhove J and Rasmussen HH Human 2-D PAGE databases for proteome analysis in health and disease: http://biobase.dk/cgi-bin/celis. FEBS Lett 1996;398:129-34.
3. Wilkins MR, Pasquali C, Appel RD, Ou K, Golaz O, Sanchez JC, Yan JX, Gooley AA, Hughes G, Humphery-Smith I, Williams KL and Hochstrasser DF From proteins to proteomes: large scale protein identification by two- dimensional electrophoresis and amino acid analysis. Biotechnology 1996;14:61-5.
4. Pandey A and Mann M Proteomics to study genes and genomes. Nature 2000;405:837-46.
5. O'Farrell PH High resolution two-dimensional electrophoresis of proteins. J Biol Chem 1975;250:4007-21.
6. Klose J and Spielmann H Gel isoelectric focusing of mouse lactate dehydrogenase: heterogeneity of the isoenzymes A4 and X4. Biochem Genet 1975;13:707-20.
7. Mann M, Hendrickson RC and Pandey A Analysis of Proteins and Proteomes by Mass Spectrometry. Annu Rev Biochem 2001;70:437-73.
8. Jungblut P and Thiede B Protein identification from 2-DE gels by MALDI mass spectrometry. Mass Spectrom Rev 1997;16:145-62.
9. Pandey A, Podtelejnikov AV, Blagoev B, Bustelo XR, Mann M and Lodish HF Analysis of receptor signaling pathways by mass spectrometry: identification of vav-2 as a substrate of the epidermal and platelet- derived growth factor receptors. Proc Natl Acad Sci 2000;97:179-84.
10. Ahn NG and Resing KA Toward the phosphoproteome. Nat Biotechnol 2001;19:317-8.
11. Arrell DK, Neverova I and Van Eyk JE Cardiovascular proteomics: evolution and potential. Circ Res 2001;88:763-73.
12. Hwang JJ, Dzau VJ and Liew CC Genomics and the pathophysiology of heart failure. Curr Cardiol Rep 2001;3:198-207.
13. Vondriska TM, Klein JB and Ping P Use of functional proteomics to investigate PKC epsilon-mediated cardioprotection: the signaling module hypothesis. Am J Physiol 2001;280:H1434-H41.
14. Michiels C, Arnould T and Remacle J Endothelial cell responses to hypoxia: initiation of a cascade of cellular interactions. Biochim Biophys Acta 2000;1497:1-10.
15. Semenza GL HIF-1: mediator of physiological and pathophysiological responses to hypoxia. J Appl Physiol 2000;88:1474-80.
16. Semenza GL Regulation of mammalian O2 homeostasis by hypoxia-inducible factor 1. Annu Rev Cell Dev Biol 1999;15:551-78.
17. Gorlach A, Camenisch G, Kvietikova I, Vogt L, Wenger RH and Gassmann M Efficient translation of mouse hypoxia-inducible factor-1alpha under normoxic and hypoxic conditions. Biochim Biophys Acta 2000;1493:125-34.
18. Jaakkola P, Mole DR, Tian YM, Wilson MI, Gielbert J, Gaskell SJ, Kriegsheim A, Hebestreit HF, Mukherji M, Schofield CJ, Maxwell PH, Pugh CW and Ratcliffe PJ Targeting of HIF-alpha to the von Hippel-Lindau ubiquitylation complex by O2-regulated prolyl hydroxylation. Science 2001;292:468-72.
19. Ivan M, Kondo K, Yang H, Kim W, Valiando J, Ohh M, Salic A, Asara JM, Lane WS and Kaelin WG HIFalpha targeted for VHL-mediated destruction by proline hydroxylation: implications for O2 sensing. Science 2001;292:464-8.
20. Steiner S and Witzmann FA Proteomics: applications and opportunities in preclinical drug development. Electrophoresis 2000;21:2099-104.

4. USING COMPARATIVE GENOME ANALYSIS TO FIND INTERACTION PARTNERS FOR FRATAXIN

M.A. Huynen

Introduction

The sequencing of complete genomes has provided the opportunity, not only to interpret the function of a protein within its proteomic context, but also to predict new functional interactions between proteins using comparative genome analysis [1]. Various methods have been proposed and demonstrated to predict functional interaction between proteins based on the genomic context of their genes [2,3,4,5]. These methods are all based on variations of the idea that genes that are somehow associated with each other on the genome tend to encode proteins that functionally interact. The types of genomic association that they use are either a) the fusion of genes; b) the conservation of gene order, e.g. when genes are located in operons; c) the co-occurrence of genes in genomes (also called 'phylogenetic profiles'). A systematic analysis of the correlation between on the one hand the type of genomic context and on the other hand the type of functional interaction shows that conservation of genomic context can indeed be a reliable indication of a functional interaction (figure 1). The functional interactions that are reflected in the conservation of genomic context include a wide variety of relations between proteins, including direct physical interactions but also less direct ones, like being part of the same metabolic or regulatory pathway. When there is prior knowledge about a protein's involvement in a process, yet the exact function of the protein is not known, the co-occurrence of genes in genomes can more specifically pinpoint in which sub-process the protein plays a role [6, 7]. Here we use genome comparisons to predict functional interactions for frataxin, a mitochondrial protein that has no detectable homologs with known function and that presently has a unique fold [8, 9]. Severely reduced levels of frataxin cause the disease Friedreich's ataxia [10], which is characterized by degeneration of large sensory neurons and spinocerebellar tracts, cardiomyopathy and increased likelihood of diabetes [11]. In mitochondria, reduced levels of frataxin result in the absence of iron-sulfur (Fe-S) cluster dependent enzymes, accumulation of iron deposits, DNA damage and oxidative stress [12]. Based on such observations the main hypothesis about frataxin's function is that it is directly involved in iron homeostasis of the mitochondria. Alternatively it has been proposed that frataxin is involved in Fe-S cluster assembly on Fe-S proteins [13]. Recent findings that the yeast ortholog of frataxin precipitates with iron support the first hypothesis [14], they could however not be reproduced with purified human frataxin itself [8]. Here we show that the frataxin gene and its orthologs (*cyaY* in Bacteria) have the same phylogenetic distribution as the chaperones *hscA* and *hscB/JAC1*, supporting a direct role in the assembly of Fe-S cluster proteins, rather than in iron homeostasis.

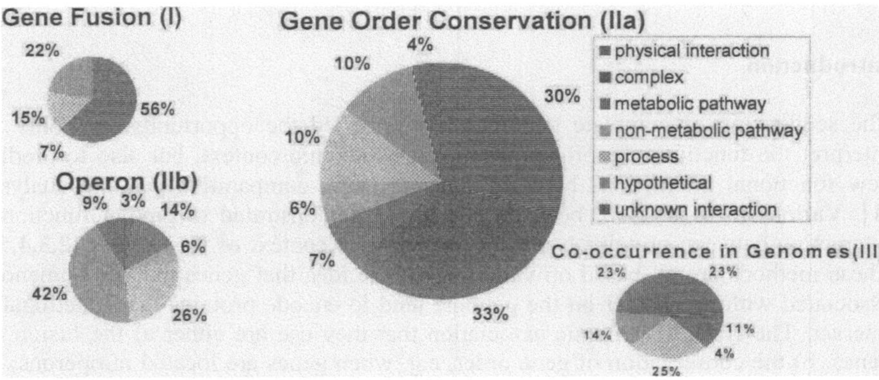

Figure 1 *Correlations between various types of functional interactions and types of genomic context.*

For the minimal genome of the gram-positive Bacterium *Mycoplasma genitalium* (480 genes) we determined the usefullness of the conservation of genomic context for the prediction of functional interactions. The surface areas of the circles are proportional to the number of genes to which the various types of genomic context apply. We studied four types of genomic context : I) Gene fusion : two or more *M. genitalium* genes are fused in any of the other published genomes, IIa) Gene Order Conservation : two or more *M. genitalium* genes are neighbors of each other in at least two genomes IIb) Operon : two or more *M. genitalium* genes are conserved in potential operons with each other in at least two genomes, but they are not conserved as neighbors. III) Co-occurrence in Genomes : two or more *M. genitalium* genes have a similar phylogenetic distribution (see [5] for further technical details). Classification was done by manual inspection, allowing detection of all possible described functional interactions between proteins. Subsequently the functional interactions were divided along the following, hierarchical, classification scheme.

1: direct physical interaction between the proteins.

2: indirect physical interaction: i.e. the proteins are part of the same protein complex, but there is no evidence that they interact directly with each other.

3: the proteins are part of a single metabolic pathway.

4: the proteins are part of a non-metabolic pathway, either regulatory or otherwise.

5: the proteins take part in the same process.

6: pairs of proteins of which at least one is hypothetical.

7: proteins with known functions between which no functional interactions are known (an estimate of the fraction of false positives).
Class 5 was only considered if the functional interactions between the proteins did not fall in class 1-4.

Materials and Methods

The determination of orthology relations between the genes in *Buchnera* and the other 55 sequenced genomes was done in two steps. First we compared all the predicted protein sequences from *Buchnera* with those from the other genomes using the Smith-Waterman algorithm [43] to detect homology relations. From these we selected bidirectional best, not-overlapping (to include the possibility of fission/fusion of genes), homologs (E<0.01) [2]. This procedure gives a first-order approximation of the orthology relations between genomes. It is however far from perfect, the sequence similarity between some orthologous genes has eroded too far for their homology to be detected by the Smith-Waterman algorithm and highly variable rates of evolution can lead to situations in which bidirectional best hits do not reflect orthology relations. Thus in the second step we improved the quality of the orthology predictions for the 20 genes that had the most similar phylogenetic distribution to frataxin in the following manner. We performed iterative PSI-Blast searches (5 iterations, E<0.001) [44] to select all family members of these genes. They were aligned using clustalx [45] and Neighbor-joining trees were constructed [46]. Subsequently we made high quality orthology predictions by manual inspection of these phylogenetic trees and the groups of orthologous genes were selected that had the most similar distribution to the frataxin gene. This procedure revealed only two genes with the same distribution as frataxin, *hscA* and *hscB/JAC1*. Other genes from *Buchnera* have at least a discrepancy of seven in their phylogenetic distribution: i. e. there are seven genomes where either they are present and frataxin is not, or vice versa. In quantitative analyses of the fraction of false positives of functional interactions that are predicted by the co-occurrence of genes we observed none at the co-occurrence level of frataxin *hscA* and *hscB/JAC1* (data not shown).
All sequenced prokaryotic genomes (for an overview see www.tigr.org/tdb/mdb/mdbcomplete.html) and those of *S.cerevisiae*, *D.melanogaster* and *A.thaliana* were obtained from genbank (ftp.ncbi.nlm.nih.gov/pub/genomes). The *S. pombe* genome was obtained from the European *Schizosaccharomyces* genome sequencing project, http://www.sanger.ac.uk/Projects/S_pombe), *C. albicans* from the Stanford Genome Technology Center, http://www-sequence.stanford.edu/group/candida/search.html, and the *H.sapiens* genome from Ensembl, http://www.ensembl.org.
Species names in full: *Homo sapiens, Drosophila melanogaster, Caenorhabditis elegans, Arabidopsis thaliana, Saccharomyces cerevisiae, Candida albicans, Schizosaccharomyces pombe, Rickettsia prowazekii, Caulobacter crescentus, Melorhizobium loti, Neisseria meningitidis, Xylella fastidiosa, Pseudomonas aeruginosa, Haemophilus influenzae, Pasteurella multocida, Vibrio cholerae, Escherichia coli, Campylobacter jejuni, Helicobacter pylori, Deinococcus*

radiodurans, Mycobacterium tuberculosis, Mycoplasma genitalium, Bacillus subtilis, Aquifex aeolicus, Methanococcus jannaschii, Aeropyrum pernix.

Results

Co-occurrence of frataxin with proteins involved in iron-sulfur cluster assembly in the Bacteria

Frataxin (Cyay in Bacteria) is not fused with any gene with known function in any of the published genomes. A search for conservation of gene order with STRING(http://www.bork.embl-heidelberg.de/STRING) revealed that it is not conserved with any genes in potential operons in any of the published genomes either. We therefore resorted to the third type of genomic context, the co-occurrence of genes, to find potential interaction partners for frataxin. Orthologs of the human frataxin gene were found in all sequenced eukaryotic genomes, and in most proteobacteria (purple bacteria), specifically in all but one of the sequenced γ-proteobacteria, in all of the sequenced β-proteobacteria and in one of the sequenced α-proteobacteria: *Rickettsia prowazekii*, the closest fully sequenced relative to the ancestor of the mitochondria. That frataxin is only present in the proteobacterial clade led to the proposal that in eukaryotes the protein is targeted to the mitochondrion [15], which was substantiated by subsequent experimental evidence [16,17,18,10].

To find possible interaction partners for frataxin we determined the phylogenetic distribution of all genes of the smallest genome that contains the frataxin gene: the intracellular symbiont proteobacterium *Buchnera* (see Materials and Methods). As summarized in Figure 2, only two genes have a phylogenetic distribution that is identical to that of frataxin: The chaperone pair *hscA* and *hscB/JAC1*. In Bacteria this pair is part of the so-called iron sulfur cluster (isc) assembly operon [16] that in *Escherichia coli* encodes 9 proteins: A hypothetical RNA methylase (*E. coli* gene number EC2532), a hypothetical helix-turn-helix containing transcriptional regulatory protein *iscR*, *iscS*, *iscU*, *iscA*, *hscB*, *hscA*, *fdx* (a 2Fe-2S ferredoxin), and a third hypothetical protein (EC2524). A number of the encoded proteins appear involved in the generation of iron-sulfur clusters on ferredoxin in *Escherichia coli*: IscS, IscA, HscA, HscB and EC2524 [20]. The clustering of these genes into one operon (actually two operons in *Rickettsia* and *Neisseria*) has, within the bacteria, the same phylogenetic distribution as the frataxin gene. At the level of the individual genes this however only applies to *hscB* and *hscA*. Other genes from the cluster are either more widespread in the Bacteria (EC2532, EC2531, *iscS*, *iscU*, *iscA* and *fdx*) or less widespread (EC2524).

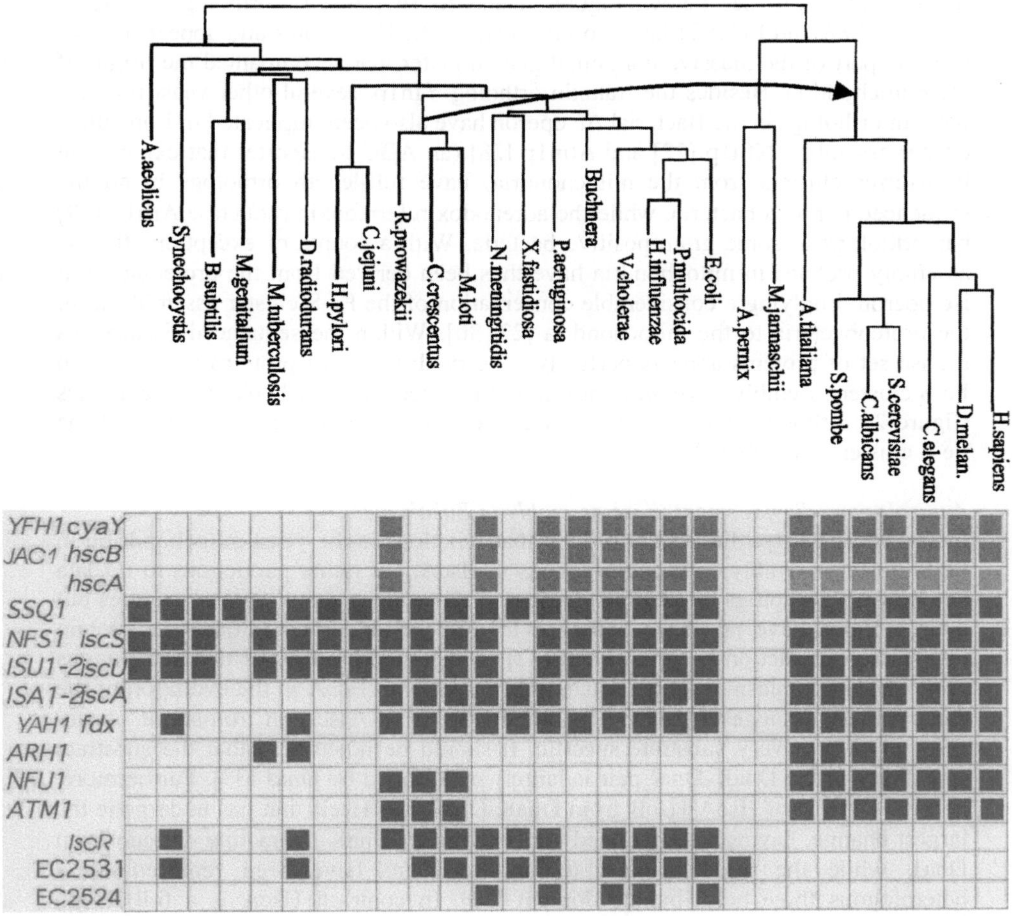

***Figure 2** The phylogenetic distribution of genes involved in iron-sulfur cluster assembly in proteobacteria and Eukaryotes. Several other prokaryotic species, including the Archaea M.jannaschii and A.pernix) have been included for reference. The first two columns give the gene names in S. cerevisiae and prokaryotes. The genes for frataxin (YFH1 in S. cerevisiae, cyaY in bacteria), hscA and hscB have identical distributions and are indicated in gray. Proteins encoded by orthologs of hscA in the eukaryotes are not located in the mitochondria and are therefore indicated in lighter gray. Black squares indicate the presence in the various genomes of full-length orthologs of the other genes implicated in iron-sulfur cluster assembly.*

Evolution of the Eukaryotic isc assembly in mitochondria.

Eukaryotic orthologs, and in one case a paralog, of most of the Bacterial isc proteins have been implicated in Fe-S cluster generation in yeast. These are : Nfs1p [21] (ortholog of IscS), Isu1p and Isu2p [22] (orthologs of IscU), Isa1p and Isa2p [23, 24] (orthologs of IscA), Yah1p [25] (ortholog of the fdx gene product), Ssq1p [26] (paralog of HscA, see below, Figure 2) and Jac1p [26, 27] (ortholog of HscB). EC2532, *iscR* and EC2524 have no orthologs in the Eukaryotes and appear to have been no part of the massive horizontal gene transfer that accompanied the origin of the mitochondria. Besides the frataxin ortholog Yfh1p, several other yeast proteins without orthologs in the Bacterial *isc* operon have also been implicated in iron-sulfur cluster assembly. Nfu1p [22] and Atm1p [28] (an ABC transporter that exports the iron-sulfur clusters from the mitochondria) have full-length orthologs in all the sequenced α-proteobacteria, while the adrenodoxin/ferredoxin reductase Arh1p [29] has orthologs in some gram-positive bacteria. With a couple of exceptions the isc assembly proteins in mitochondria have thus been derived from the proteobacterial isc operon, implying a considerable conservation of the Fe-S cluster assembly from the proteobacteria to the mitochondria [22, 30]. Within the sequenced Eukaryotes the isc set of proteins appears perfectly conserved: the yeast proteins implicated in Fe-S cluster assembly have orthologs in all the other sequenced eukaryotic genomes (figure 2), although some variation might exist in the compartmentalization of the Fe-S cluster assembly [31].

A paralogous displacement in isc assembly in Eukaryotes.
Interestingly Ssq1p, the HSP70 protein that functions in the yeast mitochondrion in Fe-S cluster assembly, is not orthologous to HscA, but rather paralogous to it (data not shown). Orthologs of *hscA* are present in all sequenced eukaryotic genomes but their proteins have not been observed in mitochondria. In addition, the protein localization prediction program Psort [32] predicts the proteins of this orthologous group to be cytoplasmic. This switching of DnaK with HscA in the evolution of the eukaryotic cell suggests that the functioning of DnaK/HscA in iron-sulfur cluster assembly is not very substrate specific. It should be noted here that the substrate specificity of the DnaK-DnaJ pair is largely determined by DnaJ [33]. Furthermore, in the evolution of HscA-HscB from DnaK-DnaJ, it is HscB that has undergone the largest change, having only retained the N-terminal, DnaK interacting domain from DnaJ, while the middle and C-terminal domains have been replaced by a heterologous three helix bundle domain [34]. In contrast, HscA is a full length homolog of DnaK. It has retained functionality as a chaperone for standard substrates such as rhodanese or citrate synthase [35]. Relative to HscB, HscA has thus retained more of the structure and function of its ancestral protein. HscA might thus easier be replaced by DnaK in the evolution of the Fe-S cluster assembly in the mitochondrion from that in its ancestral α-proteobacterium, than HscB by DnaJ.

Co-Evolution of the HscA, HscB and frataxin genes
The patterns in the occurrence of genes can be combined with the evolutionary history of the genomes to reveal whether genes have not only been invented at the same time, but also have been lost together (figure 3). Such co-loss of genes increases the statistical significance of their co-occurrence and the likelihood that the proteins they encode are functionally linked [6,7]. Given their widespread phylogenetic distribution, EC2532, *iscR*, *iscS*, *iscU* and *fdx* likely existed in the bacteria before the *dnaK-dnaJ* duplication gave rise to *hscA-hscB*, apparently in the

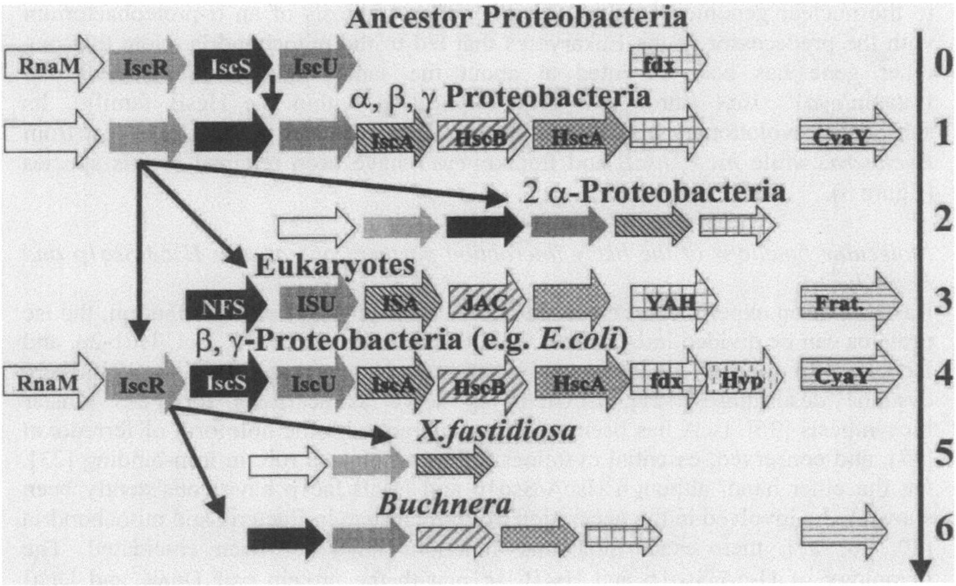

***Figure 3** A history of iron-sulfur cluster assembly in proteobacteria and eukaryotes.*

The figure depicts the evolutionary history of the main elements of the iron-sulfur cluster assembly machinery as it is currently present in mitochondria and of frataxin (cyaY in Bacteria). As shown, two elements of the cluster, HscA and HscB, are not only invented at more or less the same time as frataxin, they are also lost at least twice with it.

(0) Some parts of the iron-sulfur cluster (isc) assembly operon predate the proteobacteria.

(1) At the onset of the α, β and γ proteobacteria, a) CyaY/frataxin is invented, possibly by gene duplication from a yet unknown source, b) IscA is invented by a gene duplication from a member of the HesB gene family, c) HscB and HscA are invented by gene duplications from DnaJ and DnaK respectively.

(2) HscB, HscA and frataxin are subsequently lost from two a-proteobacteria : *M. loti* and *C. crescentus.*

(3) The complete gene cluster, except for RnaM and IscR, and frataxin are transferred from the a-proteobacterial ancestor of the mitochondria to the nuclear genome.

(4) The β and γ proteobacteria inherit the complete isc operon + frataxin, they add a hypothetical protein to the isc cluster

(5) The γ proteobacterium *X. fastidiosa* loses parts of the isc cluster, including HscB and HscA, and frataxin.

(6) The minimal genome of *Buchnera* loses parts of the isc operon but maintains HscA and HscB and it maintains frataxin

proteobacterial clade. At about the same time that frataxin was invented, this chaperone pair has thus been added to a preexisting set of isc proteins that likely already functioned in iron-sulfur cluster assembly. Subsequently the *hscA-hscB* gene pair and frataxin have been lost in *Xylella fastidiosa* and in the lineage leading *to Melorhizobium loti* and *Caulobacter crescentus* (figure 3). Furthermore, a large fraction of the set of isc genes have been transferred together with the frataxin gene to the nuclear genome of eukaryotes after the symbiosis of an α-proteobacterium with the predecessor of the Eukaryotes that led to the mitochondria. Note that one other gene has been invented at about the same time as *hscA, hscB* and frataxin/*cyaY*: *iscA* (through a gene duplication within the HesB family). Its subsequent evolutionary histories is however different. *IscA* has e.g. been lost from *Buchnera*, while *hscA*, *hscB* and frataxin/*cyaY* have been retained in this species (figure 3).

Molecular functions of the likely interaction partners of frataxin: HscA/Ssq1p and HscB/Jac1p

Based both on experimental evidence and on their phylogenetic distribution, the isc proteins can be divided into subsets. On the one hand IscS/Nfs1p, IscU/Isu1-2p, and IscA/Isa1-2p function in the assembly of iron-clusters themselves. IscS/Nfs1 are cysteine desulfurases [21]. IscU/Isu1-2p serve as scaffolds for Fe-S cluster biosynthesis [36]. IscA has been shown to interact with the holoform of ferredoxin [37], and conserved, essential cysteines in Isa1p hint at a role in iron-binding [23]. On the other hand, although HscA/Ssq1p and HscB/Jac1p have consistently been shown to be involved in the generation of Fe-S clusters in Bacteria and mitochondria [20, 26, 27], their exact molecular functions have not been elucidated. The homology of HscA/Ssq1p and HscB/Jac1p with the protein pair DnaK and DnaJ suggests that they function as chaperones, possibly facilitating the folding of Fe-S proteins. One substrate of the chaperone pair is IscU. Physical interaction between IscU and HscA/HscB has been observed in *E.coli* [38]. HscA/HscB could serve as specialized chaperones for IscU [35], but might also facilitate the transfer of Fe-S clusters formed in IscU to apoacceptor proteins [35]. While HscA is a full-length homolog of DnaK, HscB shares only the N-terminal, DnaK-binding domain with the classical DnaJ. In HscB, the zinc-finger middle domain and the C-terminal domain that in DnaJ are involved in substrate binding [39, 40] have been replaced with a three helix bundle coiled-coil like structure [34], of which the function is not yet clear.

Discussion

Comparative genome analysis provides a new means to predict protein function: genes that tend to be associated with each other on the genome, or that merely tend to occur with each other on genomes encode proteins that functionally interact. Here we have used these concepts to find interaction partners for frataxin, a protein whose function is unknown. In fifty-six available genomes we identified two genes with identical phylogenetic distributions to the frataxin/*cyaY* gene: *hscA* and *hscB*/*JAC1*. These genes have not only emerged in the same evolutionary lineage as the frataxin gene, they have also been lost at least twice with it, and they have been horizontally transferred with it in the evolution of the mitochondria. The proteins encoded by *hscA* and *hscB*, the chaperone HSP66 and the co-chaperone HSP20, have been shown to be required for the synthesis of 2Fe-2S clusters on ferredoxin in proteobacteria. *JAC1*, an ortholog of *hscB*, and *SSQ1*, a paralog of *hscA*, have been shown to be required for Fe-S cluster assembly in mitochondria of *Saccharomyces cerevisiae*. The identical phylogenetic distribution of the frataxin gene with *hscA/SSQ1* and *hscB/JAC1* suggest that frataxin plays a role in the same stage of the process of Fe-S cluster protein assembly as the HscA-HscB chaperone system, possibly as co-chaperone, or in protecting the sulfhydryl groups of Fe-S apoproteins. One possibility is that frataxin plays a role in the selection of the substrate. It could replace the substrate-selecting function of the middle and C-terminal domains of DnaJ, that are missing in HscB/Jac1p. The conservation of a string of negatively charged residues on the surface of the protein [8], supports the hypothesis of a role in peptide binding. Alternatively it could interact with HscA/Ssq1p. The paralogous displacement of HscA with Ssq1p in the evolution of Fe-S cluster assembly makes this interaction however less likely to be specific. Nevertheless, Ssq1p has been shown to be involved, together with Ssc1, in the maturation of the yeast frataxin ortholog, Yfh1p [41]. It should be noted that the types of functional interaction that correlate with the co-occurrence of genes in genomes include a large variation of functional interactions, and do not necessarily reflect physical interaction, but rather that the proteins are involved in the same (sub)process.

In any case, the strict co-occurrence of the frataxin gene with *hscA/SSQ1* and *hscB/JAC1* strongly supports a direct role of frataxin in Fe-S cluster protein assembly. The hypothesis that frataxin is directly involved in Fe-S cluster assembly and that the accumulation of iron in frataxin deficient cells is only a secondary effect is not new [13]. Recent evidence from yeast supports a direct role of frataxin in Fe-S cluster assembly [26], while in frataxin-deficient mouse cells the accumulation of iron is secondary to deficiency of Fe-S cluster proteins [42]. Based on the co-occurrence of genes in genomes and their encoded proteins having the same distribution in the eukaryotic cell, we can however be more specific in our predictions : Frataxin should function in conjunction with HscB/Jac1p.

Acknowledgements

This research was supported by BMBF. I thank Peer Bork, Toby Gibson and Berend Snel for their useful comments on the manuscript.

References

1. Huynen, M.A. and Snel, B. Gene and context: integrative approaches to genome analysis. Adv. Protein Chem. 2000;54;345-79.
2. Huynen, M.A. and Bork, P. Measuring genome evolution. Proc. Natl. Acad. Sci. USA. 1998;95:5849-56.
3. Pellegrini, M., Marcotte, E.M., Thompson, M.J., Eisenberg, D. and Yeates, T.O. Assigning protein functions by comparative genome analysis: protein phylogenetic profiles. Proc. Natl. Acad. Sci. USA 1999;96:42885-8.
4. Marcotte, E.M., Pellegrini, M., Ng, H.L., Rice, D.W., Yeates, T.O. and Eisenberg, D. Detecting protein function and protein-protein interactions from genome sequences. Science 2000;285:751-3.
5. Huynen, M.A., Snel, B., Lathe, W. 3rd and Bork, P. Predicting protein function by genomic context: quantitative evaluation and qualitative inferences. Genome Res. 2000;10:1204-10.
6. Aravind, L., Watanabe, H., Lipman, D.J. and Koonin, E.V. Lineage-specific loss and divergence of functionally linked genes in eukaryotes. Proc. Natl. Acad. Sci. USA 2000;97:11319-24.
7. Ettema, T., v.d. Oost, J. and Huynen, M.A. Modularity in the gain and loss of genes : Applications for function prediction. Trends Genet.2001;in press.
8. Musco, G., Stier, G., Kolmerer, B., Adinolfi, S., Martin, S., Frenkiel, T., Gibson T., and Pastore, A. Towards a structural understanding of Friedreich's ataxia : the solution structure of frataxin. Structure Fold. Des. 2000 ;257 :507-11.
9. Cho, S-J., Lee, M.G., Yang, J.K., Lee, J.-Y., Song, H.K. and Suh, S.W. Crystal structure of Escherichia coli CyaY protein reveals a previously unidentified fold for the evolutionarily conserved frataxin family. Proc. Natl. Acad. Sc. USA 2000;97:8932-7.
10. Campuzano, V., Montermini, L., Lutz, Y., Cova, L., Hindelang, C., Jiralersprong, S., Trottier, Y., Kish, S.J., Faucheux, B., Trouillas, P. et al. Frataxin is reduced in Friedreich ataxia patients and is associated with mitochondrial membranes. Hum. Mol. Genet.1997:6:1771-80.
11. Durr, A., Cossee, M., Agid, Y., Campuzano, V., Mignard, C., Penet, C., Mandel, J.L., Brice, A. and Koenig, M. Clinical and genetic abnormalities in patients with Friedreich's ataxia. N. Engl. J. Med. 1996;335:1169-75.
12. Puccio, H. and Koenig, M. Recent advances in the molecular pathogenesis of Friedreich's ataxia. Hum. Mol. Genet. 2000;9:887-92.
13. Foury F. Low iron concentrations and aconitase deficiency in a yeast frataxin homologue deficient strain. FEBS Lett., 1999;456:281-4.
14. Adamec, J., Rusnak, F., Owen, W.G., Naylor, S., Benson, L.M., Gacy, A.M. and Isaya, G. Iron-dependent self-assembly of the recombinant yeast frataxin: implications for Friedreich's ataxia. Am. J. Hum. Genet. 2000;67:549-62.
15. Gibson, T.J., Koonin, E.V., Musco, G., Pastore, A. and Bork, P. Friedreich's ataxia protein: phylogenetic evidence for mitochondrial dysfunction. Trends Neurosci. 1996;19:465-8.
16. Babcock, M., de Silva, D., Oaks, R., Davis-Kaplan, S., Jiralersprong, S., Montermini, L., Pandolfo, M. and Kaplan, J. Regulation of mitochondrial accumulation by Yfh1p, a putative homolog of frataxin. Science 1997;276:1709-12.
17. Foury, F. and Cazzalini, O. Deletion of the yeast homologue of the human gene associated with Friedreich's ataxia elicits iron accumulation in mitochondria FEBS Lett. 1997;411:373-7.
18. Wilson R.B. and Roof D.M. Respiratory deficiency due to loss of mitochondrial DNA in yeast lacking the frataxin homologue. Nat. Genet. 1997;16:352-7.
19. Zheng, L., Cash, V.L., Flint, D.H., Dean, D.R. Assembly of iron-sulfur clusters. Identification of an iscSUA-hscBA-fdx gene cluster from Azotobacter vinelandii. J Biol. Chem. 1998;273:13264-72.

20. Takahashi, Y. and Nakamura, M. Functional assignment of the ORF2-iscS-iscU-iscA-hscB-hscA-fdx-ORF3 gene cluster involved in the assembly of Fe-S clusters in Escherichia coli. J. Biochem. (Tokyo) 1999;126:917-26.
21. Schwartz, C.J., Djaman, O., Imlay, J.A. and Kiley, P.J. The cysteine desulfurase, IscS, has a major role in in vivo Fe-S cluster formation in Escherichia coli. Proc. Natl. Acad. Sci. USA 2000;97:9009-14.
22. Schilke, B., Voisine, C., Beinert, H. and Craig, E. Evidence for a conserved system for iron metabolism in the mitochondria of Saccharomyces cerevisiae. Proc. Natl. Acad. Sci. USA, 1999;96:10206-11.
23. Kaut, A., Lange, H., Diekert, K., Kispal, G. and Lill, R. Isa1p is a component of the mitochondrial machinery for maturation of cellular iron-sulfur proteins and requires conserved cysteine residues for function. J. Biol. Chem. 2000;275:15955-61.
24. Jensen, L.T. and Culotta, V.C. Role of Saccharomyces cerevisiae ISA1 and ISA2 in iron homeostatis. Mol. Cell. Biol. 2000;20:3918-27.
25. Lange, H., Kaut, A., Kispal, G. and Lill, R. A mitochondrial ferredoxin is essential for the biogenesis of cellular iron-sulfur proteins. Proc. Natl. Acad. Sci. USA 2000;97:1050-5.
26. Lutz, T., Westermann, B., Neupert, W. and Herrmann J.M. The mitochondrial proteins Ssq1 and Jac1 are required for the assembly of iron sulfur clusters in mitochondria. J. Mol. Biol. 2001;307:815-25.
27. Voisine, C., Cheng, Y.C., Ohlson, M., Schilke, B., Hoff, K., Beinert, H., Marszalek, J. and Craig, E. A. Jac1, a mitochondrial J-type chaperone, is involved in the biogenesis of Fe/S clusters in Saccharomyces cerevisiae. Proc. Natl. Acad. Sci. USA 2001;98:1483-8.
28. Kispal, G., Csere, P., Prohl, C. and Lill, R. The mitochondrial proteins Atm1p and Nfs1p are essential for biogenesis of cytosolic Fe/S proteins. EMBO J. 1999;18:3981-9.
29. Manzella, L., Barros, M.H. and Nobrega, F.G. ARH1 of Saccharomyces cerevisiae : a new essential gene that codes for a protein homologous to the human adrenodoxin reductase. Yeast 1998;14:839-46.
30. Lill, R. and Kispal, G. Maturation of cellular Fe-S proteins: an essential function of mitochondria. Trends Bioch. Sci. 2000;25:352-6.
31. Tong, W. H. and Roualt, T. Distinct iron-sulfur cluster assembly proteins exist in the cytosol and mitochondria of human cells. EMBO J. 2000;19:5692-700.
32. Nakai, K. and Horton, P. PSORT : a program for detecting sorting signals in proteins and predicting their subcellular localization. Trends Biochem. Sci. 1999;24:34-6.
33. Rudiger, S., Schneider-Mergener, J. and Bukau, B. Its substrate specificity characterizes the DnaJ co-chaperone as a scanning factor for the DnaK chaperone. EMBO J., 2001;20:1042-50.
34. Cupp-Vickery, J. R. and Vickery, L. E. Crystal structure of Hsc20, a J-type Co-chaperone from Escherichia coli. J. Mol. Biol. 2000;304:835-45.
35. Silberg J. J., Hoff K. G., Tapley T. L. and Vickery L. E. The Fe/S assembly protein IscU behaves as a substrate for the molecular chaperone Hsc66 from Escherichia coli. J. Biol. Chem. 2001;276:1696-700.
36. Agar, J. N., Krebs, C., Frazzon, J., Huynh, B. H., Dean, D. R. and Johnson, M. K. IscU as a scaffold for iron-sulfur cluster biosynthesis : sequential assembly if [2Fe-2S] and [4Fe-4S] clusters in IscU. Biochemistry 2000;27:7856-62.
37. Ollagnier-De-Choudens, S., Mattioli, T., Takahashi, Y. and Fontecave, M. Iron sulfur cluster assembly: characterization of IscA and evidence for a functional complex with ferredoxin. J. Biol. Chem. 2001;276:22604-7.
38. Hoff, K. G., Silberg, J. J. and Vickery, L. E. Interaction of the iron-sulfur cluster assembly protein IscU with the Hsc66/Hsc20 molecular chaperone system of Escherichia coli. Proc. Natl. Acad. Sci. USA, 2001;97:7790-5.

39. Banecki, B., Liberek, K., Wall, D., Wawrzynow, A., Georgopoulis, C., Bertoli, E., Tanafi, F. and Zylicz, M. Structure-function analysis of the zinc-finger region of the DnaJ molecular chaperone. J. Biol. Chem.1996; 271:14840-8.
40. Zsabo, A., Korzun, R., Hartl, F. U. and Flanagan, J. A zinc finger-like domain of the molecular chaperone DnaJ is involved in binding to denatured protein substrates. EMBO J., 1996;5:408-17.
41. Voisine, C., Schilke, B., Ohlson, M., Beinert, H., Marszalek, J. and Craig, E. A. Role of the Mitochondrial Hsp70s, Ssc1 and Ssq1, in the Maturation of Yfh1. Mol. Cel. Biol 2000;20:3677-84.
42. Puccio, H., Simon, D., Cossee, M., Criqui-Filipe, P., Tiziano, F., Meiki, J., Hindelang, C., Matyas, R., Rustin, P. and Koenig, M. Mouse models for Friedreich ataxia exhibit cardiomyopathy, sensory nerve defect and Fe-S enzyme deficiency followed by intramitochondrial iron deposits. Nat. Genet.2001;27:181-6.
43. Smith, T. and Waterman, M.S.Identification of common molecular subsequences. J. Mol. Biol.1981;147:195-7.
44. Altschul, S.F., Madden, T.L., Schaffer, A.A., Zhang, G., Zhang, Z., Miller, W. and Lipman, D.J. Gapped BLAST and PSI-BLAST : a new generation of protein database search programs. Nucl. Acids Res. 1997;25:3389-302.
45. Jeanmougin F., Thompson J. D., Gouy M., Higgins, D. G. and Gibson T. J. Multiple sequence alignment with ClustalX. Trends Biochem. Sci.1998;23:403-5.
46. Saitou, N. and Nei, M. The neighbor-joing method : a new method for reconstructing phylogenetic trees. Mol. Biol. Evol. 1987;4:406-25.
47. Snel, B., Bork, P. and Huynen, M.A. Genome phylogeny based on gene content. Nat. Genet. 1999;21:108-10.

ESF workshop Maastricht 2001: Session 2

Genomics and pathophysiology of atherosclerosis

5. IN VITRO-IN VIVO GENE EXPRESSION ANALYSIS IN ATHEROSCLEROSIS

A.J.G.Horrevoets, R.J.Dekker, R.D. Fontijn, S. van Soest and H. Pannekoek

Introduction

Atherosclerosis, the pathologic inflammatory response to injury of the human vessel wall, has been long recognized for its complexity of initiation, progression and ultimate appearance of clinical symptoms [1]. Many proteins and other compounds have been implicated in atherogenesis, and this list is now growing exponentially with the recent advances in high-throughput gene expression profiling [1,2,3,4,5,6,7,8,9]. Indeed, a plethora of individual genes show altered expression during atherosclerosis, but the development of intervention strategies based on such individual genes in animal models has been rather challenging. The translation into treatment of atherosclerosis in man has proven even more difficult. A clear gene-environment interaction, most notably Western-type diet and life-style, lies at the basis of disease development. This indicates that disturbed patterns of gene-expression rather than single culprit genes form the basis for the widespread penetrance of the disease in the elderly Western population. We are applying functional Genomics to the study of atherosclerosis, with the goal of characterizing healthy and diseased gene expression profiles. While our immediate objective is to characterize those genes that are differentially expressed during atherogenesis, our long-term goal is to determine how a healthy gene expression profile can be induced in the cells of the vascular wall. This implies not only to identify differentially expressed genes but also to determine their function and, most importantly, to analyze the integrated pathways and mechanisms through which their expression is regulated. In this report we will describe the use of differential display RT-PCR and cDNA microarray expression analysis to determine changes in gene expression profiles in cultured vascular endothelial cells in response to pro- and anti-atherogenic stimuli. We will briefly explore the computational analysis of such gene expression profiles as detected by a custom cardiovascular microarray. Finally, we show that insights that were gained in vitro can be extended to the in vivo (atherosclerotic) vascular wall.

Results

Cytokine-stimulated endothelial cell gene expression
Inflammation is one of the basic initiating events in atherosclerosis [1]. Crucial first steps in atherogenesis are the infiltration of the vessel wall by monocytes, the differentiation of the invading monocytes into macrophages, and the eventual conversion of macrophages into lipid-laden foam cells. We sought to determine the impact of the initial invasion of the vessel wall on the expression of genes in endothelial cells. We activated cultured endothelial cells from the human umbilical vein (HUVEC) with either macrophages supernatant containing a variety of cytokines or with purified tumor necrosis factor (TNF-alpha), a dominant secretion product of activated monocytes. The subsequently performed extensive differential display RT-PCR analysis, using 144 different primer combinations enabling the visualization of > 10,000 mRNA species in 10 samples, resulted in the identification of 106 genes that were differentially expressed in response to cytokine stimulation [2]. Several novel genes were identified and cloned [2,10,11] and in vivo relevance was shown by mRNA in situ hybridization on human vascular tissue [2]. The gene products stimulated by cytokines were involved in virtually all aspects of endothelial cell behavior, including leukocyte trafficking (cellular interactions), proliferation and apoptosis, protection against oxidative damage, signaling, the cytoskeleton and even vesicular transport, as exemplified in Table 1 [2,10,11].
The main new insight that we learned from this unbiased inventory of changes in endothelial cell gene expression profile was that even a single stimulus has an immense influence on the mRNA repertoire, covering several percent of the genes in the human genome [2,9]. Furthermore, the majority of the genes that we identified are presently without established in vivo function. Both pro- and anti inflammatory (protective) genes were induced, which precludes the direct identification of single suitable targets for future intervention studies.

Shear stress-stimulated endothelial cell gene expression

The risk factors for atherosclerosis are of a systemic nature, being high plasma cholesterol levels, age, smoking and metabolic disorders like diabetes mellitus and hyperhomocysteinemia. It is therefore remarkable that the initial localization of lesions is confined to specific and reproducible positions in the arterial tree. The hypothesis that this focality of atherosclerosis is caused by local blood flow turbulence at vessel bifurcations and curvatures has been firmly supported over the last decades [5,12,13]. Lowering of the shear stress on the vascular endothelium at sites of flow turbulence, creating steep shear stress gradients is now believed to be one of the initiating factors in atherogenesis [13]. Consequently, the ability of the atheroprotective force of shear stress to modulate the expression of endothelial genes has triggered ample research effort on the identification of shear-regulated genes. Figure 1 shows a direct comparison of the expression levels of a large part of the human genome in human cultured endothelial cells grown under either static or shear-stressed conditions.

Gene	*GB*	*Affected process*
MCP-1	X14768	cellular interactions
GM-CSF	M11220	cellular interactions
IL-8	M28130	cellular interactions
RANTES	M21121	cellular interactions
PCTA-1	L78132	cellular interactions
A20	M59465	apoptosis
CA2-1*= IAP-1	AF070674	apoptosis
AG8-2*= stannin	AF070673	apoptosis
GG2-1*= SCC-S2	AF070671	apoptosis
BTG-1	H70177	proliferation
GSPT1	X17644	proliferation
GG10-2*= RB6K	AF070672	proliferation
RGS-5	AB008109	proliferation
Mn-SOD	X07834	anti-oxidation
ferritin heavy chain	M97164	anti-oxidation
GBP-2	M55542	signaling
SUPT4H	U43923	transcription
fibrillarin	X56597	transcription
TRIP7	L40357	transcription
myosin light chain	M22919	cell shape
GRAVIN	AB003476	cell shape
annexin-XI	L19605	vesicular transport
RAC1	M29970	vesicular transport
COX-2	D28235	prostaglandin synthesis.
RP- S11	X06617	translation
CG12-1*=apoL3	AF070675	lipid metabolism
36 ESTs	-	?

Table 1 Cytokine-induced changes in endothelial cell gene expression as revealed by differential display RT-PCR. Cumulative data on affected genes, the corresponding Genbank (GB) accession number and affected processes of cytokine-stimulated endothelial cells. These data have been taken from Horrevoets et al.[1] and were up-dated by consulting a recent version of Unigene (Build 116, http:www.ncbi.nlm.nih.gov/UniGene). *Full length cloned [2,10,11]

A striking outcome of such a micro-array experiment is that many genes show an altered level of expression, with an over two-fold induction or repression for 230 genes amounting to ~5.7% of the human genome [18]. Thus, like with cytokine stimulation, hundreds of genes with quite diverse biochemical functions are modulated by a single stimulus [2,5,18]. A comparison of these datasets revealed that although several genes were specific for one or the other stimulus, there was a large overlap of genes that were induced by either stimulus, suggesting that most genes are more general stress-responses [18]. A further analysis of all these genes required an integrated high-throughput approach as explained in the next section.

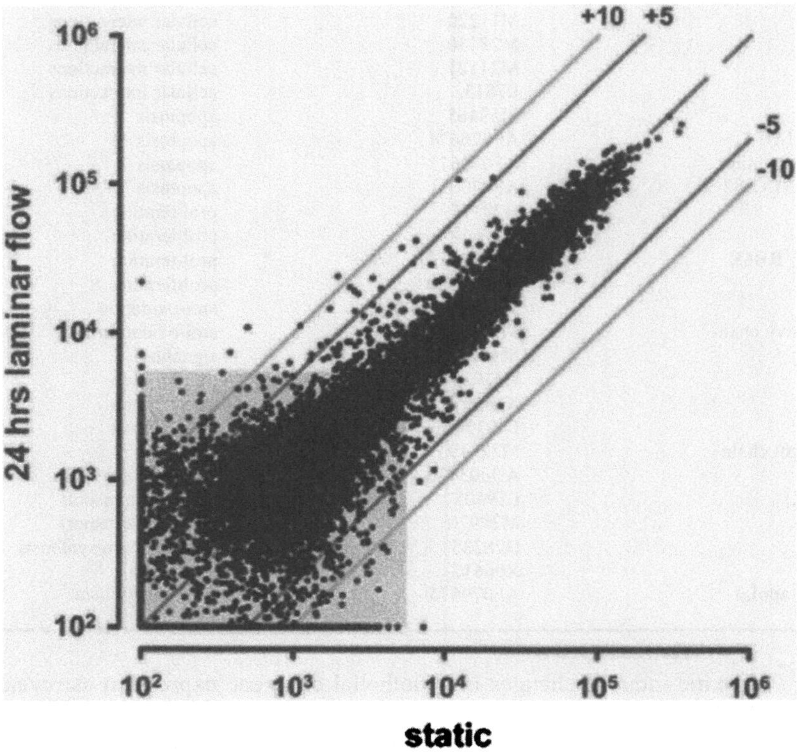

Figure 1 *Comparison of gene expression profiles of static versus shear stressed cultures of vascular endothelial cells as probed by a micro-array containing 18,000 human cDNAs. The lines represent a 5- or 10-fold up or down regulation, respectively.*

Integration and analysis of expression data by micro-array and clustering algorithms

A custom glass-based DNA micro-array was constructed, composed of 320 genes (in triplicate) that were identified in our studies so far, based on their modulation of expression by either pro- or anti-atherogenic stimuli [9,18]. These arrays were used to study the effect of a set of established modulators of endothelial gene expression, including shear stress, nitric oxide (NO), the cytokines/growth factors TNF-α, interleukin-1β (IL-1β), transforming growth factor-β (TGF-β), vascular endothelial growth factor (VEGF), as well as thrombin, on the expression of the selected set of genes. To that end, HUVEC were cultured in the continuous presence of these agents for 2, 6 and 24 hours to allow identification of early and late induced genes, whereas non-stimulated controls were taken at 0 and 24 hours. Other human cell-lines like HL-60 (myelomonocytic) and HeLa (cervical carcinoma) were included in the analysis, to provide additional information on cell-type specific expression [18]. Data were acquired and analyzed by two-way hierchical clustering analysis, allowing the clustering or grouping of both genes (y-axis) and stimuli (x-axis) based on similarity of expression profiles [14,15,16]. An example of two of such clusters is shown in figure 2.

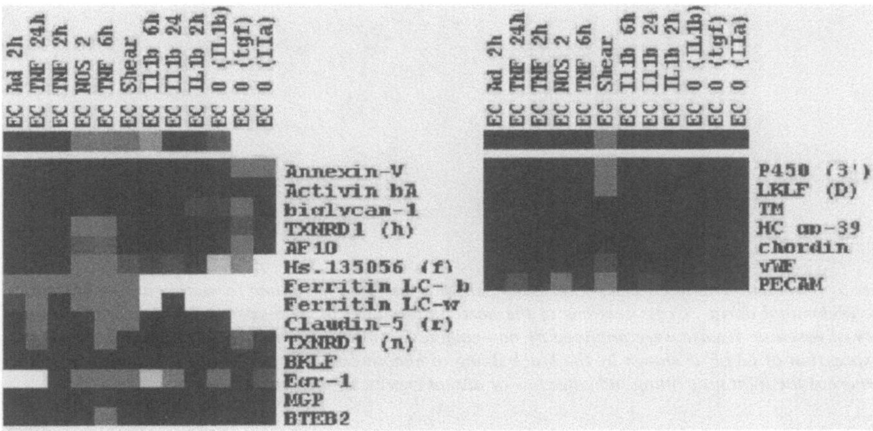

Figure 2 *Application of two-way hierarchical clustering to a customized cardiovascular micro-array. This algorithm allows genes to be arranged in different panels according to their expression profiles. Red (dark grey) indicates high expression, green (light grey) indicates decreased expression.*

The genes expressed in the left panel were induced by shear stress as well as by other inducers like TNF-alpha. The common theme linking all of the genes grouped here is that they are general stress response genes and protect cells against damage. None of these genes are specific for endothelial cells. In contrast, the genes expressed in the right panel were specifically induced by shear stress and not by any other stimulus, indicating they are not general stress-response genes. As most of these genes are considered to be predominantly specific for endothelial cells, a correlation between expression in response to shear stress and endothelial cell specificity is suggested. The striking outcome of these experiments was that of the >200 genes that were modulated by shear stress, <5 were specific for this stimulus, with the transcription factor Lung Krueppel like factor (LKLF) as most specific example.

In situ hybridization reveals in vivo gene expression profiles

The endothelial flow-specific expression of LKLF in vitro prompted the study of its expression pattern in human vascular tissue specimens, in relation to diseased versus healthy vessels and their described flow patterns [19], an example of which is shown in Figure 3.

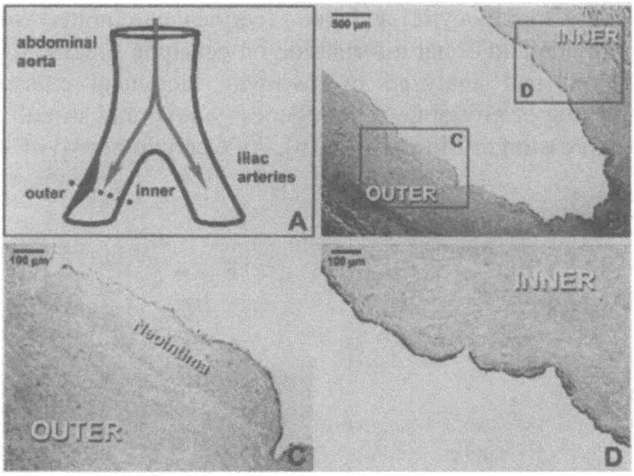

Figure 3 *Correlation between LKLF expression, shear stress, and neointima formation at the bifurcation of the abdominal aorta. Cross sections of the aortic bifurcation of a 68-year old female with no prior history of vascular disease were analyzed by non-radioactive mRNA in situ hybridization. Positive signal for expression of LKLF is shown by the black lining of endothelial cells covering the vessel wall (panel D), whereas the light grey lining indicates low or absent expression (panel C).*

At the inner wall of the bifurcation, where shear stress is high, LKLF expression is also high and the intima is normal (neointima is absent). In contrast, at the outer bifurcation, where shear stress is low, LKLF expression is almost absent and a pronounced neointima is apparent. These experiments illustrate the positive correlation between the in vitro deduced specificity of LKLF gene expression in relation to shear stress and its in vivo gene expression as determined by blood flow and vessel architecture. Gene disruption of this gene in embryonic mice has indeed shown that the function of this endothelial gene is essential for the formation of a normal, quiescent tunica media [17]. We are currently attempting to elucidate the role of the transcription factor LKLF in the induction and regression of atherosclerosis.

The enormous amount of expression data that emerges from gene expression profiling directly on diseased human tissue is both overwhelming and puzzling. Genes that are induced in diseased versus normal tissue are frequently non-specific and often represent a general protective stress response rather than being causative to the disease. Thus, a large-scale micro-array analysis identified the prominent induction of genes encoding ribosomal proteins in cancerous tissue, which are naturally not causative to the disease and hardly constitute suitable targets for future intervention [20]. Similarly, gene expression in tumor endothelium turned out to be very reminiscent of profiles found during natural processes like wound healing and corpus luteum formation [21]. The extensive analysis of gene expression during several types of inflammatory diseases in comparison to gene expression in control unaffected tissues will ultimately yield information on the specific expression of groups of genes that are causally and uniquely related to the process of atherosclerosis. We have set out to dissect out the already enormous and complex impact of isolated model stimuli on isolated model cells [9]. The global patterns of gene expression provide information about the relationships between different stimuli or disease states and allow cell-specific responses to be distinguished from more general cellular responses to stress. Thus, we are using high throughput gene expression profiling in combination with mathematical algorithms to identify clusters of genes and their relationships [14,15,16]. Examination of individual clusters of genes permits the identification of genes that are specifically induced by a particular stimulus and not by other stimuli. Basically, a gene expression cluster is no more then a coordinately expressed set of genes, not necessarily of closely related biochemical function. It is assumed that single key transcription factors govern the coordinate expression of such clusters [22]. As transcription factors lie at the basis of altered steady state mRNA levels, they are the most direct targets for modulation of deregulated transcriptional programs. The transcription factor family of proteins has vastly expanded in vertebrates as evidenced by the now completed sequence of the human genome [23,24]. Most family members have no direct counterpart in lower life forms used as "model systems" like C.elegans and S. cerevisiae, and therefore still remain largely unexplored [24]. Comparing the genomes of different species allows the identification of common regulatory motifs that often remained relatively unaffected during evolution [25]. Comparison of promoter sequences with expression profiles allows the detection of common sequence motifs and their corresponding transcription factors [22,26]. Intervention strategies aimed at derived pathways of gene-expression in cultured human cells and mouse models for atherosclerosis can next be performed using DNA-decoys derived from such analyses. Such decoys can be used both to analyze and intervene in disturbed patterns of gene expression. The use of decoys has been shown to be a powerful approach. For example the use of an NF-κB decoy was shown to inhibit CAD after cardiac transplantation in rodents [27]. The final identification of the key transcription factors will then allow the direct intervention in gene expression, preferably by small molecules. One example of the latter approach, which is already in clinical use, is the class of antihyperlipidemic drugs known as fibrates, which lower plasma triglyceride levels by targeting the nuclear receptor (transcription factor) Peroxisome Proliferator-Activated Receptor (PPAR-alpha).

Conclusion

Hundreds of genes are differentially expressed in atherosclerosis. However, the challenge is to analyze, integrate, and reduce this information to a manageable number of parameters that can be potentially influenced by therapeutic interventions. High-throughput technologies, the sequencing of the human and mouse genomes, and advanced bioinformatics techniques now permit the identification of genes that are transcribed in coordinated fashion to produce a healthy or diseased gene expression profile. The ultimate goal will be to target the relevant transcription factors and thereby cause the diseased tissue to revert to a healthy pattern of gene expression.

Acknowledgements

This work has been supported by grants from the Netherlands Heart Foundation: NHS M93.007 (Molecular Cardiology Program), NHS 96.094, NHS 97.209 and NHS 2000.144

References

1. Ross R. Atherosclerosis-an inflammatory disease. N. Engl. J. Med. 1999;340:115-26.
2. Horrevoets AJG, Fontijn RD, van Zonneveld A, de Vries CJM, ten Cate JW, and Pannekoek H. Vascular endothelial genes that are responsive to tumor necrosis factor-alpha in vitro are expressed in atherosclerotic lesions, including Inhibitor of Apoptosis Protein-1, stannin and two novel genes. BLOOD 1999;93:3418-31.
3. de Vries CJM, van Achterberg TAE, Horrevoets AJG, ten Cate JW, and Pannekoek H. Differential display identification of 40 genes with altered expression in activated human smooth muscle cells: local expression in atherosclerotic lesions of smags, smooth muscle activation-specific genes. J. Biol. Chem. 2000;275:23939-47.
4. Faber BC, Cleutjens KB, Niessen RL, Aarts PL, Boon W, Greenberg AS, Kitslaar PJ, Tordoir JH, and Daemen MJ. Identification of genes potentially involved in rupture of human atherosclerotic plaques. Circ Res. 2001;89:547-54.
5. Garcia-Cardena G, Comander J, Anderson KR, Blackman BR, Gimbrone MA Jr. Biomechanical activation of vascular endothelium as a determinant of its functional phenotype. Proc Natl Acad Sci USA 2001;98:4478-85.
6. Adams LD, Geary RL, McManus B, Schwartz SM. A comparison of aorta and vena cava medial message expression by cDNA array analysis identifies a set of 68 consistently differentially expressed genes, all in aortic media. Circ Res 2000;87:623-31.
7. Shiffman D, Mikita T, Tai JT, Wade DP, Porter JG, Seilhamer JJ, Somogyi R, Liang S, Lawn RM.
8. Large-scale gene expression analysis of cholesterol-loaded macrophages. J Biol Chem. 2000;275:37324-32.
9. Hashimoto S, Suzuki T, Dong HY, Yamazaki N, Matsushima K. Serial analysis of gene expression in human monocytes and macrophages. Blood 1999;94:837-44.
10. van Soest S, Horrevoets AJG, Beauchamp NJ, and Pannekoek H. Current technologies in gene expression profiling: applications to cardiovascular research. Fibrinolysis and Proteolysis 2000;14:73-81.
11. Fontijn RD, Goud B, Echard A, Jollivet F, van Marle J, Pannekoek H, and Horrevoets AJG. The human kinesin-like protein RB6K is under tight cell cycle control and is essential for cytokinesis. *Mol. Cell. Biol.* 2001;21:2944-55.
12. Monajemi H, Fontijn RD, Pannekoek H, and Horrevoets AJG. The Apolipoprotein L gene cluster has emerged recently in evolution and is expressed in human vascular tissue. Submitted.
13. Asakura T, Karino, T. Flow patterns and spatial distribution of atherosclerotic lesions in human coronary arteries Circ Res 1990;66:1045-66.
14. Hajra L, Evans AI, Chen M, Hyduk SJ, Collins T, and Cybulsky MI. The NFkB signal transduction pathway in aortic endothelial cells is primed for activation in regions predisposed to atherosclerotic lesion formation. Proc. Natl. Acad. Sci. USA 2000;97:9052-7.
15. Shalon D, Smith SJ, Brown PO. A DNA microarray system for analyzing complex DNA samples using two-color fluorescent probe hybridization. Genome Res. 1996;6:639-45.
16. Eisen MB, Spellman PT, Brown PO, Botstein D. Cluster analysis and display of genome-wide expression patterns. Proc Natl Acad Sci USA. 1998;95:14863-8.
17. Tamayo P, Slonim D, Mesitow J, Zhu Q, Kitareewan S, Dmitrovsky E, Lander ES, and Golub TR. Interpreting patterns of gene expression with self-organizing maps: Methods and application to hematopoietic differentiation. Proc. Natl. Acad. Sci. 1999;96:2907-12.
18. Kuo CT, Veselits ML, Barton KP, et al. The LKLF transcription factor is required for normal tunica media formation and blood vessel stabilization during murine embryogenesis. Genes Dev. 1997;11:2996-3006.

19. Dekker RJ, van Soest S, Pannekoek H, and Horrevoets AJG. A micro-array analysis of fluid shear-stress modulated genes in endothelial cells discriminates NFkB-dependent and independent pathways. Submitted.
20. Dekker RJ, Pannekoek H, and Horrevoets AJG. Lung Kruppel-like Factor is specifically upregulated by fluid shear-stress in endothelial cells in vitro and absent from atherosclerotic lesions in vivo. Submitted.
21. Alon U, Barkai N, Notterman DA, Gish K, Ybarra S, Mack D, and Levine AJ. Broad patterns of gene expression revealed by clustering analysis of tumor and normal colon tissues probed by oligonucleotide arrays. Proc. Natl. Acad. Sci. USA. 1999;96:6745-50.
22. St. Croix B, Rago C, Velculescu V, et al. Genes expressed in human tumor endothelium. Science 2000;289:1197-202.
23. Zhu G, Spellman PT, Volpe T, Brown PO, Botstein D, Davis TN, and Futcher B. Two yeast forkhead genes regulate the cell cycle and pseudohyphal growth. Nature 2000;406:90-4.
24. Lander ES, et al (International human genome sequencing consortium). Initial sequencing and analysis of the human genome. Nature 2001;409:860-921
25. Venter JC. et al. The sequence of the human genome. Science 2001;291:1304-51.
26. Rubin, E.M., and Tall, A. Perspectives for vascular genomics.. *Nature* 407 (2000), 265-9
27. Bussemaker HJ, Li H, Siggia ED. Regulatory element detection using correlation with expression. Nature Genetics 27(2001), 167-71
28. Feeley BT, Miniati DN, Park AK, Hoyt EG, and Robbins RC. Nuclear factor-kappaB transcription factor decoy treatment inhibits graft coronary artery disease after cardiac transplantation in rodents. Transplantation. 2000;70:1560-8.

ESF workshop Maastricht 2001: Session 3

Genomics in hypertension

6. HUMAN ESSENTIAL HYPERTENSION: ROLE OF THE GENES OF THE RENIN-ALDOSTERONE SYSTEM.

X. Jeunemaitre, S. Disse-Nicodème, A. Gimenez-Roqueplo, P. Corvol.

Introduction

A number of epidemiologic studies have shown that individual blood pressure levels result from both genetic predisposition and environmental factors. It is generally accepted that approximately 30% of the variance of blood pressure is attributable to genetic heritability and 50% to environmental influences [1].

With the notable exception of few autosomal forms of hypertension for which the molecular basis has been recently elucidated, there is no indication on the number of genetic loci involved in the regulation of blood pressure, the frequency of deleterious alleles, their mode of transmission, and the quantitative effect of any single allele on blood pressure. The unimodal distribution of blood pressure within each age group and in each sex strongly suggests, but does not definitively prove, that several loci are involved. Because of the likely etiologic heterogeneity of the disease, it is difficult to expect that a single biochemical or DNA genetic marker will help the clinician in the management of most hypertensive patients. However, genetic markers alone and in combination will probably be useful indicators for elucidating the various genetic loci linked to high blood pressure through intermediate phenotypes, and for predicting the sensitivity to antihypertensive agents. This review deals with the possible involvement of the genes of the renin aldosterone system in human essential hypertension.

The genes of the renin angiotensin aldosterone system are a good illustration of a "candidate gene" approach since this system is well known to be involved in the control of blood pressure and in the pathogenesis of several forms of experimental and human hypertension. This system consists of four main proteins: renin, angiotensinogen, angiotensin I-converting enzyme (ACE) and angiotensin II receptors. During the past years, considerable progress has been achieved since all these genes have been cloned in humans, and informative genetic markers identified. Since very complete reviews have already been published on that matter from our group [2] and others [3-5], we will only summarize the main findings obtained that concern essential hypertension and/or blood pressure regulation.

The renin gene

Activation of the renin system depends on the renin angiotensinogen reaction which is the first and rate-limiting step leading to angiotensin II production. Quite interesting was the pioneering observation by Rapp *et al.* [6] of a cosegregation between a renin gene polymorphism and blood pressure level in a F2 population generated from crosses between inbred salt sensitive and salt resistant Dahl rats. In humans, renin levels seem at least partly heritable, as observed by Grim and colleagues in twins submitted to well standardized conditions of posture and sodium diet [7].

The human renin gene is located on the short arm of chromosome 1 (1q32-1q42). Several restriction fragment length polymorphisms (RFLPs) have been located throughout the renin gene. Soubrier *et al.* [8] reported a study comparing the frequency of renin RFLPs in a large and contrasted population of 120 normotensive and 120 rigorously selected hypertensive subjects. Renin gene allele and haplotype frequencies were similar in the hypertensive and the normotensive groups.

To explore further the potential role of the renin gene as a genetic determinant of hypertension, we used the hypertensive sib pairs approach [9]. No linkage was found between the renin gene and hypertension, suggesting again that the renin gene does not have a frequent and/or important role in the pathogenesis of essential hypertension. However, the definitive exclusion of a contribution of the renin gene in the heritability of essential hypertension would require more powerful linkage studies, such as the use of a reliable renin intermediate phenotype, and a more polymorphic marker of the renin locus. It is interesting to observe that some of the genome scans conducted since then, have found the chromosome 1q32 region as a possible locus for hypertension.

The angiotensin I-converting enzyme (ACE)

Angiotensin I-converting enzyme (ACE) is a zinc metalloprotease whose main functions are to convert angiotensin I into angiotensin II, and to inactivate bradykinin. It is assumed that this step of the renin angiotensin system is not limiting in plasma, but the local generation of angiotensin I and the degradation of a bradykinin might depend on the level of ACE expressed in tissues.

Relationship between plasma ACE levels and genotype.

From a geneticist point of view, plasma ACE concentration is an interesting marker as it varies markedly between individuals, a variance which is due, in large part, to a major genetic effect [10]. The most studied ACE gene polymorphism is an insertion/deletion (I/D) of a 287 base pair DNA fragment in the intron 16 of the gene, corresponding to an Alu sequence [11]. In the original study by Rigat *et al.* [11], the ACE I/D polymorphism accounted for 47% of the total variance of serum ACE. Another study combining segregation and linkage analysis in 98 healthy nuclear families showed that the ACE I/D polymorphism is only a neutral marker in strong linkage disequilibrium with the putative functional variant [12]. Recently, Rieder and colleagues performed the complete genomic sequence of the ACE gene from 11 individuals, representing the longest contiguous scan (24 kb) for sequence variation in human DNA [13]. They identified 78 varying sites in 22 chromosomes that resolved into 13 distinct haplotypes, 17 being in absolute linkage disequilibrium with the ACE I/D polymorphism, producing two distinct and distantly related clades. The authors suggested that the causal variant should be located within the 3' part of the gene. Despite these efforts, the causative variant(s) responsible for the increase in ACE has(ve) yet to be found, which might foreshadow the difficulty of identifying causal molecular variants in complex traits.

ACE gene polymorphism and hypertension.

The observation that plasma ACE levels are under the direct control of an ACE gene variant made the ACE I/D polymorphism one of the most popular markers tested in cardiovascular diseases. The first genome scan performed in a F2 rat population generated from stroke-prone spontaneously hypertensive rats (SHR/SP) and normotensive (WKY) in genetically hypertensive rats rendered this hypothesis even more attractive: two groups of investigators found a significant linkage between NaCl-increased blood pressure and a gene locus on rat chromosome 10 [14,15].

Several association studies in different clinical settings have been reported and suggest that the ACE gene does not play a major role on blood pressure variance in these populations (X; Jeunemaitre *et al.*, Genetics of Human Hypertension, in Comprehensive Cardiovascular Medicine, ed. E..J. Topol, in press).

Some positive results suggest, however, that the ACE locus might influence blood pressure variability in a sex-specific manner. In a logistic regression analysis of 3095 participants in the Framingham Heart Study [16], the adjusted odds ratios for hypertension among men for the DD and DI versus II genotypes were 1.59 and 1.18, respectively, whereas no effect was observed in women. Positive results were also reported by Fornage *et al.* [17] in the analysis of a large population-based sample of 1488 siblings having a mean age of 15 years and belonging to the youngest generation of 583 randomly ascertained three-generation pedigrees from Rochester, Minnesota. In sex-specific analyses, genetic variation in the region of the ACE gene explained as much as 35% of the interindividual blood pressure variation, again in males but not in females. Finally, Julier *et al.* [18] conducted an affected sib-pair analysis in French and U.K. families and explored the region of chromosome 17q 23-32, based on the location of the ACE locus and on the QTL observed in rat on the homologous region of chromosome 10. Significant evidence of linkage was found near two closely linked microsatellite markers, D17S183 and D17S934, that reside 18 cM proximal to the ACE locus in the homology region.

Taken together, these results suggest that the ACE gene does not play a crucial role on blood pressure variance in the overall population. It may influence blood pressure, however, in a sex-specific manner and in combination with other polymorphisms [19].

The angiotensinogen gene.

Angiotensinogen (AGT) , the renin substrate, is mainly synthesized by the liver and is the unique substrate for renin. In humans, AGT plasma concentration is within a range where variations of its concentration directly affects the angiotensin I production rate.

Angiotensinogen and essential hypertension

The human angiotensinogen gene belongs to the superfamily of serpins and is localized to chromosome 1q42.3, in the same region as human renin. With the group of JM Lalouel, we reported the first molecular elements suggesting a role of the AGT gene in human essential hypertension [20]. An extensive study was performed in two large series of hypertensive sibships yielding a total of 379 sib pairs (Salt Lake City, Utah, USA and Paris, France) and using a highly polymorphic microsatellite marker at the AGT locus. An excess of AGT allele sharing was found mainly in severely hypertensive sibpairs and in men. Other linkage studies have since been reported with controversial results. Caulfield et al. showed a strong linkage and an association of the AGT gene locus to essential hypertension in a set of British families [21,22]. In contrast, no evidence for linkage was found in a large European study involving 630 affected sibling pairs, either in the whole panel or in family subsets selected for severity or early onset of disease [23]. Linkage of the AGT gene to essential hypertension was also found in 63 affected African Caribbean sibling pairs [24]. Similarly, positive albeit modest significant excess of AGT allele sharing, was found in 46 extended Mexican American [25]. No linkage was found between the AGT locus and hypertension in 310 hypertensive Chinese sibling pairs [26]. Data must be analyzed in the context of the ethnicity as there is a strong effect of race on the allele frequency of most of the polymorphisms found at the AGT locus. For example, the 235T allele frequency varies from 40% in Caucasians to 80% in the Asian and African-American population and even 93% in Nigerians [27].

The association between the M235T polymorphism and essential hypertension has been tested in a large number of case-control studies that we reviewed recently [28]. A meta-analysis of case-control studies representing 5493 Caucasian patients showed that the 235T allele was significantly but mildly associated with hypertension (OR: 1.20; 95% [CI]: 1.11 to 1.29; P<.0001), association which increased in studies with positive family history (OR: 1.42; 95% CI: 1.25 to 1.61, P<.0001) [29]. Another meta-analysis that included 10720 whites showed a 32% increase in the risk of elevated blood pressure associated with the 235T allele [30]. More recently, Sethi et al. studied 9100 men and women from the Danish general population [31]. On multifactorial logistic regression analysis, women homozygous for the 235T allele versus non carriers had an odds ratio for elevated blood pressure of 1.29 which increased to 1.50 if they were also homozygous for the T174 allele. No significant association was found in men. Altogether, these results probably highlight the modest effect of the AGT locus in the overall population and the difficulty of identifying susceptibility genes by linkage analysis in complex diseases.

Angiotensinogen gene and plasma angiotensinogen
Among the 15 polymorphisms initially identified [20], two of them leading to aminoacid changes, 174M and M235T, were found to be associated with hypertension and to plasma AGT concentration. This association between plasma angiotensinogen level and the M235T genotype was further confirmed in white children [32], in the Monica Augsburg cohort [33] and in the Danish general population [31]. Such a relation has also been found in African American young individuals [34]. Because of the intra and interassay variability of the plasma angiotensinogen measurement and the mild association with the M235T polymorphism, a large number of individuals is required to detect this relation.
Confirmation of the impact of an increased expression of angiotensinogen on blood pressure was obtained by Smithies and colleagues [35]: mice bearing one, two, three or four functional angiotensinogen copies displayed a gene-dosage effect on plasma angiotensinogen concentration and blood pressure level, paralleling the M235T variant effect observed in humans. The functional variant may be located in the promoter region, in strong linkage disequilibrium with the M235T variant. This promoter variant may be responsible for an increase in AGT gene transcription [36].

Angiotensinogen and pregnancy-induced hypertension
One of the possibility is that molecular variants of the AGT might represent also a susceptible allele for pregnancy-induced hypertension, for which a familial tendency has been documented. Two reports have suggested that the angiotensinogen locus could play an important role in the occurrence of pregnancy-induced hypertension [37,38]. However, other studies found no indication of association or linkage between pre-eclampsia and the AGT gene. Some of the contradictory results of the case-control studies could be explained by the fact that the angiotensinogen gene predisposes to only a given subset of patients.

The angiotensin II type 1 receptor.

Angiotensin II receptors which mediate all the biological and physiological effects of the renin angiotensin system are also candidate genes for essential hypertension. The AT_1 subtype is a G-coupled receptor inserted into the plasma membrane of angiotensin II target cells, vascular smooth muscle cells, renal vasculature and mesangial cells, adrenal and brain. The human gene has been cloned and is located on chromosome 3q21-3q25. There was no evidence of linkage between a microsatellite marker of the AT_1 receptor gene and hypertension in a hypertensive sib pair study but an informative diallelic marker A1166C present in the 3' untranslated region of the AT1R gene was found significantly more frequently in 206 hypertensive subjects than in 298 normotensive control population, suggesting that a variant of AT_1 receptor exerts a small effect on blood pressure [39]. Other polymorphisms have been described, especially at the promoter region of the AT1R gene. None of these newly characterized polymorphisms has shown evidence for association with hypertension in a large Caucasian population-based sample [40]. However, recent studies performed in Finland are in favour of linkage between hypertension and the AT1R locus and of an association with the A1166C polymorphism [41,42]. In 218 Caucasian hypertensive patients, we found a positive relationship between the acute blood pressure response to angiotensin II and the AT1R locus [43].

Aldosterone synthase (CYP11B2)

The CYP11B2 gene coding for the aldosterone synthase gene is also an attractive candidate gene. It has been tested for its possible association with essential hypertension [44-46] and primary aldosteronism [47]. A polymorphism in the promoter region, C-344T, has been associated with variations in plasma aldosterone [44,48,49], suggesting that it could favour sodium retention and high blood pressure. Contradictory results have been obtained according to the populations and the clinical and biochemical parameters studied [49-53].

Conclusion

It is important to note that most of the studies performed so far on candidate genes, such as those of the renin aldosterone system, have been conducted in very simple designs, by testing one by one the role of each candidate. However, it is recognized that several genes control blood pressure level in an additive or interactive manner, together with environmental factors. A more integrated approach needs to be designed in which several genes and some environmental factors could be tested with or without *a priori* specific hypothesis [54]. In that regard, the recent study performed by Staessen and colleagues [19] showed the combined effect of the ACE I/D, the alpha-adducin Gly460Trp and the aldosterone synthase C-344T polymorphisms on blood pressure level.

Acknowledgments

This work was supported by Grants form INSERM, Collège de France, Bristol-Myers Squibb, Association Claude Bernard and Association Naturalia and Biologia

References

1. Ward R. Familial aggregation and genetic epidemiology of blood pressure. In "Hypertension: Pathophysiology, Diagnosis and Management". *Laragh JH and Brenner BM, Eds, Raven Press, Ltd, New York* 1990:81-100.
2. Corvol P, Soubrier F, Jeunemaitre X. Molecular genetics of the renin-angiotensin-aldosterone system in human hypertension. Pathol Biol (Paris*)* 1997;45:229-39.
3. Rieder MJ, Nickerson DA. Hypertension and single nucleotide polymorphisms. Curr Hypertens Rep 2000;2:44-9.
4. Danser AH, Schunkert H. Renin-angiotensin system gene polymorphisms: potential mechanisms for their association with cardiovascular diseases. Eur J Pharmacol 2000;410:303-16.
5. Wang JG, Staessen JA. Genetic polymorphisms in the renin-angiotensin system: relevance for susceptibility to cardiovascular disease. Eur J Pharmacol 2000;410:289-302.
6. Rapp JP, Wang SM, Dene H. A genetic polymorphism in the renin gene of Dahl rats cosegregates with blood pressure. Science 1989;243:542-4.
7. Grim CE, Luft FC, Miller JZ, et al. An approach to the evaluation of genetic influences on factors that regulate arterial blood pressure in man. Hypertension 1980;2[suppl2]:I-34-42.
8. Soubrier F, Jeunemaitre X, Rigat B, et al. Similar frequencies of renin gene restriction fragment length polymorphisms in hypertensive and normotensive subjects. Hypertension 1990;16:712-7.
9. Jeunemaitre X, Rigat B, Charru A, et al. Sib pair linkage analysis of renin gene haplotypes in human essential hypertension. Hum Genet 1992;88:301-6.
10. Cambien F, Alhenc-Gelas F, Herbeth B, et al. Familial resemblance of plasma angiotensin-converting enzyme level: the Nancy Study. Am J Hum Genet 1988;43:774-80.
11. Rigat B, Hubert C, Alhenc-Gelas F, et al. An insertion/deletion polymorphism in the angiotensin I-converting enzyme gene accounting for half the variance of serum enzyme levels. J Clin Invest 1990;86:1343-6.
12. Tiret L, Rigat B, Visvikis S, et al. Evidence, from combined segregation and linkage analysis, that a variant of the angiotensin I-converting enzyme (ACE) gene controls plasma ACE levels. Am J Hum Genet 1992;51:197-205.
13. Rieder MJ, Taylor SL, Clark AG, et al. Sequence variation in the human angiotensin converting enzyme. Nat Genet 1999;22:59-62.
14. Hilbert P, Lindpaintner K, Beckmann JS, et al. Chromosomal mapping of two genetic loci associated with blood-pressure regulation in hereditary hypertensive rats. Nature 1991;353:521-9.
15. Jacob HJ, Lindpaintner K, Lincoln SE, et al. Genetic mapping of a gene causing hypertension in the stroke-prone spontaneously hypertensive rat. Cell 1991;67:213-24.
16. O'Donnell CJ, Lindpaintner K, Larson MG, et al. Evidence for association and genetic linkage of the angiotensin- converting enzyme locus with hypertension and blood pressure in men but not women in the Framingham Heart Study. Circulation 1998;97:1766-72.
17. Fornage M, Amos CI, Kardia S, et al. Variation in the region of the angiotensin-converting enzyme gene influences interindividual differences in blood pressure levels in young white males. Circulation 1998;97:1773-9.
18. Julier C, Delepine M, Keavney B, et al. Genetic susceptibility for human familial essential hypertension in a region of homology with blood pressure linkage on rat chromosome 10. Hum Mol Genet 1997;6:2077-85.
19. Staessen JA, Wang JG, Brand E, et al. Effects of three candidate genes on prevalence and incidence of hypertension in a Caucasian population. J Hypertens 2001;19:1349-58.
20. Jeunemaitre X, Soubrier F, Kotelevtsev YV, et al. Molecular basis of human hypertension: role of angiotensinogen. Cell 1992;71:169-80.

21. Caulfield M, Lavender P, Farrall M, et al. Linkage of the angiotensinogen gene to essential hypertension. N Engl J Med 1994;330:1629-33.
22. Caulfield M, Lavender P, Newell-Price J, et al. Angiotensinogen in human essential hypertension. Hypertension 1996;28:1123-5.
23. Brand E, Chatelain N, Keavney B, et al. Evaluation of the angiotensinogen locus in human essential hypertension: a European study. Hypertension 1998;31:725-9.
24. Caulfield M, Lavender P, Newell-Price J, et al. Linkage of the angiotensinogen gene locus to human essential hypertension in African Caribbeans. J Clin Invest 1995;96:687-92.
25. Atwood LD, Kammerer CM, Samollow PB, et al. Linkage of essential hypertension to the angiotensinogen locus in Mexican Americans. Hypertension 1997;30:326-30.
26. Niu T, Xu X, Rogus J, et al. Angiotensinogen gene and hypertension in Chinese. J Clin Invest 1998;101:188-94.
27. Rotimi C, Morrison L, Cooper R, et al. Angiotensinogen gene in human hypertension. Lack of an association of the 235T allele among African Americans. *Hypertension* 1994;24:591-4.
28. Jeunemaitre X, Gimenez-Roqueplo AP, Celerier J, et al. Angiotensinogen variants and human hypertension. Curr Hypertens Rep 1999;1:31-41.
29. Kunz R, Kreutz R, Beige J, et al. Association between the angiotensinogen 235T-variant and essential hypertension in whites: a systematic review and methodological appraisal. Hypertension 1997;30:1331-7.
30. Staessen JA, Kuznetsova T, Wang JG, et al. M235T angiotensinogen gene polymorphism and cardiovascular renal risk. J Hypertens 1999;17:9-17.
31. Sethi AA, Tybjaerg-Hansen A, Gronholdt ML, et al. Angiotensinogen mutations and risk for ischemic heart disease, myocardial infarction, and ischemic cerebrovascular disease. Six case- control studies from the Copenhagen City Heart Study. Ann Intern Med 2001;134:941-54.
32. Bloem LJ, Manatunga AK, Tewksbury DA, et al. The serum angiotensinogen concentration and variants of the angiotensinogen gene in white and black children. J Clin Invest 1995;95:948-53.
33. Schunkert H, Hense HW, Gimenez-Roqueplo AP, et al. The angiotensinogen T235 variant and the use of antihypertensive drugs in a population-based cohort. Hypertension 1997;29:628-33.
34. Bloem LJ, Foroud TM, Ambrosius WT, et al. Association of the angiotensinogen gene to serum angiotensinogen in blacks and whites. Hypertension 1997;29:1078-82.
35. Smithies O, Kim HS. Targeted gene duplication and disruption for analyzing quantitative genetic traits in mice. Proc Natl Acad Sci U S A 1994;91:3612-5.
36. Inoue I, Nakajima T, Williams CS, et al. A nucleotide substitution in the promoter of human angiotensinogen is associated with essential hypertension and affects basal transcription in vitro. J Clin Invest 1997;99:1786-97.
37. Ward K, Hata A, Jeunemaitre X, et al. A molecular variant of angiotensinogen associated with preeclampsia. Nat Genet 1993;4:59-61.
38. Arngrimsson R, Purandare S, Connor M, et al. Angiotensinogen: a candidate gene involved in preeclampsia? Nat Genet 1993;4:114-5.
39. Bonnardeaux A, Davies E, Jeunemaitre X, et al. Angiotensin II type 1 receptor gene polymorphisms in human essential hypertension. Hypertension 1994;24:63-9.
40. Zhang X, Erdmann J, Regitz-Zagrosek V, et al. Evaluation of three polymorphisms in the promoter region of the angiotensin II type I receptor gene. J Hypertens 2000;18:267-72.
41. Kainulainen K, Perola M, Terwilliger J, et al. Evidence for involvement of the type 1 angiotensin II receptor locus in essential hypertension. Hypertension 1999;33:844-9.
42. Perola M, Kainulainen K, Pajukanta P, et al. Genome-wide scan of predisposing loci for increased diastolic blood pressure in Finnish siblings. J Hypertens 2000;18:1579-85.
43. Vuagnat A, Giacche M, Hopkins PN, et al. Plasma LDL cholesterol is a strong predictor of the blood pressure increase following an acute infusion of angiotensin II. J Mol Med

2001; 79:175-183
44. Brand E, Chatelain N, Mulatero P, et al. Structural analysis and evaluation of the aldosterone synthase gene in hypertension. Hypertension 1998;32:198-204.
45. Davies E, Holloway CD, Ingram MC, et al. Aldosterone excretion rate and blood pressure in essential hypertension are related to polymorphic differences in the aldosterone synthase gene CYP11B2. Hypertension 1999;33:703-7.
46. Komiya I, Yamada T, Takara M, et al. Lys(173)Arg and -344T/C variants of CYP11B2 in Japanese patients with low-renin hypertension. Hypertension 2000;35:699-703.
47. Mulatero P, Schiavone D, Fallo F, et al. CYP11B2 gene polymorphisms in idiopathic hyperaldosteronism. Hypertension 2000;35:694-8.
48. Clyne CD, Zhang Y, Slutsker L, et al. Angiotensin II and potassium regulate human CYP11B2 transcription through common cis-elements. Mol Endocrinol 1997;11:638-49.
49. Hautanena A, Lankinen L, Kupari M, et al. Associations between aldosterone synthase gene polymorphism and the adrenocortical function in males. J Intern Med 1998;244:11-8.
50. Kupari M, Hautanen A, Lankinen L, et al. Associations between human aldosterone synthase (CYP11B2) gene polymorphisms and left ventricular size, mass, and function. Circulation 1998;97:569-75.
51. Schunkert H, Hengstenberg C, Holmer SR, et al. Lack of association between a polymorphism of the aldosterone synthase gene and left ventricular structure. Circulation 1999;99:2255-60.
52. Hautanen A, Toivanen P, Manttari M, et al. Joint effects of an aldosterone synthase (CYP11B2) gene polymorphism and classic risk factors on risk of myocardial infarction. Circulation 1999;100:2213-8.
53. Tamaki S, Iwai N, Tsujita Y, et al. Genetic polymorphism of CYP11B2 gene and hypertension in Japanese. Hypertension 1999;33:266-70.
54. Williams SM, Addy JH, Phillips JA, et al. Combinations of variations in multiple genes are associated with hypertension. Hypertension 2000;36:2-6.

7. GENE THERAPY FOR HYPERTENSION: FUTURE OR FICTION?

J.P. Fennell, M.J. Brosnan, A.J. Frater, A.H. Baker, and A.F. Dominiczak

Introduction

Initial therapeutic applications of gene therapy focused on cancer or severe single gene disorders such as Duchenne's muscular dystrophy and cystic fibrosis [1]. As our understanding of disease pathology, molecular biology and vectors improves, gene transfer for complex diseases is becoming a realistic prospect. At present, the vast majority of vascular gene transfer studies serve as an investigational, rather than a therapeutic tool. Sizeable obstacles remain, notably in terms of safe and efficient gene-delivery vectors. However, the high mortality rate in the developed world (> 250,000/year in the UK [2]) from vascular disease demonstrates a clear need for new methods of treatment. Essential hypertension is associated with endothelial dysfunction [3] and contributes significantly to cardiovascular risk. The goals of gene therapy for hypertension would be long-term reduction in blood pressure and prevention or reversal of end-organ damage with few side effects. This work will review current vascular gene transfer for hypertension with particular emphasis on *in vivo* gene transfer in models of essential hypertension.

Overexpression of Vasodilator Genes

There are two major gene therapy approaches used to alleviate hypertension: augmenting vasodilation by genes that result in expression of vasodilatory products and the transfer of antisense molecules to inhibit vasoconstriction [1].

Kallikrein

Kallikrein cleaves kininogen producing vasodilator kinin peptides. By binding to bradykinin receptors, kinins stimulate the release of the potent vasodilators prostacyclin, endothelium-derived hyperpolarising factor and nitric oxide. One of the first successful *in vivo* gene transfer studies to reduce blood pressure was done by Dr Chao's group [4]. They produced impressive results by infusing DNA plasmids containing the human kallikrein gene into the SHR model of hypertension. The reduction in blood pressure compared to controls was sustained for 6 weeks and reached a maximal reduction of 46 mmHg. In subsequent work, *in vivo* kallikrein gene transfer using adenoviral and DNA plasmid vectors was compared [5]. Both Adenoviral-mediated gene expression, as expected was only transient, and lasted 4 weeks but unexpectedly, plasmid kallikrein gene transfer reduced blood pressure for up to 8 weeks. These results have not been replicated by other groups thus far. Adenoviral gene transfer of kallikrein has also been tested in other models of hypertension and showed a 26 mm Hg reduction in the Goldblatt model and a 31-day delay in reaching the blood pressure of sham-operated controls [6]. In a deoxycorticosterone acetate (DOCA) salt rat model, this vector produced a maximal reduction of 50 mm Hg and the antihypertensive effect lasted for more than 23 days [7]. In a model of chronic renal failure adenoviral delivery of kallikrein improved hypertension, cardiac hypertrophy and renal injury [8].

Adrenomedullin
Adrenomedullin causes vasodilation, natriuresis and reduces extracellular matrix formation. Adenoviral-mediated gene transfer of adrenomedullin in the DOCA salt hypertensive rat produced a maximal reduction in blood pressure of 41 mm Hg 9 days after tail vein injection [9]. This reduction lasted nearly 20 days and was accompanied by a reduction in cardiac hypertrophy and fibrosis. Similar effects in reducing hypertension, cardiac fibrosis and hypertrophy were shown in the Dahl salt sensitive [10] and the Goldblatt rat models [11] of hypertension.

Atrial Natriuretic Peptide
Atrial natriuretic peptide (ANP) has powerful natriuretic and hypotensive effects, making it a candidate for antihypertensive gene therapy. Systemic plasmid-mediated ANP gene transfer in the SHR resulted in a 7-week reduction in blood pressure in young animals (4 weeks old) but had no effect in older animals (12 weeks old) [12]. Adenoviral ANP gene transfer in the Dahl salt-sensitive rat resulted in a maximal blood pressure reduction of 28 mm Hg that lasted more than 3 weeks and a significant reduction in mortality and aortic hypertrophy [13].
These studies have shown proof of the principle of systemic gene therapy for hypertension but much work needs to be done to improve the duration of expression and safety of the vectors used. Potential long-term side effects of both the vector and the transgene need to be fully evaluated before clinical trials could begin.

Nitric Oxide Synthase and Superoxide Dismutase
Hypertension, atherosclerosis and diabetic vascular disease all exhibit endothelial dysfunction, which is characterised by reduced nitric oxide (NO) and NO-mediated vasodilation. Continuous basal NO production is responsible for maintaining vasodilation and inhibition of platelet aggregation and adhesion. NO also has antioxidant effects and it inhibits extracellular matrix synthesis, vascular smooth muscle cell proliferation and migration. NO plays an anti-inflammatory role by inhibiting nuclear factor-κB activation, adhesion molecule expression and neutrophil activation and adhesion. The importance of NO in normal endothelial function

makes it a very attractive and logical choice for gene transfer in vascular disease. A further advantage of nitric oxide synthase (NOS) is that its product, NO is diffusible and therefore able to act beyond the infected cells. NO availability can be increased either by increasing NO production by increasing NOS activity or by preventing NO degradation by superoxide (O_2^-) using superoxide dismutase (SOD) gene transfer. Increasing NOS activity alone may not be sufficient because under conditions of reduced tetrahydrobiopterin[14] or L-arginine[15] availability, NOS production of NO is uncoupled and the detrimental reactive oxidative species O_2^- [16] is produced instead.

Systemic injections of endothelial NOS (eNOS) DNA plasmids resulted in a 21 mm Hg reduction in blood pressure in the SHR model of hypertension which lasted for 12 weeks [17]. However, the biodistribution of the transgene was not assessed. Ex vivo infection of carotid rings from an angiotensin-II rabbit model of hypertension with an adenovirus encoding eNOS restored vasomotor function [18]. Data from our laboratory demonstrated that in vivo infection of carotid arteries in the spontaneously hypertensive stroke-prone rat (SHRSP) restored NO availability to the level seen in the normotensive Wistar-Kyoto strain (figure 1) [19]. The SHRSP model is an excellent model of human hypertension because it develops progressive hypertension, left ventricular hypertrophy, haemorrhagic and ischaemic strokes and a more moderate phenotype is seen in females. CuZnSOD gene transfer was unable to improve NO availability in either angiotensin II-infusion or SHRSP hypertensive models [18,20]. Recently we have shown that in vivo gene transfer of the extracellular SOD but not the manganese SOD isoform significantly increased NO availability in the SHRSP (unpublished results).

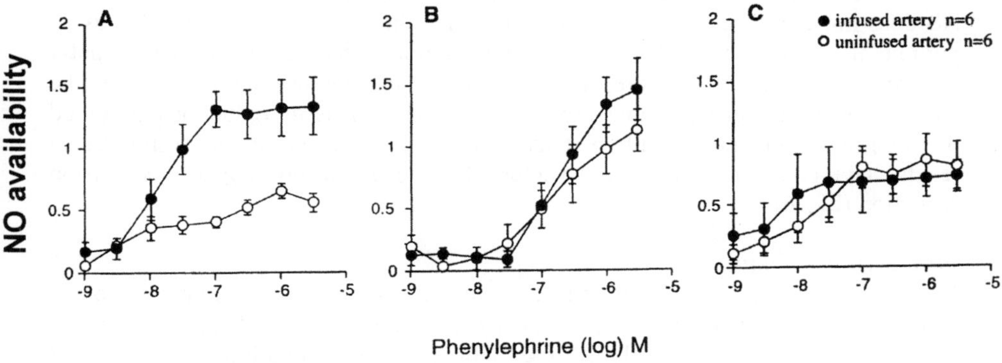

Figure 1 *Assessment of NO availability 24 hours after in vivo eNOS gene transfer in the SHRSP carotid artery. Ex vivo contractility studies revealed a significant increase in NO availability relative to the uninfused contralateral artery after eNOS gene transfer in SHRSP animals (A); P=0.007. Contractility studies revealed no significant increase in NO availability after eNOS gene transfer in normotensive WKY rats (B) or after β-gal gene transfer in SHRSP rats (C); P=0.2 and 0.9 respectively. All results are expressed as a mean ± SEM. From reference [19], with permission.*

Patients with essential hypertension frequently have other co-existing vascular problems, especially type II diabetes, with its potential to increase oxidative stress and aggravate endothelial dysfunction. Gene transfer methods aimed at increasing NO availability, in addition to reducing blood pressure, may also act directly to improve endothelial function in these conditions. A gene transfer method that increases NO availability could theoretically be of benefit to many vascular diseases through its anti-thrombotic, anti-inflammatory and anti-proliferatory effects. Local delivery of NOS has been shown to improve endothelial function in atherosclerotic arteries [21] and prevent intimal hyperplasia in balloon-injured arteries [22]. Human clinical trials using local vascular gene therapy for endothelial dysfunction are beginning to produce successful results for coronary and peripheral vascular disease (V.Dzau, personal communication).

Inhibition of Vasoconstrictor Genes

Antisense oligonucleotides

There is little clinical experience of increasing ANP or kallikrein to treat hypertension. Several groups are using gene transfer methods based on the better-understood physiological pathways of the pharmacological treatment of hypertension. This allows direct comparisons to be made between conventional therapy and gene therapy for hypertension. One such approach uses antisense (AS) oligonucleotides to block the renin-angiotensin system, e.g. by producing a genetic beta-blocker or angiotensin type 1 receptor (AT_1-R) inhibition. Antisense oligonucleotides are single-stranded sequences of nucleotides that are complementary to specific mRNA. They bind mRNA preventing translation and thus transiently suppress target protein synthesis. The high specificity results in fewer side effects compared to pharmacological treatment. Another advantage is that AS does not cross the blood brain barrier or the placenta. Wielbo et al. [24] using peripheral administration of liposomes containing angiotensinogen AS significantly lowered mean arterial pressure, angiotensinogen and angiotensin II levels in the adult SHR. Other groups have produced similar short-term reductions in blood pressure in various models of hypertension using AS alone, but the most impressive results have coupled AS to a viral vector which results in prolonged hypertension reduction [25,26].

Viral delivery of antisense

Viral delivery of AS has been used both to prevent and reverse hypertension in animal models. Retroviral transfer of AT_1-R AS in 5-day-old WKY and SHR rats reduced blood pressure solely in the SHR [27]. The addition of the AT_1-R antagonist, losartan was unable to further reduce blood pressure in the SHR. Losartan alone produced a very similar reduction in blood pressure to the AT_1-R AS vector but its effect lasted less than 24 hours whereas AS treatment lasted 90 days.
Retroviral delivery of AT_1-R AS in neonatal SHR animals prevented the development of ventricular hypertrophy, perivascular fibrosis and hypertension for at least 120 days [26]. Similar studies in WKY animals did not affect basal blood

pressure but gave complete protection against angiotensin II-induced hypertension [28]. Intracardiac injection of retroviral AS for AT_1-R [29] and ACE [30] in 5-day-

Table 1 Vasodilator Gene Transfer Studies in Hypertension

Vector	Gene	Delivery method	Disease model	Duration of study	Max. Δ in BP/other phenotype	Ref.
Plasmid	Kallikrein	*In vivo*	SHR	6 weeks	- 46 mm Hg	Wang [4]
Plasmid Adenovirus	Kallikrein	*In vivo*	SHR	8 weeks 4 weeks	- 20 mm Hg - 12 mm Hg	Chao [5]
Adenovirus	Kallikrein	*In vivo*	Goldblatt hypertensive rat	24 days	- 26 mm Hg	Yayama [6]
Adenovirus	Kallikrein	*In vivo*	DOCA salt rat	23 days	- 50 mm Hg	Dobrzynski [7]
Adenovirus	Kallikrein	*In vivo*	Rat Chronic renal failure	5 weeks	- 37 mm Hg	Wolf [8]
Plasmid	ANP	*In vivo*	SHR	7 weeks	- 21 mm Hg	Lin [12]
Adenovirus	ANP	*In vivo*	Dahl salt sensitive rat	3weeks	- 28 mm Hg	Lin [13]
Retrovirus	Haem oxygenase	*In vivo*	SHR	20 weeks	- 26 mm Hg	Sabaawy [23]
Adenovirus	Adreno-medullin	*In vivo*	DOCA salt rat	20 days	- 41 mm Hg	Dobrzynski [9]
Adenovirus	Adreno-medullin	*In vivo*	Dahl salt sensitive rat	4 weeks	- 31 mm Hg	Zhang [10]
Adenovirus	Adreno-medullin	*In vivo*	Goldblatt hypertensive rat	3 weeks	- 28 mm Hg	Wang [11]
Plasmid	ENOS	*In vivo x 2*	SHR	12 weeks	- 21 mm Hg	Lin [17]
Adenovirus	ENOS	*In vivo*	SHRSP	24 hours	↑ NO availability to level of WKY	Alexander [19]
Adenovirus	ENOS CuZnSOD	*In vivo*	SHRSP	24 hours	eNOS but not CuZnSOD ↑ NO availability	Alexander [20]
Adenovirus	ENOS CuZnSOD ECSOD	*Ex vivo*	Ang-II infusion	24 hours	eNOS but not SODs ↑ NO availability	Nakane [18]
Adenovirus	MnSOD ECSOD	*In vivo*	SHRSP	24 hours	ECSOD but not MnSOD ↑ NO availability	Unpublished Results

old SHR both resulted in a reduction in hypertension and prevention of the deterioration in many of the cardiac and renal parameters associated with hypertensive disease. The offspring of the infected animals were also found to have similar improvements in cardiovascular physiology. PCR or Southern blotting demonstrated the presence of the AS in the genome. However, ethical issues surrounding the potential for insertional mutagenesis and germline transmission would preclude the use of retroviruses in clinical trials for hypertension.

One of the first successful, long-term, *in vivo* reductions of hypertension was by Phillips et al.[25]. They used an adeno-associated virus (AAV) encoding AS for AT_1-R in adult SHRs. Relative to controls, a single intracerebroventricular injection of vector produced a maximal reduction in blood pressure of 23 mm Hg and the reduction lasted for more than 9 weeks. In the first study [31] to use a retroviral vector for the same purpose, adult SHR were injected for 6 consecutive days with a AT_1-R AS retrovirus, resulting in a 30 to 60 mm Hg reduction in blood pressure which was significant for up to 36 days. The intracerebroventricular or intracardiac routes of injection used in many of the above studies would not be acceptable in the clinic, but tail vein injection of angiotensinogen AS in an AAV-based plasmid also reduced blood pressure by up to 22 mm Hg [32].

Future Developments

Improved vectors

Current gene therapy vectors are not ideal for the treatment of hypertension [33]. The requirements for an *in vivo* gene therapy vector for hypertension would include *in vivo* stability, long-term expression without harmful side effects such as immunogenicity or induction of an inflammatory response and efficient target cell transduction [1]. Ideally, it would be easily produced to the high titres required for systemic administration [34]. Because hypertension is a diffuse, progressive disease the ideal treatment method would be long lasting, systemic *in vivo* gene therapy [35]. Localised short-term gene expression would, however, be appropriate for some of the complications of hypertension and atherosclerosis such as vein graft failure and post-angioplasty restenosis. Viral vectors are currently much more effective than non-viral vectors for vascular gene transfer [1]. Safety remains a major concern, especially in light of the much-publicised patient death in a recent clinical gene therapy trial [36] and the increasing evidence of the inflammatory contribution of virus infection to atherosclerosis, myocarditis and cardiomyopathies. Reversion of a replication-deficient vector to a replication-competent wild type virus is another concern. Finally, control and localisation of gene expression needs to be considered because uncontrolled or ectopic gene expression could be detrimental.

New vectors designed to improve the duration of gene expression and reduce the immune response have been developed. These include the 'gutless' or third generation adenovirus [37], lentivirus and new hybrid vectors that could potentially combine the best features of two vectors. Hybrid vectors such as liposomes coupled to Sendai virus fusion proteins [38], Adenoviral/retroviral hybrids [39] and Adenoviral/AAV vectors [40] have shown promise but need to be evaluated fully in a vascular setting.

Table 2 Vasoconstrictor Antisense Gene Transfer in Hypertension

Vector	Gene	Delivery method	Disease model	Duration of study	Δ in BP	Ref.
Liposomes	Angiotensinogen AS	*In vivo*	SHR	NA	-25 mm Hg	Wielbo [24]
AAV	AT_1-R AS	*In vivo*	SHR	9 weeks	- 23 mm Hg	Phillips [25]
Retrovirus	AT_1-R AS	*In vivo*	SHR	90 days	-25 mm Hg	Lu [27]
Retrovirus	AT_1-R AS	*In vivo*	SHR	36 days	- 60 mm Hg	Katovich [31]
Retrovirus	AT_1-R AS	*In vivo*	SHR	120 days	- 47 mm Hg	Martens [26]
Retrovirus	AT_1-R AS	*In vivo*	SHR	120 days	- 35 mm Hg also ↓ in F_1 and F_2	Reaves [29]
Retrovirus	ACE AS	*In vivo*	SHR	120 days	- 17 mm Hg also ↓ in F_1	Wang [30]
AAV-based plasmid	Angiotensinogen AS	*In vivo*	SHR	6 days	- 22 mm Hg	Tang [32]

Current vector systems are extremely inefficient for delivery of therapeutic genes to cells of the vasculature and are equally non-selective. Although adenoviral vectors can be delivered locally to the vessel wall, very high titres are required. Systemic administration of adenoviruses leads to sequestration of the vast majority of the virus to the liver and spleen with little or no uptake by endothelial cells lining the blood vessels. This limitation is also true for other vectors such as AAV. Furthermore, the use of viral promoters results in high-level expression of the transgene in non-vascular cells. A major requirement for development of vectors for future clinical use will be the design of disease-specific modified vectors systems. While some components of the viral delivery system are efficient (such as nuclear trafficking), the vectors require modification at the level of vector-cell entry and at the level of transgene expression. For many cardiovascular applications including hypertension, targeting vectors selectively and efficiently to the vascular endothelium would be beneficial. To achieve this it is possible to exploit the vascular "address system" [41] and utilise receptors that are exclusively expressed on vascular cells. Essentially, ligands that mediate binding to the selected receptor can be incorporated into vector systems (e.g. viral capsid proteins) to mediate endothelial cell-specific binding of the vector and hence cell-specific gene delivery [42]. Additional genetic modification to provide endothelial cell-specific gene expression will provide additional selectivity and enhanced safety. The FLT-1 promoter [43,44] looks particularly promising in this context. The development of inducible promoters is another area of intensive research. The ability to turn off gene expression would be a crucial safety feature in the event of unforeseen clinical side effects. Several inducible gene expression systems have been developed using exogenous ligands such as tetracycline or rapamycin. Preliminary studies using the tetracycline system in a viral vector encoding AT_1-R AS showed that the administration of a tetracycline resulted in a significant reduction in blood pressure in the SHR [33]. The ultimate aim would be the development of disease-specific inducible promoters, where gene expression would be induced by local pathology such as shear stress, hypoxia, hypoglycaemia or the release of inflammatory cytokines. Thus, modifications of viral and non-viral vector systems at the level of vector-cell interactions and transgene expression will improve the safety and efficiency of vascular gene transfer and will result in increased efficacy and safety for future clinical vascular gene therapy.

Other Developments
As genetic linkage and association studies for essential hypertension progress, other suitable candidates for gene transfer will be discovered [45]. Technical improvements in intravascular delivery and catheter development will aid gene transfer to the area of choice and changes in the vessel wall will be more easily assessed by improvements such as intravascular ultrasound. As our understanding of vectors and disease pathogenesis improves, and technical improvements in vector targeting, delivery and assessment are developed, clinical trials for hypertension will one day become possible.

Conclusion

Gene therapy presents a tremendous opportunity to improve the clinician's arsenal against a large number of diseases where treatment is inadequate or cures are not available. There are now a large number of different genes, vectors and approaches that require further assessment in vascular disease models. Vector systems in particular need to be improved such that gene delivery is safe, effective and occurs in the correct location and for the correct duration for the selected disease. Gene therapy for hypertension will not reach the clinic in the short term because of the availability of effective pharmacological treatment. Optimised, long-term gene therapy methods without the side effects of current pharmacological interventions would be of great benefit both in terms of clinical outcome and improved patient compliance in hypertension and other cardiovascular diseases.

References

1. Phillips MI: Is gene therapy for hypertension possible? Hypertension 1999;33:8-13.
2. Petersen S, Rayner M, Press V: Coronary heart disease statistics. British Heart Foundation, 2000.
3. Panza JA, Quyyumi AA, Brush JEJ, et al: Abnormal endothelium-dependent vascular relaxation in patients with essential hypertension. N. Engl. J. Med. 1990;323:22-7.
4. Wang C, Chao L, Chao J: Direct gene delivery of human tissue kallikrein reduces blood pressure in spontaneously hypertensive rats. J Clin Invest 1995;95:1710-6.
5. Chao J, Chao L: Kallikrein gene therapy: a new strategy for hypertensive diseases. Immunopharmacology 1997;36:229-36.
6. Yayama K, Wang C, Chao L, et al: Kallikrein gene delivery attenuates hypertension and cardiac hypertrophy and enhances renal function in Goldblatt hypertensive rats. Hypertension 1998;31:1104-10.
7. Dobrzynski E, Yoshida H, Chao J, et al: Adenovirus-mediated kallikrein gene delivery attenuates hypertension and protects against renal injury in deoxycorticosterone-salt rats. Immunopharmacology 1999;44:57-65.
8. Wolf WC, Yoshida H, Agata J, et al: Human tissue kallikrein gene delivery attenuates hypertension, renal injury, and cardiac remodeling in chronic renal failure. Kidney Int 2000;58:730-9.
9. Dobrzynski E, Wang C, Chao J, et al: Adrenomedullin gene delivery attenuates hypertension, cardiac remodeling, and renal injury in deoxycorticosterone acetate-salt hypertensive rats. Hypertension 2000;36:995-1001.
10. Zhang JJ, Yoshida H, Chao L, et al: Human adrenomedullin gene delivery protects against cardiac hypertrophy, fibrosis, and renal damage in hypertensive dahl salt-sensitive rats. Hum Gene Ther 2000;11:1817-27.
11. Wang C, Dobrzynski E, Chao J, et al: Adrenomedullin gene delivery attenuates renal damage and cardiac hypertrophy in Goldblatt hypertensive rats. Am J Physiol Renal Physiol 2001;280:F964-71
12. Lin KF, Chao J, Chao L: Human atrial natriuretic peptide gene delivery reduces blood pressure in hypertensive rats. Hypertension 1995;26:847-53.
13. Lin KF, Chao J, Chao L: Atrial natriuretic peptide gene delivery reduces stroke-induced mortality rate in Dahl salt-sensitive rats. Hypertension 1999;33:219-24.
14. Vásquez-Vivar J, Kalyanaraman B, Martasek P, et al: Superoxide generation by endothelial nitric oxide synthase: the influence of cofactors. Proceedings of the National Academy of Sciences of the United States of America 1998;95:9220-5.
15. Vergnani L, Hatrik S, Ricci F, et al: Effect of native and oxidized low-density lipoprotein on endothelial nitric oxide and superoxide production : key role of L-arginine availability. Circulation 2000;101:1261-6.
16. McIntyre M, Bohr DF, Dominiczak AF: Endothelial function in hypertension: the role of superoxide anion. Hypertension 1999;34:539-45.
17. Lin KF, Chao L, Chao J: Prolonged reduction of high blood pressure with human nitric oxide synthase gene delivery. Hypertension 1997;30:307-13.
18. Nakane H, Miller FJJ, Faraci FM, et al: Gene transfer of endothelial nitric oxide synthase reduces angiotensin II-induced endothelial dysfunction. Hypertension 2000;35:595-601.
19. Alexander MY, Brosnan MJ, Hamilton CA, et al: Gene transfer of endothelial nitric oxide synthase improves nitric oxide-dependent endothelial function in a hypertensive rat model. Cardiovascular Research 1999;43:798-807.
20. Alexander MY, Brosnan MJ, Hamilton CA, et al: Gene transfer of endothelial nitric oxide synthase but not Cu/Zn superoxide dismutase restores nitric oxide availability in the SHRSP. Cardiovascular Research 2000;47:609-17.

21. Ooboshi H, Toyoda K, Faraci FM, et al: Improvement of relaxation in an atherosclerotic artery by gene transfer of endothelial nitric oxide synthase. Arteriosclerosis, Thrombosis & Vascular Biology 1998;18:1752-8.
22. Janssens S, Flaherty D, Nong Z, et al: Human endothelial nitric oxide synthase gene transfer inhibits vascular smooth muscle cell proliferation and neointima formation after balloon injury in rats. Circulation 1998;97:1274-81.
23. Sabaawy HE, Zhang F, Nguyen X, et al: Human heme oxygenase-1 gene transfer lowers blood pressure and promotes growth in spontaneously hypertensive rats. Hypertension 2001;38:210-5.
24. Wielbo D, Simon A, Phillips MI, et al: Inhibition of hypertension by peripheral administration of antisense oligodeoxynucleotides. Hypertension 1996;28:147-51.
25. Phillips MI, Mohuczy-Dominiak D, Coffey M, et al: Prolonged reduction of high blood pressure with an in vivo, nonpathogenic, adeno-associated viral vector delivery of AT1-R mRNA antisense. Hypertension 1997;29:374-80.
26. Martens JR, Reaves PY, Lu D, et al: Prevention of renovascular and cardiac pathophysiological changes in hypertension by angiotensin II type 1 receptor antisense gene therapy. Proc Natl Acad Sci USA 1998;95:2664-9.
27. Lu D, Raizada MK, Iyer S, et al: Losartan versus gene therapy: chronic control of high blood pressure in spontaneously hypertensive rats. Hypertension 1997;30:363-70.
28. Pachori AS, Wang H, Gelband CH, et al: Inability to induce hypertension in normotensive rat expressing AT(1) receptor antisense. Circ Res 2000;86:1167-72.
29. Reaves PY, Gelband CH, Wang H, et al: Permanent cardiovascular protection from hypertension by the AT(1) receptor antisense gene therapy in hypertensive rat offspring. Circ Res 1999;85:e44-50
30. Wang H, Reaves PY, Gardon ML, et al: Angiotensin I-converting enzyme antisense gene therapy causes permanent antihypertensive effects in the SHR.Hypertension 2000;35: 202-8.
31. Katovich MJ, Gelband CH, Reaves P, et al: Reversal of hypertension by angiotensin II type 1 receptor antisense gene therapy in the adult SHR. Am J Physiol 1999;277:H1260-4
32. Tang X, Mohuczy D, Zhang YC, et al: Intravenous angiotensinogen antisense in AAV-based vector decreases hypertension. Am J Physiol 1999;277:H2392-9
33. Pachori AS, Huentelman MJ, Francis SC, et al: The future of hypertension therapy: sense, antisense, or nonsense? Hypertension 2001;37:357-64.
34. Anderson WF: Human gene therapy. Nature 1998;392:25-30.
35. Ponder KP: Systemic gene therapy for cardiovascular disease. Trends Cardiovasc Med 1999;9:158-62.
36. Somia N, Verma IM: Gene therapy: trials and tribulations. Nat Rev Genet 2000;1:91-9.
37. Schiedner G, Morral N, Parks RJ, et al: Genomic DNA transfer with a high-capacity adenovirus vector results in improved in vivo gene expression and decreased toxicity. Nat Genet 1998;18:180-3.
38. Dzau VJ, Mann MJ, Morishita R, et al: Fusigenic viral liposome for gene therapy in cardiovascular diseases. Proc Natl Acad Sci USA 1996;93:11421-5.
39. Feng M, Jackson WHJ, Goldman CK, et al: Stable in vivo gene transduction via a novel adenoviral/retroviral chimeric vector. Nat Biotechnol 1997;15:866-70
40. Recchia A, Parks RJ, Lamartina S, et al: Site-specific integration mediated by a hybrid adenovirus/adeno-associated virus vector. Proc Natl Acad Sci USA 1999;96:2615-20.
41. Rajotte D, Arap W, Hagedorn M, et al: Molecular heterogeneity of the vascular endothelium revealed by in vivo phage display. J Clin Invest 1998;102:430-7.
42. Nicklin SA, Buening H, Dishart KL, et al: Efficient and Selective AAV2-Mediated Gene Transfer Directed to Human Vascular Endothelial Cells. Mol Ther 2001;4:In press
43. Nicklin SA, Reynolds PN, Brosnan MJ, et al: Analysis of cell-specific promoters for viral gene therapy targeted at the vascular endothelium. Hypertension 2001;38:65-70.

44. Reynolds PN, Nicklin SA, Kaliberova L, et al: Combined transductional and transcriptional targeting improves the specificity of transgene expression in vivo. Nat Biotechnol 2001;19: (in press)
45. Dominiczak AF, Negrin DC, Clark JS, et al: Genes and hypertension: from gene mapping in experimental models to vascular gene transfer strategies. Hypertension 2000;35:164-72.

8. HORMONES AND SIGNALLING PATHWAYS

G. Lembo

Introduction

Several hormones are known to participate in blood pressure homeostasis. In particular, the renin-angiotensin system influences blood volume and vascular tone, while the adrenergic system affects vascular tone, cardiac contraction, and heart rate. Thus, it is no surprise that changes in hormonal status can have a strong influence on blood pressure. In fact, most forms of secondary hypertension are determined by an alteration in secretion of specific hormones. As examples, we could mention renovascular hypertension, where the elevation in blood pressure is mainly due to increases in the activity of the renin-angiotensin system with both systemic and renal effects; adrenal hypertension, where systemic vasoconstriction is mediated by a potentiation of sympathoadrenal and renin-angiotensin activity; pheochromocytoma, where an excess of circulating catecholamines can cause serious cardiovascular complications; and several others, like hypothyroidism, hyperparathyroidism, and acromegaly.
Furthermore, also essential hypertension has been extensively correlated with abnormalities in hormone regulation, as sympathetic hyperactivation and insulin resistance. This latter dysfunction is a common clinical finding that represents an important risk factor for the development of hypertension and cardiovascular diseases, since insulin resistance occurs in about 50% of hypertensive subjects. Moreover, it is often found in association with other metabolic abnormalities, such as glucose intolerance, type II diabetes mellitus, dyslipidemia, and obesity. The association of these cardiovascular risk factors, called 'insulin resistance syndrome' or 'syndrome X', is responsible for a 2- to 3-fold increase in cardiovascular morbidity and mortality [1]. However, little is known about the mechanisms linking insulin resistance and hypertension.

Insulin and hypertension:

It has been reported that insulin, like other hormones whose main action is not related to the regulation of blood pressure, such as oestrogens and leptin, have secondary effects on cardiac output and on vascular resistance, thus contributing to blood pressure variations. In particular, insulin modulates various factors with opposing effects on blood pressure homeostasis. In fact, hyperinsulinemia evokes a net reflex increase in sympathetic nervous activity, whereas it is able to attenuate the vasoconstrictive effects of the reflex sympathetic activation. Based on these findings a hypothetical sequence may be formulated in which, on the one hand, resistance to insulin-induced glucose disposal leads to compensatory hyperinsulinaemia and then to sympathetic overactivity and, to the other hand, resistance to the vasodilator action of insulin may result in an imbalance of the sympathetic control on peripheral vascular tone. To test this hypothesis we investigated whether insulin was able to exert such an interference with the sympathetic vasoconstrictor effect in hypertensive patients who showed a reduced sensitivity to insulin metabolic action. For this reason, we applied a graded negative pressure to the lower body (LBNP) to patients with mild to moderate essential hypertension both in control conditions and during infusion of insulin in brachial artery [2]. In these patients we failed to observe the insulin-modulating effect on the forearm sympathetic-evoked vasoconstrictor response induced by LBNP. This observation further supports the hypothesis that in hypertensive patients a loss of the insulin-induced attenuation of the sympathetic vasoconstrictor response may act synergistically with sympathetic over-responsiveness to lead to increased peripheral resistance, resulting in hypertension. However it was still unclear whether the abnormal vascular action of insulin is a primary event in essential hypertension or a feature acquired in conjunction with high blood pressure levels. To clarify this issue, we evaluated the effects of insulin on contractile responses of aortic rings to graded doses of norepinephrine, the major sympathetic neurotransmitter, in spontaneously hypertensive rats before and after the establishment of hypertension, as well as in age and sex-matched Wistar Kyoto rats, the normotensive reference strain [3]. The results of this study clearly indicate that when at early stages systolic blood pressure does not appear to be different between the two rat strains, the defect in vascular response to insulin is already present in the rats that will develop hypertension. This observation suggests that the vascular insulin resistance could precede the appearance of a stable hypertensive condition, such as demonstrated for hyperinsulinemia and insulin resistance for human offspring of hypertensive parents [4]. Thus, insulin resistance may play a causative or permissive role in the development of hypertension. To elucidate the pathophysiological mechanisms that account for insulin resistance, we tried to unravel the molecular pathway through which insulin realizes its vascular effects. Therefore, we explored in the forearm of normal subjects whether an endothelial mediator may play a role in insulin modulation of sympathetic-evoked vascular responses [5]. Insulin modulated the vascular effect elicited by α_2 and β-adrenergic stimulation, but it had no effect on α_1-adrenergic-evoked vasoconstriction, suggesting that there is a specific crosstalk between insulin and the signal transduction pathways activated in response to α_2 and β-adrenergic agonists. Moreover, the vascular effects of insulin were abolished by the nitric oxide synthase inhibitor L-NMMA and by endothelium removal. This finding clearly indicates that an intracellular pathway at the endothelial level is modulated by the hormone, leading to enhanced production of nitric oxide, the main endothelial vasodilator. Stimulation of α_2 receptors is known to activate nitric oxide release through two distinct pathways, one of them involving receptor coupling with G_i proteins. To identify which pathway was altered by insulin, we stimulated rat aortic rings concomitantly with an α_2 agonist and with insulin in presence of a toxin able to uncouple G_i from the receptor, specifically disrupting the signal transduction originated by the G_i proteins [6]. The toxin completely abolished insulin-induced potentiation of α_2-adrenergic vasorelaxation, thus showing the exact pathway involved in insulin endothelial action. Interestingly, very high doses of insulin, 100 times more than physiological levels, have been shown to induce a direct nitric oxide release, through a PI3kinase-Akt-eNOS pathway [7]. This dose-dependent vascular response to insulin is likely related to a different activation threshold of this pathway. In other words, while very high levels of insulin clearly activate the production of nitric oxide, lower physiological levels of the hormone are able only to sensitize the nitric oxide production induced by other heterologous stimuli. Therefore, the vascular action of insulin is realized through the enhancement of a physiological pathway modulating positively eNOS activity.

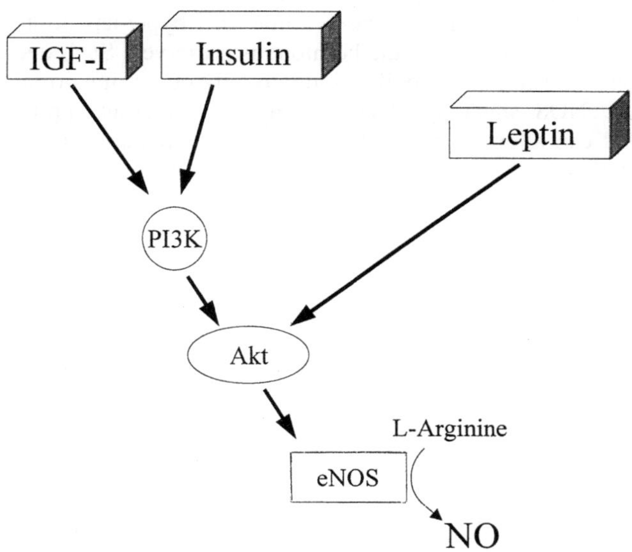

Figure 1. *Signalling pathways connecting several hormones with nitric oxide production in endothelial cells.*

Significantly, we observed the same effects, eliciting a direct stimulation of nitric oxide at high doses and a sensitization of the α_2-adrenergic-pathway at low doses, with another hormone, Insulin-like Growth Factor-I, that shares a great similarity with insulin [8], suggesting that the same pathway could underlie the vascular responses elicited by both substances. On the other hand, we have demonstrated that leptin, an adipocyte hormone strictly linked to obesity and hypertension, realizes a

direct activation of eNOS through Akt phosphorylation as insulin, but in a PI3kinase-independent manner [9], indicating that different hormones are able to elicit similar endothelial responses through different signal transduction pathways. However, the signaling pathways connecting insulin or leptin to eNOS intersect in a downstream protein kinase, Akt, suggesting that a vascular interaction between the two hormones is possible. To test this hypothesis, we evaluated the effects of insulin on vasodilator responses of rat aortic rings to graded doses of leptin [10].
Insulin preincubation potentiated leptin-evoked vasodilatations, and this effect was inhibited by a PI3kinase inhibitor, thus suggesting that insulin can sensitize nitric oxide production through an enhanced activation of Akt. In this regard, a recent study has shown that mice genetically ablated for an Akt specific subtype present insulin resistance in several organs [11], confirming the importance of Akt on insulin action.

Conclusion

The impact of insulin resistance on hypertension depends on a decreased insulin-evoked sensitization of nitric oxide release, while the hyperactivation of the sympathetic nervous system induced by the hormone is preserved. Such decrease in the facilitating action of insulin on vasodilatation may rely on an impairment in the insulin/PI3kinase/Akt/eNOS signaling. The application of genetic approaches to insulin resistance and can allow to better delineate this molecular pathway and to identify new therapeutic targets.

References

1. Haffner SM, Valdez RA, Hazuda HP, et al. Prospective analysis of the insulin-resistance syndrome (syndrome X). Diabetes 1992;41:715-22
2. Lembo G, Rendina V, Iaccarino G, et al. Insulin does not modulate reflex forearm sympathetic vasoconstriction in patients with essential hypertension. J Hypertens 1993;11:S272-3
3. Lembo G, Iaccarino G, Vecchione C, et al. Insulin modulation of vascular reactivity is already impaired in prehypertensive spontaneously hypertensive rats. Hypertension 1995;26:290-3
4. Ferrari P, Weidmann P, Shaw S. Alterd insulin sensitivity, hyperinsulinemia, and dyslipidemia in individuals with an hypertensive parent. Am J Med 1991;91:589-96
5. Lembo G, Iaccarino G, Vecchione C, et al. Insulin modulation of an endothelial nitric oxide component present in the α_2- and β-adrenergic responses in human forearm. J Clin Invest 1997;100:2007-14
6. Lembo G, Iaccarino G, Vecchione C, et al. Insulin enhances endothelial α_2 adrenergic vasorelaxation by a pertussis toxin mechanism. Hypertension 1997;30:1128-34
7. Zeng G, Nystrom FH, Ravichandran LV, et al. Roles for insulin receptor, PI3-kinase, and Akt in insulin-signaling pathways related to production of nitric oxide in human vascular endothelial cells. Circulation 2000;101:1539-45
8. Vecchione C, Colella S, Fratta L, et al. Impaired insulin-like growth factor I vasorelaxant effects in hypertension. Hypertension 2001;37:1480-5
9. Vecchione C, Colella S, Maffei A, et al. Leptin effect on endothelial nitric oxide is mediated by a PI3kinase-independent Akt phosphorylation. Diabetes 2001 (in press)
10. Vecchione C, Marino G, Maffei A, et al. Insulin potentiates leptin vasodilation through a PI3K-dependent mechanism. Circulation 2001, abstract
11. Cho H, Mu J, Kim JK, et al. Insulin resistance and a diabetes mellitus-like syndrome in mice lacking the protein kinase Akt2 (PKBβ). Science 2001;292:1728-31

9. ENDOTHELIAL CHANGES IN HYPERTENSION

C. Zaragoza and S. Lamas

Introduction

Hypertension is the second cause of death in developed countries. Human hypertension and several animal models of hypertension are associated with increased peripheral vascular resistance and changes at the endothelial layer of the vessels. The endothelium is a dynamic organ which responds to several stimuli triggered by a wide plethora of effectors in order to restore the loss of function [1]. There is accumulating evidence pointing towards a dysregulation of endothelial function as key in the development of cardiovascular diseases like hypertension. As a consequence of such dysregulation, the endothelium suffers changes in structure and function, which under certain circumstances contribute to the genesis of different cardiovascular diseases. Here we will summarize the most important changes occurring within the vascular endothelium in response to hypertension.

Structural changes in hypertension

As a result of their location, endothelial cells are exposed to mechanical forces: pressure, tension, and shear stress. Of these forces, shear stress appears to be very important in mediating endothelial changes both morphological and metabolic. Cells exposed to positive shear stress exhibit a reorientation in respect to the blood flow. It is of particular interest that regions experiencing laminar blood flow tend to be protected against atherogenesis. By contrast there is a strict correlation between areas suffering high shear stress forces (such as those in the bifurcation of coronary arteries) with the formation of atherosclerotic lesions [2], although the mechanism by which shear stress is transduced into metabolic pathways remains poorly understood. Mechanical forces activate the cascade of the Mitogen Activated Protein Kinases (MAPK) ERK1/2 (Extracellular signal Regulated Kinase), inducing gene expression [3]. However, the link between shear stress and ERK activation is not well documented.

Arterial pressure plays a pivotal role in the maintenance of the vessel architecture. Hypertension is associated with alterations in the structure and number of arteries, and these modifications affect the whole artery layer system from the endothelium to the adventitia. The endothelial response to hypertension is correlated with a re-orientation, polarity loosening, and a dramatic change in cell morphology turning the cells into a polyedric-like shape. As a result of these changes, endothelial uniformity is lost and protrusion of the cells into the lumen of the vessel wall takes place [4]. One of the causes by which hypertensive patients acquire atherosclerotic lesions is thought to be mediated by the loss of the ability of endothelial cells to respond to shear stress forces, although it is not the sole cause of this vascular pathology.
Hypertension-mediated phenotypic changes of the endothelium wall are also associated with the modification in the composition of the extracellular matrix (ECM) at the intima layer, and the balance between the expression and activity of metalloproteinases, ECM degrading enzymes, and their specific inhibitors TIMPs [5]. As we will further discuss, hypertension is strongly associated with severe cardiovascular pathological entities like atherosclerosis.

Metabolic changes in hypertension

The endothelial wall is the main sensor of dynamic stress, arterial pressure, and chemicals, and it is the signal transducer of the cardiovascular system from the blood flow to the vessels. Endothelial cells are in charge of maintaining vascular tone and of regulating cell adhesion and platelet aggregation to the vascular wall.
Endothelial changes in response to hypertension are mediated by the synergistic effect of mechanical and chemical stimuli, and the consequence is the release of vasoactive factors, which are intended to restore the functionality of the endothelium. We can classify these factors into the following categories: relaxing factors, contracting factors, reactive oxygen species, growth factors, and coagulation and fibrinolysis factors. A brief description of the physiologic functions of these factors follows.

Relaxing factors

In response to different stimuli, endothelial cells synthesize the relaxing factors nitric oxide, prostacycline (PGI_2), and the endothelium derived hyperpolarizing factor (EDHF) [1](figure 1).

Nitric Oxide
Nitric oxide (NO) is a free radical gas molecule produced by the activity of the enzymes nitric oxide synthases (NOSs). Three NOS isoforms have been described. Two constitutive isoforms are expressed in endothelial cells and central nervous

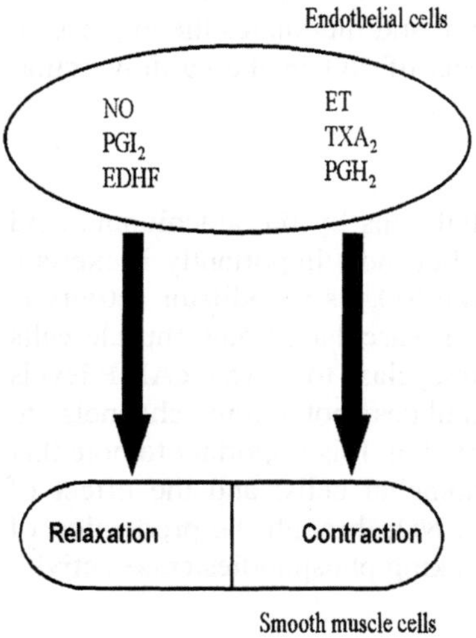

Figure 1 *Relaxing and contracting factors produced by endothelial cells.*

system (eNOS and nNOS, respectively) among other tissues, and an inducible isoform is synthesized by several cell types (iNOS) [6]. Both eNOS and nNOS are calcium dependent enzymes and produce NO constantly at low levels. By contrast, iNOS is a calcium independent isoform and in response to pro-inflammatory stimuli it generates NO very rapidly and in amounts which are several orders of magnitude higher than the constitutive isoforms. NO plays different roles depending on the tissue at which it is produced. In general it is a signaling molecule, in the central nervous system it is a neurotransmitter [7], immune cells generate NO during inflammation, and it may also act as an immune effector against several microorganisms [8]. In the vascular endothelium it is synthesized by the constitutive calcium dependent NOS isoform (eNOS). It acts as a potent vasodilator and thus is key for the maintenance of the vascular tone [9]. The main target of NO produced by the endothelial cells resides in the smooth muscle cell layer, and is the soluble guanylate cyclase, which produces cGMP and thus induces vasorelaxation [10]. Increased blood pressure induces the production of NO therefore altering the balance between vasodilatation and constriction [11].

In recent years NO has been shown to regulate gene expression. This may occur through its natural effector, cGMP, or by the post-translational modification of proteins. The latter include nitrosylation, nitration, or glutathionylation, even when solid examples for in vivo regulation have only been provided with nitrosylation [12]. Besides, the interaction of NO with several signaling cascades has been reported .

Beyond its effect in the vascular tone, NO plays an important role by controlling platelet aggregation, immune cell adhesion, and ECM remodeling. NO blocks adhesion, activation, secretion and platelet aggregation in endothelial cells, and these effects are in part cGMP-dependent [13], although results from recent studies are starting to raise the question about direct protein modifications [14]. NO attenuates monocyte and leukocyte adhesion to the endothelium and modulates the expression of extracellular matrix components [15]. The effects of NO in the cardiovascular system are summarized in Table 1 [16].

Prostacycline

The prostacycline PGI_2 is synthesized in endothelial cells by the arachidonic acid route and is also a smooth muscle relaxing factor, but most importantly it exerts a powerful effect blocking platelet aggregation. Unlike NO, its vasodilator activity is determined by the expression of specific receptors in vascular smooth muscle cells [17]. Prostacyclin receptors are coupled to adenylate cyclase to elevate cAMP levels in vascular smooth muscle. This in turn stimulates potassium channels to hyperpolarize cell membranes and inhibit cell contraction. It is important to note that prostacyclin facilitates the release of NO by endothelial cells, and the effect of prostacyclin in smooth muscle is increased by NO. NO, through the production of cGMP, increases the half-life of prostacyclin by blocking phosphodiesterase activity which breaks down cAMP [18].

Table 1. Steps at which NO is involved in the vascular system

-Relaxation
-Vascular tone
-Platelet aggregation
-Extracellular matrix remodeling
-Immunity
-Cell signaling

Endothelium-derived hyperpolarizing factor

Smooth muscle relaxation is mediated in the arteries by polarization changes mediated by potassium channels. As we mentioned above this is how PGI_2 mediates muscle relaxation and is ATP dependent. However, ATP-independent polarization channels exist, and they are used by EDHF to produce smooth muscle relaxation. EDHF composition is unknown and it is also possible that more than one EDHF exists in nature [19]. It was postulated that it might be an arachidonic acid derivative. Others claim that it may be just potassium released by endothelial cells [20].

The relaxing properties of EDHF are restricted to the microvasculature [21]. In fact, in human coronary arteries, as well as in animal blood vessels, the importance of EDHF increases as the arterial diameter decreases, and thus EDHF probably plays a significant role in the regulation of peripheral vascular resistance and local hemodynamics .

There is in addition a relationship between NO and EDHF. Elevated NO levels induce an impairment of the EDHF effect, thus providing a basis for the argument that EDHF is just a back up mechanism of relaxation in situations of low NO production [22].

Contracting factors

Endothelial cells synthesize factors which mediate smooth muscle contractility. These factors are endothelins and arachidonic acid derivatives like PGH_2 and thromboxane A2 (TXA_2, figure1).

Endothelin

Endothelin is a potent vasoconstrictor peptide composed by 21 amino acids. Endothelin is synthesized from a variety of tissues, and among the 3 isoforms of endothelin found (endothelin-1, -2, and –3), endothelial cells synthesize endothelin-1 [23].

Endothelin-1 (ET-1) is produced in the vascular endothelium by means of different stimuli including shear stress, increased blood pressure, hypoxia, growth factors, cytokines, angiotensin, thrombin, or adrenaline. In humans ET-1 gene expression yields a 201 amino acid polypeptide precursor which is processed to form an intermediate known as big ET-1. Big ET-1 will be further cleaved by the endothelin converting enzyme-1 protease, to render the active form of ET-1.

The mechanism by which ET-1 induces vasoconstriction is by binding to specific receptors located at the surface of a wide variety of cell types. In mammalian tissues two types of ET receptors ET_a, and ET_b receptors have been reported. ET_a is preferentially located in vascular smooth muscle and promotes contractility , while the ET_b receptor is present at the endothelial layer, and mediates the release of NO and prostacyclin [24]. Thus, as mentioned above in relation with other factors, there is a cross talk between ET and NO, as also increased cGMP levels inhibit the release of ET-1. Hence, under physiological conditions, blood flow and pressure regulate the production of ET-1 and NO. Increased blood flow induces the production of NO and ET-1 release is inhibited by cGMP, therefore promoting vasorelaxation. In addition recent studies revealed an inhibitory effect of NO by binding to the ET_b receptor [25].

Arachidonic acid derivatives

Under different pathophysiological conditions, endothelial cells synthesize other vasoconstrictive factors which are most of them derived from the arachidonic acid cascade. The two best-known factors synthesized by endothelial cells are the cyclic endoperoxide PGH_2 and thromboxane A_2. Both factors share in common the same receptor which makes them very similar in function: vasoconstriction and platelet aggregation. During physiological conditions they do not play a dominant role, however, under high shear stress, hypertension, and hypercholesterolemia their synthesis is increased [26], and it is believed that both work as a negative feedback mechanism of the effects mediated by NO.

Reactive oxygen species (ROS)

ROS are produced in the cardiovascular system by endothelial cells and vascular smooth muscle cells, and they participate in several signaling cascades associated with cardiovascular pathology. During physiological conditions, oxygen is reduced to produce superoxide anion (O_2^-), hydrogen peroxide, and water. In the vascular wall, oxidative stress produces an excess of O_2^-, which is metabolized to synthesize hydroxyl radicals and peroxynitrite ($OONO^-$) when it reacts with NO. ROS contribute to the alteration of blood flow, platelet aggregation, cell adhesion, and cell growth. Although several enzymatic systems are potentially able to generate ROS in vascular cells, the main contributors are NADH/NADPH oxidase, xanthine oxidase, and eNOS [27].

Under hypertensive conditions O_2^- is produced by the enzymes mentioned above, and is in part mediated by the activation of the renin-angiotensin system, which increases the levels of the p22phox subunit of the NADPH oxidase, and O_2^-, and is partially reverted by the treatment with superoxide dismutase (SOD).

The reaction which takes place in endothelial cells between O_2^- and NO to form the species $OONO^-$ is of particular importance [28]. The reaction takes place at a rate which is 3-4 times higher than the dismutation of O_2^- by SOD. This implies an inhibitory effect of NO function by the production of O_2^-. In addition $OONO^-$ is a strong oxidant, capable of inducing tyrosine nitration, and of inhibiting several enzymatic activities. Besides, nitrogen dioxide a metabolite produced as a result of its degradation is also highly oxidant and cytotoxic [29].

Growth factors

Endothelial cells synthesize growth factors during several physiological and pathological conditions. Changes in the endothelium as a result of growth factor expression are produced during angiogenesis, new vessel formation and hypoxia by the synthesis of vascular endothelial growth factor (VEGF), which promotes endothelial cell growth and migration [30]. Endothelial cells also synthesize another potent mitogenic effector, PDGF, which induces smooth muscle cell proliferation, and regulates the activity of VEGF and bFGF (fibroblast growth factor), also synthesized by endothelial cells in response to tissue damage, promoting smooth muscle cell proliferation [31]. Another type of growth factors produced by the endothelium are insulin-like growth factors (IGF-I, and IGF-II), which have similar effects as insulin, also promoting vascular smooth muscle cell proliferation [32].

Coagulation and fibrinolysis

Endothelial cells provide to the cardiovascular system an antithrombotic and profibrinolytic surface where platelet aggregation and clotting are prevented. Hypertension is one of the conditions where this protective mechanism is altered, mostly mediated as a consequence of mechanical and biochemical forces.

The critical step at which endothelial cells become prothrombotic involves the activation of tissue factor (TF), which induces the formation of fibrin located at the cell surface [33]. TF also leads to the expression of several surface receptors of coagulation including PAR-1, -2 (Protease Activated Receptor –1, and –2), and integrins involved in coagulation [34].

Plasminogen activators like t-Pa, are expressed in endothelial cell sub-populations mostly located at the microvasculature, thus its effect is restricted to those locations [35]. Another plasminogen activator, u-PA seems to be expressed during wound repair and angiogenesis [36]. Both activators bind to the endothelial cell surface through receptors. In the case of t-PA it binds to annexin-II while u-PA binds to endothelial cells through the u-PA receptor (u-PAR).

On the other hand, endothelial cells also express plasminogen activator inhibitors (PAIs), which prevent plasminogen activation [37]. They also express the Von Willebrand factor which is procoagulant, prothrombotic and antifibrinolytic [38]. Therefore, in order to keep vessels in a status where aggregation and clotting are inhibited, endothelial cells stay in constant activity, and any alteration affecting this balance, such as hypertension may dramatically affect coagulation and fibrinolysis.

Functional changes in hypertension: endothelial dysfunction

The vascular endothelium has developed mechanisms to sense any potential aggressive situation in order to promptly response by the synthesis and cross talk of vasoactive factors as described above. However, chronic increased blood pressure, hypertension, and other pathological conditions lead to a dysregulation of endothelial homeostasis, which finally results in the so-called endothelial dysfunction.

Endothelial dysfunction is a term which defines the consequence of the imbalance in the synthesis of vasoactive factors by the endothelium: impaired vasorelaxation, loss of vascular tone, change to a pro-adhesion phenotype, and increased platelet aggregation. The impairment of endothelium-dependent functions may lead to localized inflammation and ultimately to severe vascular disorders such as atherosclerosis, thrombosis, and stroke [39].

During hypertension endothelial dysfunction is mostly promoted by changes in blood pressure and blood flow across the vessels, and it is associated with an impairment in the ability to synthesize and release endothelial vasoactive factors [40].

The impairment in the ability to produce NO is the most common manifestation which leads to hypertensive endothelial dysfunction. In fact, null mice for eNOS do exhibit a hypertensive phenotype [9]. There are several mechanisms proposed to explain the reduction observed in the amount of NO during hypertension: inhibition of eNOS expression, alterations in eNOS activity, reduction in the amount of L-Arginine (the substrate of NO synthesis) and a diminished availability of co-factors required for the synthesis of NO [6]. Another mechanism involved in the reduction of NO associated to hypertension is an increased oxidative stress by the activation of several enzymes including NADPH oxidase, cycloxygenase, or cytochrome P450

reductase, which lead to an increase in the amount of O_2 . As result of these enzymatic activities O_2^- anions synthesized react with NO to produce $OONO^-$, thus diminishing vasorelaxation (figure 2). Another side effect associated with the increased oxidative stress is the cytotoxicity which may follow the generation of $OONO^-$ (protein nitration, lipid peroxidation, impairment of gene expression) [41].
As discussed above, ET-1 is the main vasoconstrictive effector produced by endothelial cells. In some studies in humans with severe hypertension, ET-1 is present at high levels in plasma. However, the renal function of these patients is seriously impaired. In fact, hypertensive patients with normal renal function have no differences in circulating ET-1 levels with respect to healthy subjects [42]. Another evidence supporting the idea that ET-1 is important for the course of endothelial dysfunction arose from recent studies in which blockage of the ET_a receptor during hypertension improved endothelial dysfunction of hypertensive rats [43]. On the other hand, in experimental models of hypertension the reduction of vasorelaxation induced by acetylcholine is thought to be mediated by the release of a prostanoid-like vasoconstrictor effector, presumably TXA_2 as TXA_2 agonist receptors have been shown to increase acetylcholine relaxation.

Figure 2
Antagonistic effects of NO and O_2^- on smooth muscle relaxation are mediated through the formation of peroxynitrite

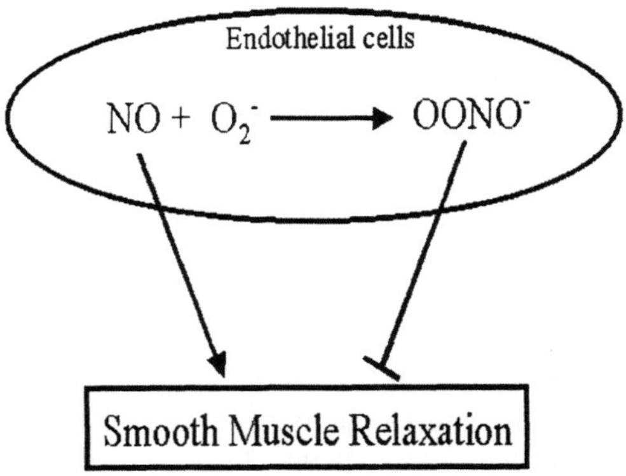

Endothelial dysfunction is not only associated with alterations in the vascular tone. It is also correlated with structural variations in the vascular wall, and the involvement of cardiovascular pathologies such as atherosclerosis. The magnitude of vascular remodeling associated to hypertension is different depending on the vessel. In small arteries there is a reduction in the diameter of the lumen and the reduction in the number of vessels, while in the large arteries there is a decrease in the elasticity which promotes hypertrophy of the left ventricle. In addition, these structural alterations will be the cause of hypoperfusion and ischemia, with the risk of angina and cardiac arrest.
Endothelial dysfunction is also involved in atherogenesis, and this is due mainly to the capacity of native and oxidized LDL to modify the expression and activity of endothelial vasoactive factors [44]. Hypertensive patients are more susceptible to develop atherosclerotic lesions than normal subjects. In this case scenario it has been shown that hypertension correlates with an increase of adhesion molecules like

ICAM and VCAM [45, 46]. How hypertension increases adhesion molecule levels is not well understood. Pro-inflammatory cytokines, through the activation of the transcription factor NF□B, tend to increase the expression of adhesion molecules and monocyte chemoattractant proteins which in the end promote the adhesion of monocytes and leucocytes to the endothelium of arteries. Hypertension induces the activation of pro-inflammatory cytokines like IL-1 and tumor necrosis factor alpha (TNF-α) [47]. Under these circumstances, reduced NO availability facilitates platelet aggregation, monocyte adhesion, and smooth muscle cell proliferation in the vessels, finally culminating in the development of the atherosclerotic lesion.

Conclusion

The former concept of the endothelium as a static barrier of the cardiovascular system is no longer accepted. Endothelial cells are located in the interface between the blood flow and specialized cells which in response to stimuli from both sides can transmit signals bidirectionally. Indeed, their location and plasticity make endothelial cells key mediators in the regulation of vascular function. Cardiovascular pathologies like hypertension are promoters of endothelial dysfunction, thus altering the endothelial homeostatic properties conferred to the vascular system. Among the different factors which contribute to endothelial dysfunction, failure in NO bioavailabilty is of particular relevance. Inhibition of NO production within the endothelial layer exacerbates atherosclerosis in animal models, whereas L-Arginine diminishes lesions in animals and in hypercholesterolemic young humans. The inhibition of NO production associated to hypertension is in part responsible of atherosclerotic lesions, and this is in part due to changes experimented by endothelial cells which in turn are accompanied by alterations in function: lack of relaxation with loss of homeostasis, transformation into a pro-adhesive phenotype, and facilitation of platelet aggregation. A profound understanding of the mechanisms of endothelial dysfunction associated to changes in the vascular system and its potential reversion will lead to better-targeted therapeutic strategies for normalization of endothelial function in cardiovascular diseases.

References

1. Luscher TF, Barton M. Biology of the endothelium. Clin Cardiol 1997;20:II-3-10.
2. Traub O, Berk BC. Laminar shear stress: Mechanisms by which endothelial cells transduce an atheroprotective force. Arterioscler Thromb Vasc Biol 1998;18:677-85.
3. Surapisitchat J, Hoefen RJ, Pi X, et al. Fluid shear stress inhibits tnf-alpha activation of jnk but not erk1/2 or p38 in human umbilical vein endothelial cells: Inhibitory crosstalk among mapk family members. Proc Natl Acad Sci U S A 2001;98:6476-81.
4. Prewitt RL. Teaching vascular adaptations to mechanical stress. Am J Physiol 1999;277:S211-3.
5. Li-Saw-Hee FL, Edmunds E, Blann AD, et al. Matrix metalloproteinase-9 and tissue inhibitor metalloproteinase-1 levels in essential hypertension. Relationship to left ventricular mass and anti-hypertensive therapy. Int J Cardiol 2000;75:43-7.
6. Cooke JP, Dzau VJ. Nitric oxide synthase: Role in the genesis of vascular disease. Annu Rev Med 1997;48:489-509.
7. Bredt DS. Endogenous nitric oxide synthesis: Biological functions and pathophysiology. Free Radic Res 1999;31:577-96.
8. Zaragoza C, Ocampo CJ, Saura M, et al. Inducible nitric oxide synthase protection against coxsackievirus pancreatitis. J Immunol 1999;163:5497-504.
9. Shesely EG, Maeda N, Kim HS, et al. Elevated blood pressures in mice lacking endothelial nitric oxide synthase. Proc Natl Acad Sci U S A 1996;93:13176-81.
10. Murad F. Cyclic guanosine monophosphate as a mediator of vasodilation. J Clin Invest 1986;78:1-5.
11. Raij L. Hypertension and cardiovascular risk factors: Role of the angiotensin ii-nitric oxide interaction. Hypertension 2001;37:767-73.
12. Stamler JS, Toone EJ, Lipton SA, et al. (s)no signals: Translocation, regulation, and a consensus motif. Neuron 1997;18:691-6.
13. Vaandrager AB, de Jonge HR. Signalling by cgmp-dependent protein kinases. Mol Cell Biochem 1996;157:23-30.
14. Tsikas D, Ikic M, Tewes KS, et al. Inhibition of platelet aggregation by s-nitroso-cysteine via cgmp- independent mechanisms: Evidence of inhibition of thromboxane a2 synthesis in human blood platelets. FEBS Lett 1999;442:162-6.
15. Sawicki G, Salas E, Murat J, et al. Release of gelatinase a during platelet activation mediates aggregation. Nature 1997;386:616-9.
16. Ignarro LJ, Cirino G, Casini A, et al. Nitric oxide as a signaling molecule in the vascular system: An overview. J Cardiovasc Pharmacol 1999;34:879-86.
17. Breyer RM, Bagdassarian CK, Myers SA, et al. Prostanoid receptors: Subtypes and signaling. Annu Rev Pharmacol Toxicol 2001;41:661-90.
18. Delpy E, Coste H, Gouville AC. Effects of cyclic gmp elevation on isoprenaline-induced increase in cyclic amp and relaxation in rat aortic smooth muscle: Role of phosphodiesterase 3. Br J Pharmacol 1996;119:471-8.
19. Garland CJ, Plane F, Kemp BK, et al. Endothelium-dependent hyperpolarization: A role in the control of vascular tone. Trends Pharmacol Sci 1995;16:23-30.
20. Coleman HA, Tare M, Parkington HC. Edhf is not k+ but may be due to spread of current from the endothelium in guinea pig arterioles. Am J Physiol Heart Circ Physiol 2001;280:H2478-83.
21. Coats P, Johnston F, MacDonald J, et al. Endothelium-derived hyperpolarizing factor : Identification and mechanisms of action in human subcutaneous resistance arteries. Circulation 2001;103:1702-8.
22. Campbell WB, Harder DR. Prologue: Edhf--what is it? Am J Physiol Heart Circ Physiol 2001;280:H2413-6.

23. Banasik JL. Endothelins: New players in cardiovascular physiology and disease. J Cardiovasc Nurs 1994;8:87-104.
24. Masaki T, Miwa S, Sawamura T, et al. Subcellular mechanisms of endothelin action in vascular system. Eur J Pharmacol 1999;375:133-8.
25. Wiley KE, Davenport AP. Nitric oxide-mediated modulation of the endothelin-1 signalling pathway in the human cardiovascular system. Br J Pharmacol 2001;132:213-20.
26. Mayhan WG. Role of prostaglandin h2-thromboxane a2 in responses of cerebral arterioles during chronic hypertension. Am J Physiol 1992;262:H539-43.
27. Gorlach A, Brandes RP, Nguyen K, et al. A gp91phox containing nadph oxidase selectively expressed in endothelial cells is a major source of oxygen radical generation in the arterial wall. Circ Res 2000;87:26-32.
28. Navarro-Antolin J, Lopez-Munoz MJ, Klatt P, et al. Formation of peroxynitrite in vascular endothelial cells exposed to cyclosporine a. Faseb J 2001;15:1291-3.
29. Beckman JS. $OONO^-$: Rebounding from nitric oxide. Circ Res 2001;89:295-7.
30. Faller DV. Endothelial cell responses to hypoxic stress. Clin Exp Pharmacol Physiol 1999;26:74-84.
31. Sandirasegarane L, Charles R, Bourbon N, et al. No regulates pdgf-induced activation of pkb but not erk in a7r5 cells: Implications for vascular growth arrest. Am J Physiol Cell Physiol 2000;279:C225-35.
32. Liu X, Lin CS, Spencer EM, et al. Insulin-like growth factor-i promotes proliferation and migration of cavernous smooth muscle cells. Biochem Biophys Res Commun 2001;280:1307-15.
33. Drake TA, Morrissey JH, Edgington TS. Selective cellular expression of tissue factor in human tissues. Implications for disorders of hemostasis and thrombosis. Am J Pathol 1989;134:1087-97.
34. Langer F, Morys-Wortmann C, Kusters B, et al. Endothelial protease-activated receptor-2 induces tissue factor expression and von willebrand factor release. Br J Haematol 1999;105:542-50.
35. Levin EG, Santell L, Osborn KG. The expression of endothelial tissue plasminogen activator in vivo: A function defined by vessel size and anatomic location. J Cell Sci 1997;110:139-48.
36. Bacharach E, Itin A, Keshet E. In vivo patterns of expression of urokinase and its inhibitor pai-1 suggest a concerted role in regulating physiological angiogenesis. Proc Natl Acad Sci U S A 1992;89:10686-90.
37. Higazi AA, Mazar A, Wang J, et al. Single-chain urokinase-type plasminogen activator bound to its receptor is relatively resistant to plasminogen activator inhibitor type 1. Blood 1996;87:3545-9.
38. Brinkhous KM, Reddick RL, Read MS, et al. Von willebrand factor and animal models: Contributions to gene therapy, thrombotic thrombocytopenic purpura, and coronary artery thrombosis. Mayo Clin Proc 1991;66:733-42.
39. Taddei S, Virdis A, Ghiadoni L, et al. Endothelial dysfunction in hypertension. J Nephrol 2000;13:205-10.
40. Spieker LE, Noll G, Ruschitzka FT, et al. Working under pressure: The vascular endothelium in arterial hypertension. J Hum Hypertens 2000;14:617-30.
41. Szabo C, Ferrer-Sueta G, Zingarelli B, et al. Mercaptoethylguanidine and guanidine inhibitors of nitric-oxide synthase react with peroxynitrite and protect against peroxynitrite-induced oxidative damage. J Biol Chem 1997;272:9030-6.
42. Pinto-Sietsma SJ, Paul M. A role for endothelin in the pathogenesis of hypertension: Fact or fiction? Kidney Int Suppl 1998;67:S115-21.
43. Park JB, Schiffrin EL. Et(a) receptor antagonist prevents blood pressure elevation and vascular remodeling in aldosterone-infused rats. Hypertension 2001;37:1444-9.
44. Hernandez-Perera O, Perez-Sala D, Navarro-Antolin J, et al. Effects of the 3-hydroxy-3-methylglutaryl-coa reductase inhibitors, atorvastatin and simvastatin, on the expression of

endothelin-1 and endothelial nitric oxide synthase in vascular endothelial cells. J Clin Invest 1998;101:2711-9.
45. Liu Y, Liu T, McCarron RM, et al. Evidence for activation of endothelium and monocytes in hypertensive rats. Am J Physiol 1996;270:H2125-31.
46. Tummala PE, Chen XL, Sundell CL, et al. Angiotensin ii induces vascular cell adhesion molecule-1 expression in rat vasculature: A potential link between the renin-angiotensin system and atherosclerosis. Circulation 1999;100:1223-9.
47. Dorffel Y, Latsch C, Stuhlmuller B, et al. Preactivated peripheral blood monocytes in patients with essential hypertension. Hypertension 1999;34:113-7.

ESF workshop Maastricht 2001: Session 4

Genomics in cardiac hypertrophy and failure

10. TOWARDS ELUCIDATION OF GENETIC PATHWAYS IN CARDIAC HYPERTROPHY: TECHNIQUE TO DEVELOP MICROARRAYS

B.J.C.van den Bosch, P.A.Doevendans, D.J.Lips, J.M.W.Geurts, H.J.M. Smeets

Introduction

The incidence of heart failure has increased the past two decades in the Western countries. Currently, 1.5% of the population in our society suffers from pump failure. Over the age of 75 years the prevalence rises to 10%. Genetic causes of familial dilated cardiomyopathy, covering about 20% of all patients with dilated cardiomyopathy [1], are currently being resolved at a rapid pace and include cytoskeletal, sarcomeric and metabolic genes [2]. In other families, loci have been identified but the genetic causes largely remain to be resolved. Dilated cardiomyopathy can also occur as part of specific syndromes [3]. Most patients, however, develop pump failure as end stage heart disease due to chronic cardiovascular disease, like hypertensive heart disease and coronary heart disease. Pump failure is then preceded by asymptomatic myocardial hypertrophy. During the development of hypertrophy, a foetal gene program is being restarted, with among others re-expression of atrial natriuretic factor (ANF), skeletal α-actin, atrial myosin light chain (MLC-2a) and (in rats, but not in humans) beta-myosin heavy chain [4]. Our studies aim at the identification of the molecular program underlying myocardial hypertrophy by applying the powerful strategy of gene expression microarrays. Insight in the early pathogenic events of hypertrophy may provide clues for therapeutic interventions.

Microarrays have been used to determine gene expression differences between normal, hypertrophic and failing heart [5]. Because affected human cardiac tissue will display complex expression patterns and cDNA microarrays will identify many alterations in gene expression levels, we have started with well-controlled animal models. Several mouse models for cardiac hypertrophy exist. Some of them are transgenic mice, mimicking genetic defects in human patients with familial hypertrophic cardiomyopathy (for example α-MHC: Arg403Glu, α-tropomyosin: Asp175Asn), troponin-T (arg92Gln) [6], others involve mechanically (Transverse

Aortic Banding: TAC [7]) or chemically induced hypertrophy (injections with isoproterenol or angiotensin II) and can be reversed (isoproterenol) [5]. This paper describes the TAC mouse model for hypertrophy and the development of in-house microarray technology, using both commercially available cDNA as well as oligonucleotide collections. Several technical approaches will be discussed and an overview of the pitfalls and possibilities will be presented. As the first large scale experiments are currently being performed and analysed, we will present updated results during the meeting.

Material and methods

TAC mice

By Transverse Aortic Banding (TAC) a stenosis is established in the thoracic aorta of mice, resulting in a pressure overload, which triggers a hypertrophic response in the heart [7]. The procedure is done with visual help of a micro-dissecting microscope. The aorta has to be prepared for the banding: first the tissue is dissected cranially of the aorta. This should be done between the first and second truncus of the aortic arch (e.g. truncus brachiocephalicus and the left common carotid artery). When free, the aorta is dissected caudally. After this a 7-0 silk ligature is pulled around the aorta with a modified micro-hook. This ligature is subsequently closed against a 0.4 millimetre canula. The closure should be done with a surgical knot. When the ligature is closed, the canula is withdrawn. It is essential that the closing time of the aorta does not exceed 10 seconds. Moreover the banding should be in the right place and constrict the aorta. Then, the thorax and skin are closed. The operation takes about 15 to 20 minutes. Post operation the animals receive pain relief (buprinorfine 2 mg/kg subcutaneously). It is administered at the end of the day as well as the beginning and the end of the following day. The mice are kept alive for 48 hours, 1 week, 2 weeks, 3 weeks or 6 weeks. The mice are anaesthetised (ketamine 100 mg/kg, intramuscularly; xylazine 5 mg/kg, subcutaneous) and sacrificed through cervical dislocation. The hearts are harvested and stored in RNAlater (Ambion).

cDNA and oligonucleotide collections

The 4,000 IMAGE cDNA clones (Incyte Genomics), derived from heart of adult (C57BL/6J, 4 weeks of age) [8] and foetal mice (NIH/Swiss, 13 day embryo hearts) [9] were supplied as glycerol stocks. PCR reactions (cycling conditions: 30sec 96°C, 30 rounds [45sec 94°C, 45sec 55°C, 150sec 72°C] followed by 5 min 72°C; PCR buffer 50mM KCl, 10mM Tris-HCl (pH=8), 4mM $MgCl_2$, 1mM dNTPs), universal primers (20 pmol of each primer M13 FWD [5' GTT TTC CCA GTC ACG ACG TTG 3'] and M13 REV [5' TGA GCG GAT AAC AAT TTC ACA CAG 3']) were performed using 2 μl of the glycerol stocks as template. By electrophoresis on 1% agarose gels, the quality of the PCR products was checked. A smart ladder (Eurogentec) was used for quantification and sizing of products. Products were purified and concentrated to achieve a final spot concentration of 300-400 μg/μl. Products were then dissolved in 50% DMSO prior to printing on the slides (Corning CMT-GAPS™ II slides). As controls, a number of housekeeping genes were amplified either from clones already present in the library or directly from cDNA (EF1α, Vimentin, β2 microglobulin, HPRT, UBE1, GAPDH, HPRT, β actin, Vimentin, Keratin 8, ribosomal protein S9). In addition, a number of hypertrophy markers was included as well (ANF, SERCA2, PLN, βMHC, α2-macroglobuline). Oligonucleotides (50-mers) were designed from the 3'-UTR region of genes and were tested for specificity against the mouse genome database (Celera). In addition, commercially available sets of oligonucleotides (Operon, 70-mers) were used as well.

Preparation and analysis of microarrays

The layout of microarrays was created using CloneTracker software (Version 1.4, BioDiscovery, Inc.). Adult and foetal cDNA clones or oligonucleotides were spotted with the Affymetrix 417 arrayer (four hits per dot in 50% DMSO) in duplicate. After printing, the slides were baked at 80°C for 4 hours. Total left ventricle RNA of TAC mice at different time points after induction of hypertrophy and controls was isolated using the TRIzol protocol (Life Technologies). Following a quality check using the Bioanalyser (Agilent), RNA was fluorescently labelled by direct incorporation of Cy3 and Cy5-labelled dCTP nucleotides (NEN) in a first strand cDNA synthesis reaction (Superscript II RT kit, Life Technologies). Hybridisation of the arrays was performed according to the updated TIGR protocol [10]. Several adjustments to the original protocol were tested to improve the intensity of the signals and the quality of the data. After washing the microarray (final stringency 0.1 SSC and 0.2% SDS at room temperature for 4 minutes) was scanned to quantify the signal intensity of the probe, bound to the cDNA targets or oligonucleotides on the microarray. All experiments are being performed in triplet. The resulting image and spot intensities are being analysed with Imagene (Version 4.2, BioDiscovery, Inc).

Results and Discussion

cDNA collections

Of the 4,000 cDNA clones of the library, 480 failed to amplify and 1200 showed multiple products. The poor quality of these cDNA libraries has been reported before [11]. Adjustments of PCR conditions did not lead to improvement of results. Sequence analysis of a number of clones with single bands revealed that the inserts were as expected. As abnormal gene expression profiles will be confirmed anyhow in replicate experiments or by RT-PCR, we decided to spot all fragments irrespective of the presence of multiple bands per clone.

All commercially available clone collections still have this disadvantage of being either incomplete, of poor quality, not specific, or a combination of these. This means that either all clones have to be characterised before spotting or that experiments have to be performed following the microarray experiment to validate the results. It is, however, obvious that large polluted data sets can mask specific

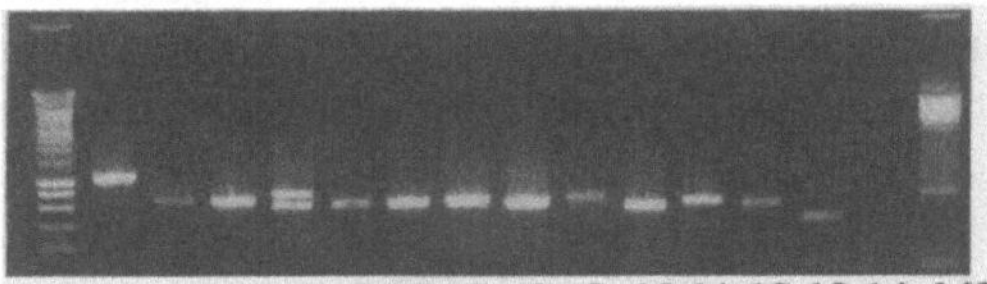

***Figure 1**: PCR fragments of cDNA clones. Lanes 1 tot 13 show 13 different cDNA clones (all, except lane 1 around 600 bp in size). Lane 4 contains a contaminated fragment. Note the differences in PCR efficiency. Lane 14 is a negative control.*

pathways. Furthermore, cDNAs can show homology to other genes as well and can trouble the detection of specific gene expression differences. A major advantage of the use of oligonucleotides is that these can be BLASTed against the respective genome sequences and potential cross-hybridisation can unambiguously be determined. Because of these inherent problems with the commercially available clone sets, it is also clear that all results from literature have to be taken with great caution.

Microarray protocol

In setting up the in-house microarray facility, we have tested every step in the protocol. Several types of glass slides (Telechem, poly-L-lysine [Sigma], aminosilane coated [Sigma, Corning]) and attachment chemistry (baking for 2-4 hours, UV crosslinking, amino-linking) were tested, both for PCR fragments and long oligonucleotides (50-, 60- and 70-mers). CMT-GAPS™ II glass slides (Corning) consistently yielded good results, both for PCR fragments and oligonucleotides. Several spotting solutions were compared, amongst which 3x SSC versus 50% DMSO. Highest signal intensities were achieved with 50% DMSO printed cDNAs. DMSO as a printing buffer has the additional advantage that the PCR fragments are denatured better, that evaporation is slower and that spot morphology is better compared to 3xSSC. Attachment to the glass surface was achieved by baking at 80°C for 4 hours. Additional UV cross-linking of the microarrays did not improve the results further and was therefore omitted from the protocol. Total RNA was labelled by a Reverse Transcription reaction using oligo(DT) priming and dCTP-Cy3 or dCTP-Cy5 incorporation. Several amounts of RNA were evaluated and labelling of 20 μg total RNA for 5 hours at 42°C turned out to be the lowest amount, yielding excellent results. After labelling the probe was purified using Autoseq™ G-50 columns (Pharmacia) and dye incorporation was checked by performing a wavescan from 200nm to 850nm. An incorporation of 2-5% was generally obtained, which was consistent with data from others [12].

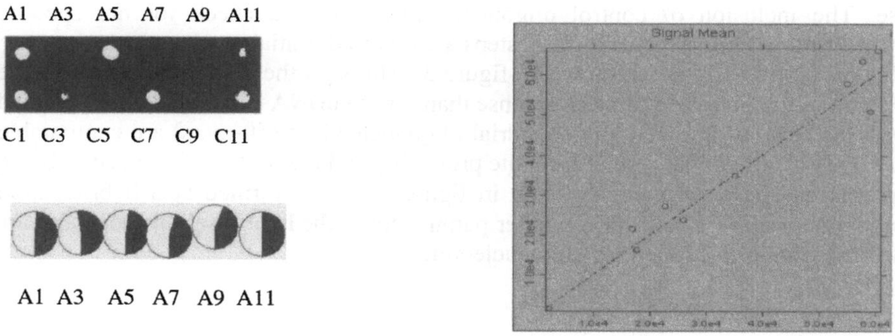

***Figure** 2: Control versus control experiment. Part of a microarray showing equal intensity of Cy3 and Cy5 signals, either by colour (white), gene pies (equal distribution of white and black) or scatter plot (linear relationship between Cy3 and Cy5 incorporation).*

Signal to noise ratios were further improved by optimising prehybridisation and hybridisation conditions. Prehybridisation was performed to block or inactivate free amine groups on the slide to prevent non-specific binding of labelled cDNA to the slide. Using 50% formamide in the prehybridisation and hybridisation solutions markedly reduced background, precluding the necessity for additional blocking agents. It should, however, be noted that 1% BSA is an essential component of the prehybridisation solution, since it was demonstrated that this greatly reduces background signal. Prehybridisation of the slides is now being performed in 5x SSC, 0.1% SDS, 1% BSA, and an end concentration of 50% formamide. Probe was added in a final concentration of 0.8μg/μl. Hybridisation (50% formamide, 5x SSC, 0.1% SDS) was performed overnight at 42°C. After hybridisation, slides were first agitated in a low stringency buffer (1x SSC, 0.2% SDS) for 4 minutes at 42°C, then in a high stringency buffer (0.1x SSC, 0.2% SDS) for 4 minutes at room temperature and finally in 0.1x SSC for 4 minutes at room temperature. Slides were scanned with Affymetrix 418 Scanner.

Control vs. control experiment

In an initial test experiment, total RNA was isolated from WT mouse heart and equivalent amounts (10μg) were labelled with Cy3 and Cy5. Subsequently, Cy3 and Cy5 labelled RNA was cohybridised on a single microarray containing several mouse cDNA clones. As expected, all spots turned yellow, indicating equal expression. The scatter plot showed good correlation between both Cy3 and Cy5 labelling signals and the gene pies also indicate equal expression.

cDNAs versus oligonucleotides

As the human, mouse and many other genome sequences have been characterised recently [13,14], it is only a matter of time before all genes and corresponding exon sequences are known. Given the drawbacks of cDNA collections, it can be expected that gene- or even exon-specific oligonucleotides will become available for general use. The inclusion of control oligonucleotides allows a check for non-specific hybridisation. Furthermore, no PCR steps are required. Initial results of the oligonucleotide arrays are shown in figure 3. Although the signal intensities of the Operon oligonucleotides are less intense than for the cDNAs, the intensities obtained with the custom-designed mitochondrial oligonucleotides (50-mers) are comparable to cDNA. These initial results are quite promising and the protocol is currently being optimised. The experiment depicted in figure 3 was performed at a hybridisation temperature of 35°C and several other parameters of the hybridisation procedure are currently being fine tuned for oligonucleotides.

Figure 3. *cDNA versus oligonucleotide microarray. The left side of the array shows 96 Operon oligonucleotides and the right upper sideshows 96 custom-made mitochondrial oligonucleotides (50 mers) in duplicate, hybridised with Cy5 labelled total RNA derived from a patient with a NDUFS7 mutation.*

Figure 4. *cDNA microarray containing 4,000 IMAGE cDNA mouse clones in duplo. As controls, a number of housekeeping genes are present (EF1α, Vimentine, β2 microglobuline, HPRT, UBE1, GAPDH, HPRT, β actine, Vimentin, Keratin 8, ribosomal protein S9). In addition, a number of hypertrophic markers is present as well (ANF, SERCA2, PLN, βMHC, α2-macroglobuline). The array was hybridised with labelled total RNA from a TAC mouse (labelled with Cy3-dCTP) 14 days after induction of hypertrophy and a sham-operated mouse (labelled with Cy5-dCTP). This picture only shows the Cy5 fluorescence.*

"Global" analysis of 4,000 heart specific clones
Finally, as all parameters have been adjusted and optimised, we have made microarrays with all 4,000 cDNA clones and controls in duplicate and we have started to test RNA from the TAC mouse model, compared to controls (figure 4). Conclusions can not be drawn at this point, because experiments are still ongoing and data still have to be analysed.

Conclusions

The establishment of an in-house microarray facility requires a major investment in advanced technology, large gene collections and bioinformatics tools. It is essential to combine all three aspects in a genetics and genomics based unit to have the necessary expertise to convert large datasets to real genomic and biological knowledge and to functional and pathogenic pathways. As gene expression profiling is the first step of globalisation of genetic and medical research, it is important to develop the expertise in-house. Otherwise, an entire new field of -omics research and technological developments will not be covered and one has to rely on commercial manufacturers. Although appealing at first sight, because part of these processes can be well-standardised in an industrial environment, it is obvious that the flexibility to design specific experiments will be less and more importantly that no genomics infrastructure will be established. This would be harmful to deal with future developments in every field of medical science. In this paper we describe the technical protocol for preparing a large cardiac-specific mouse microarray to study the process of cardiac hypertrophy. Although predominantly focussing on cDNA collections, we expect that oligonucleotides collections are far more promising and will become the gene expression platform of the years to come.

Acknowledgements

This work has been supported by grants from the Netherlands Heart Foundation (99.122).

References

1. Michels VV, Moll PP, Miller FA, Tajik AJ, Chu JS, Driscoll DJ, Burnett JC, Rodeheffer RJ, Chesebro JH, Tazelaar HD. The frequency of familial dilated cardiomyopathy in a series of patients with idiopathic dilated cardiomyopathy. N Engl J Med. 1992;326:77-2.
2. Schönberger J, Seidman CE. Many roads lead to a broken heart: The genetics of dilated cardiomyopathy. Am J Hum Genet 2001;69:249-60.
3. Santorelli FM, Tessa A , D'amati G, Casali C. The emerging concept of mitochondrial cardiomyopathies. Am Heart J. 2001;141:E1.
4. Nicol RL, Frey N, Olson EN. From the sarcomere to the nucleus. Role of genetics and signalling in structural heart disease. Annu Rev Genomics Hum Genet 2000;01:179-223
5. Friddle CJ, Koga T, Rubin EM, Bristow J. Expression profiling reveals distinct sets of genes altered during induction and regression of cardiac hypertrophy. Proc Natl Acad Sci 2000;97:6745-50.
6. Marian AJ, Roberts R. The molecular genetic basis for hypertrophic cardiomyopathy. J Mol Cell Cardiol 2001;33:655-70.
7. Rockman HA, Ross RS, Harris AN, Knowlton KU, Steinhelper ME, Field LJ, Ross J Jr, Chien KR. Segregation of atrial-specific and inducible expression of an atrial natriuretic factor transgene in an in vivo murine model of cardiac hypertrophy.Proc Natl Acad Sci U S A. 1991;88:8277-81.
8. www.ncbi.nlm.nih.gov/UniGene/lib.cgi?ORG=Mm&LID=86
9. www.ncbi.nlm.nih.gov/UniGene/lib.cgi?ORG=Mm&LID=49
10. www.tigr.org/tdb/microarray/conciseguide.html
11. Halgren RG, Fielden MR, Fong CJ, Zacharewski TR. Assessment of clone identity and sequence fidelity for 1189 IMAGE cDNA clones. Nucleic Acids Res. 2001;29:582-8.
12. GENE-ARRAYS@ITSSRV1.UCSF.EDU
13. The genome international sequencing consortium. Initial sequencing and analysis of the human genome. Nature 2001;409: 860-921
14. Venter JC, Adams MD, Myers EW, Li PW, Mural RJ, et al. The sequence of the human genome. Science 2001;291:1304-1351

11. GENE EXPRESSION IN CARDIAC HYPERTROPHY AND FAILURE: ROLE OF G PROTEIN-COUPLED RECEPTORS

G. Esposito, A. Rapacciuolo, S.V. Naga Prasad, H.A. Rockman.

Introduction

Heart failure is a clinical syndrome characterized by progressive ventricular dilatation, depressed contractile function and premature death. In the US, heart failure accounts for the majority of the cardiovascular mortality and morbidity in patients over the age of 65 [1]. Most causes of heart failure result from coronary heart disease, but a substantial number are caused by idiopathic dilated cardiomyopathy (DCM). While inherited gene defects have been found to cause DCM, they account for only 35% of cases [2], and a much smaller percentage relative to the total number of patients presenting with heart failure. Importantly, the considerable variation in the development of clinical heart failure and in the long-term survival in all patients irrespective of etiology indicates that additional unidentified genetic factors play a significant role in the phenotypic expression. Unfortunately, these modifier genes have been recalcitrant to direct identification in human populations. In this regard, genetic studies in animals models of disease can identify candidate modifier genes and provide an insight on the genetic interactions that cause phenotypic variation in the human population [3,4]. Genes that are involved in the response of the cardiac myocyte to initiate cellular processes that lead to hypertrophy are strong candidates as susceptibility genes. In this regard, recent data demonstrate an important role of G protein-coupled receptors for the induction of in vivo cardiac hypertrophy and the activation of signaling pathways, such as the mitogen activated protein kinase and phosphoinositide-3 kinase pathways. Understanding the cellular signals that initiate the hypertrophic response will be of critical importance to identify pathways that promote deterioration of the hypertrophic heart.

Role of Gq in cardiac hypertrophy

In vitro studies have suggested a pivotal role of Gq coupled receptor signaling in promoting cardiomyocyte hypertrophy [5,6]. In cardiac myocytes, G protein-coupled receptor (GPCR) agonists such as angiotensin II, endothelin 1, phenylephrine and isoproterenol can to varying levels activate mitogen activated protein kinase (MAPK) pathways [7-9]. The MAPK superfamily includes three major pathways: the extracellular regulated kinase (ERK1/2) pathway and two stress activated protein kinase pathways, c-Jun-NH2-terminal kinase (JNK) and p38 MAPK [10,11].

The important role of GPCRs and MAPK signaling in the development of *in vivo* cardiac hypertrophy has recently been shown by the generation of transgenic mice overexpressing Gαq [12], the angiotensin AT1 receptor [13], and by the adenoviral-mediated transfer of a dominant inhibitory mutant of an upstream activator of JNK [14]. However whether Gq coupled receptor stimulation is required for the induction of MAPK pathways in *in vivo* hypertrophy remains unclear. In this regard, it was recently demonstrated that inhibition of Gq-coupled receptor signaling in transgenic mice significantly reduced the hypertrophic response to in vivo pressure overload [15]. Inhibition of Gq signaling was achieved through overexpression of a 54 amino acid carboxyl-terminal peptide of Gαq (GqI) that inhibits the heterotrimeric Gq interaction with agonist-occupied receptors [15]. Transgenic mice were created with the α-myosin heavy chain to target the Gq inhibitor to cardiac myocytes [15]. Pressure overload hypertrophy was induced in these mice by transverse aortic constriction (TAC) for 7 days, which in control mice results in a reproducible and significant left ventricular (LV) hypertrophy [16]. When TAC was applied to the GqI transgenic mice, development of LV hypertrophy was significantly attenuated compared to control mice, demonstrating the critical role for Gq signaling in the activation of the hypertrophic process [15] (figure 1). Moreover we have recently shown that in the GqI transgenic mice, induction of ERK and JNK activity was abolished, whereas the induction of p38 and p38β was robust, but delayed [17]. Taken together, these data show that the induction of ERK and JNK activity in *in vivo* pressure-overload hypertrophy is mediated through the stimulation of G(q)-coupled receptors and that non-G(q)- mediated pathways are recruited to activate p38 and p38β [17].

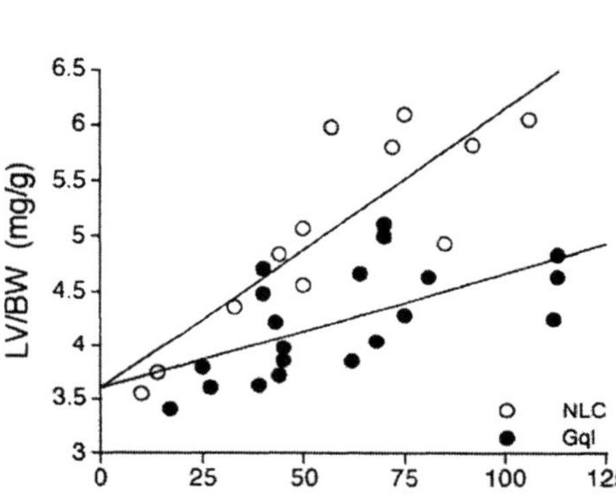

Figure 1 *Role of Gq coupled receptors in the hypertrophic response to pressure overload. Cardiac hypertrophy was measured as the ratio of left ventricular weight to body weight (LVW/BW) and is plotted against the systolic pressure gradient produced by transverse aortic constriction. for each wild type control (NLC, open circles) and transgenic mouse with myocardial-targeted expression of the Gq inhibitor (GqI) (filled circles). The slopes of the linear regression for wild type controls and GqI animals were significantly different ($P < 0.0005$, ANOVA) (from reference [15]).*

Role of catecholamines in hypertrophy and heart failure

Activation of the sympathetic nervous system is considered one of the cardinal pathophysiologic abnormalities in patients with heart failure [18] and frequently precedes the development of overt symptoms [19]. Plasma norepinephrine (NE) and renin activity are increased in patients with heart failure and are known prognostic factors for survival [18]. Elevated circulating NE and epinephrine have been implicated in contributing to the profound β-adrenergic receptor (βAR) downregulation and receptor uncoupling characteristic of end-stage human dilated cardiomyopathy [20], a process likely mediated by phosphorylation of βARs by the β-adrenergic receptor kinase (βARK1) [21,22]. Importantly, myocardial βARK1 mRNA and activity are elevated in human heart failure [23,24]. It has been postulated that chronic stimulation of myocardial βARs may adversely affect cardiomyocyte viability possibly through cAMP-mediated Ca2+ overload of the cell [25].

To further explore the role of chronic sustained adrenergic stimulation on the pathogenesis of cardiac overload, a model system in which activation of the sympathetic nervous system cannot lead to an elevation in plasma NE would be of value. In this regard, gene-targeted mice were created by homologous recombination, which lack dopamine β-hydroxylase ($Dbh^{-/-}$), the enzyme needed to convert dopamine to NE [26,27]. Although homozygous embryos die *in utero*, mortality can be prevented by treatment with L-threo-3,4-dihydroxyphenylserine (L-DOPS) administered during embryonic development. Adult $Dbh^{-/-}$ mice rescued during embryonic development have virtually no circulating NE or epinephrine [26,28]. Interestingly, $Dbh^{-/-}$ mice with absent circulating norepinephrine and epinephrine, have enhanced βAR responsiveness that is associated with a significant reduction in βARK1 activity and protein [29].

Although considerable evidence suggests that G protein-coupled receptors are involved in the hypertrophic response, it remains controversial whether catecholamines are required for the development of in vivo cardiac hypertrophy. To define the role of norepinephrine and epinephrine in the development of cardiac hypertrophy and to determine whether the absence of circulating catecholamines alters the activation of downstream myocardial signaling pathways, we performed transverse aortic constriction (TAC) in $Dbh^{-/-}$ mice [30]. Following induction of cardiac hypertrophy, the mitogen activated protein kinase signaling pathways were measured in pressure overloaded wild type and $Dbh^{-/-}$ hearts [30]. Compared to the control animals, cardiac hypertrophy was significantly blunted in $Dbh^{-/-}$ mice, which was not associated with altered cardiac function as assessed by transthoracic echocardiography in conscious mice (figure 2).

ERK, JNK and p38 MAPK pathways were all activated by 2 to 3 fold after TAC in the control animals. In contrast, induction of the three pathways (ERK1/2, JNK and p38) was completely abolished in $Dbh^{-/-}$ mice (figure 3). These data demonstrate a near complete requirement of endogenous norepinephrine and epinephrine for the induction of *in vivo* pressure overload cardiac hypertrophy and for the activation of hypertrophic signaling pathways.

Role of phosphoinositide-3 kinase in in vivo cardiac hypertrophy

As discussed above, *in vitro* and *in vivo* studies have shown that activation of Gq-coupled receptors can trigger a hypertrophic program. However, the intracellular cascades carrying the signal from Gq-coupled receptors are not fully understood. Of recent interest has been the enzyme phosphoinositide-3 kinase (PI3K), which has been implicated, in G protein- coupled signaling to stress activated protein kinases [31,32] and anti-apoptotic signaling [33].

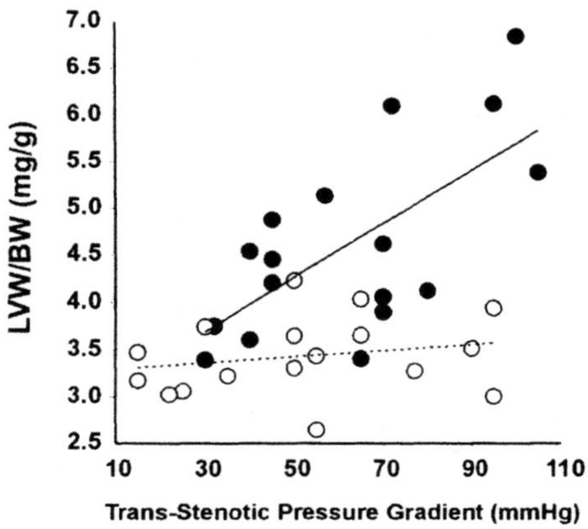

Figure 2 *Relationship between pressure load and induction of cardiac hypertrophy for all banded mice. The index of LV mass (LVW/BW) is plotted against the trans-stenotic systolic pressure gradient produced by TAC for all the banded control (n= 17, ●) and $Dbh^{-/-}$ (n= 17, o) animals. The linear regression analysis was significant for the control mice ($y= 0.0286x + 2.8325$, $r=0.66$, $p<0.01$ by ANOVA) but not for the $Dbh^{-/-}$ animals ($0.0032x + 3.2665$, $r=0.20$, p=NS). The slopes of the 2 regression lines were significantly different ($p<0.005$) by multivariate ANOVA (From [30]).*

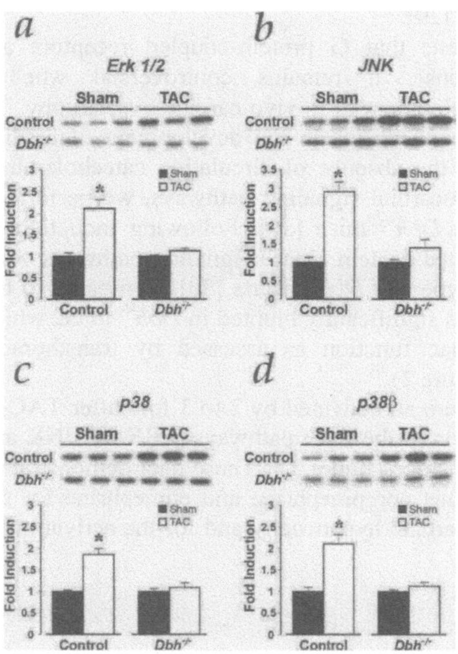

Figure 3 *Effect of pressure overload on the activation of MAPK pathways. Shown are representative autoradiograms and summary data of MAPK activity for each of the three major pathways.* ***a****, ERK1/2, *$p<0.0005$ control sham vs. control TAC.* ***b****, JNK1/2, *$p<0.0001$ control sham vs. control TAC.* ***c****, p38, *$p<0.0005$ control sham vs. control TAC.* ***d****, p38β, *$p<0.001$ control sham vs. control TAC. p=ns for $Dbh^{-/-}$ sham vs. $Dbh^{-/-}$ TAC for the above MAPKs (From [30]).*

PI3K's are a conserved family of lipid kinases that catalyze the phosphorylation in the D-3 position of the inositol ring of phosphatidylinositol (PtdIns) [33]. It is well established that PI3K's [34] and their products [33] play pivotal roles in the cell, including proliferation, differentiation, cytoskeletal organization, membrane trafficking and apoptosis [33,34]. Three different classes of PI3K's have been identified (class I, II, and III), each sharing sequence homology for the catalytic lipid kinase and PI3K protein kinase domain but show considerable diversity in activity and cellular function [34]. The best characterized are the class I PI3Ks that use PtdIns, PtdIns(4)P, PtdIns(4,5)P_2 as substrates leading to the formation of PtdIns(3)P, PtdIns(3,4)P_2 and PtdIns(3,4,5)P_3.

Given the importance of phosphoinositides in cellular function and the link between PI3K, heterotrimeric G proteins and MAPK pathways, the role of PI3K was investigated during *in vivo* cardiac pressure overload. To determine whether hypertrophic cardiac growth was accompanied by activation of PI3K, its activity was measured in hearts following induction of pressure overload by TAC [16,35]. In response to pressure overload, wild type TAC hearts had a significant increase in total PI3K activity compared to sham operated hearts (figure 4A and B) without a change in the level of PI3K protein (figure 4C). Furthermore activation of PI3K with pressure overload could be translated into activation of Protein Kinase B, a known downstream target of PI3K [35].

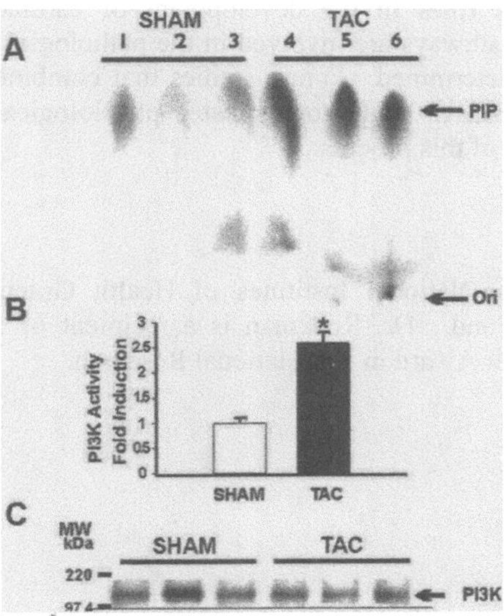

Figure 4 *Effect of in vivo pressure overload on PI3K activation in wild type mice. A) Total lipid kinase activity in cytosolic extracts from wild type sham and TAC operated hearts was measured by the ability to phosphorylate PtdIns to PtdIns(3)P. PtdIns(3)P was visualized by autoradiography following resolution on TLC plates. PIP, phosphatidylinositol-monophosphate; Ori, represents the origin of resolution. B) PI3K activity was quantified by phosphorimaging of the TLC plates in 10 sham operated mice and 10 TAC, *p <0.0001 TAC vs. sham. C) No difference in PI3K protein levels were found by immunoblotting (From [35]).*

Importantly, this study further showed that in the *in vivo* pressure overloaded heart, activation of PI3K was totally dependent on released Gβγ subunits that were derived from dissociation of the Gq class of heterotrimeric G-proteins [35]. This is important because of the role for Gq signaling in initiating the hypertrophic response during cardiac pressure overload [15].

Agonist-dependent desensitization of the βAR requires translocation and activation of the βARK1 by liberated Gβγ subunits. Subsequent internalization of agonist occupied receptors occurs as a result of the binding of β-arrestin to the phosphorylated receptor followed by interaction with the AP2 adaptor and clathrin proteins. Receptor internalization is known to require D-3 phosphoinositides that are generated by the action of PI3K [33,34], however the molecular mechanism by which PI3K acts to promote βAR internalization is not well understood. In this regard we have recently demonstrated that βARK1 and PI3K form a cytosolic complex, which leads to βARK1 mediated translocation of PI3K to the membrane in an agonist-dependent manner [36]. Furthermore, agonist induced translocation of PI3K results in rapid interaction with the receptor, which is of functional importance, since inhibition of PI3K activity attenuates βAR sequestration [36]. Therefore, agonist-dependent recruitment of PI3K to the membrane is an important step in the process of receptor internalization and links PI3K to G protein-coupled receptor activation and sequestration.

In summary, activation of a number of signaling pathways in response to G protein-coupled receptor stimulation play critical roles in the development of cardiac hypertrophy. How the activation of these pathways are involved in the pathological progression to heart failure remain to be determined. Future studies that combine the use of genetically engineered mouse models, with sophisticated physiological analysis, should enhance our understanding of this process.

Acknowledgements

This work was supported in part by the National Institutes of Health Grants HL56687 and the Burroughs Wellcome Fund. Dr. Rockman is a recipient of a Burroughs Wellcome Fund Clinical Scientist Award in Translational Research.

References

1. Braunwald E. Shattuck lecture--cardiovascular medicine at the turn of the millennium: triumphs, concerns, and opportunities. N Engl J Med 1997;337:1360-9.
2. Grunig E, Tasman JA, Kucherer H, et al. Frequency and phenotypes of familial dilated cardiomyopathy. J Am Coll Cardiol 1998;31:186-94.
3. Rozmahel R, Wilschanski M, Matin A, et al. Modulation of disease severity in cystic fibrosis transmembrane conductance regulator deficient mice by a secondary genetic factor. Nat Genet 1996;12:280-7.
4. Zielenski J, Corey M, Rozmahel R, et al. Detection of a cystic fibrosis modifier locus for meconium ileus on human chromosome 19q13. Nat Genet 1999;22:128-9.
5. Sadoshima J, Izumo S. Molecular characterization of angiotensin II--induced hypertrophy of cardiac myocytes and hyperplasia of cardiac fibroblasts. Critical role of the AT1 receptor subtype. Circ Res 1993;73:413-23.
6. Knowlton KU, Michel MC, Itani M, et al. The alpha 1A-adrenergic receptor subtype mediates biochemical, molecular, and morphologic features of cultured myocardial cell hypertrophy. J Biol Chem 1993;268:15374-80.
7. Clerk A, Bogoyevitch MA, Anderson MB, et al. Differential activation of protein kinase C isoforms by endothelin-1 and phenylephrine and subsequent stimulation of p42 and p44 mitogen- activated protein kinases in ventricular myocytes cultured from neonatal rat hearts. J Biol Chem 1994;269:32848-57.
8. Clerk A, Michael A, Sugden PH. Stimulation of the p38 mitogen-activated protein kinase pathway in neonatal rat ventricular myocytes by the G protein-coupled receptor agonists, endothelin-1 and phenylephrine: a role in cardiac myocyte hypertrophy? J Cell Biol 1998;142:523-35.
9. Kudoh S, Komuro I, Mizuno T, et al. Angiotensin II stimulates c-Jun NH2-terminal kinase in cultured cardiac myocytes of neonatal rats. Circ Res 1997;80:139-46.
10. Sugden PH, Bogoyevitch MA. Intracellular signalling through protein kinases in the heart. Cardiovasc Res 1995;30:478-92.
11. Robinson MJ, Cobb MH. Mitogen-activated protein kinase pathways. Curr Opin Cell Biol 1997;9:180-6.
12. D'Angelo DD, Sakata Y, Lorenz JN, et al. Transgenic Galphaq overexpression induces cardiac contractile failure in mice. Proc. Natl. Acad. Sci. U.S.A. 1997;94:8121-6.
13. Hein L, Stevens ME, Barsh GS, et al. Overexpression of angiotensin AT1 receptor transgene in the mouse myocardium produces a lethal phenotype associated with myocyte hyperplasia and heart block. Proc Natl Acad Sci U S A 1997;94:6391-6.
14. Choukroun G, Hajjar R, Fry S, et al. Regulation of cardiac hypertrophy in vivo by the stress-activated protein kinases/c-Jun NH(2)-terminal kinases. J Clin Invest 1999;104:391-8.
15. Akhter SA, Luttrell LM, Rockman HA, et al. Targeting the receptor-G(q) interface to inhibit in vivo pressure overload myocardial hypertrophy. Science 1998;280:574-7.
16. Rockman HA, Ross RS, Harris AN, et al. Segregation of atrial-specific and inducible expression of an atrial natriuretic factor transgene in an in vivo murine model of cardiac hypertrophy. Proc. Natl. Acad. Sci. U.S.A. 1991;88:8277-81.
17. Esposito G, Prasad SV, Rapacciuolo A, et al. Cardiac Overexpression of a G(q) Inhibitor Blocks Induction of Extracellular Signal-Regulated Kinase and c-Jun NH(2)-Terminal Kinase Activity in In Vivo Pressure Overload. Circulation 2001;103:1453-8.
18. Cohn JN, Levine TB, Olivari MT, et al. Plasma norepinephrine as a guide to prognosis in patients with chronic congestive heart failure. N. Engl. J. Med. 1984;311:819-23.

19. Francis GS, Benedict C, Johnstone DE, et al. Comparison of neuroendocrine activation in patients with left ventricular dysfunction with and without congestive heart failure. A substudy of the Studies of Left Ventricular Dysfunction (SOLVD). Circulation 1990;82:1724-9.
20. Bristow MR, Ginsburg R, Minobe W, et al. Decreased catecholamine sensitivity and beta-adrenergic-receptor density in failing human hearts. N. Engl. J. Med. 1982;307:205-11.
21. Lohse MJ, Krasel C, Winstel R, et al. G-protein-coupled receptor kinases. Kidney Int. 1996;49:1047-52.
22. Rockman HA, Chien KR, Choi DJ, et al. Expression of a β-adrenergic receptor kinase1 inhibitor prevents the development of heart failure in gene-targeted mice. Proc. Natl. Acad. Sci. U.S.A. 1998;95:7000-5.
23. Ungerer M, Bohm M, Elce JS, et al. Altered expression of beta-adrenergic receptor kinase and beta 1- adrenergic receptors in the failing human heart. Circulation 1993;87:454-63.
24. Ungerer M, Parruti G, Bohm M, et al. Expression of beta-arrestins and beta-adrenergic receptor kinases in the failing human heart. Circ. Res. 1994;74:206-13.
25. Mann DL, Kent RL, Parsons B, et al. Adrenergic effects on the biology of the adult mammalian cardiocyte. Circulation 1992;85:790-804.
26. Thomas SA, Matsumoto AM, Palmiter RD. Noradrenaline is essential for mouse fetal development. Nature 1995;374:643-6.
27. Thomas SA, Palmiter RD. Thermoregulatory and metabolic phenotypes of mice lacking noradrenaline and adrenaline. Nature 1997;387:94-7.
28. Thomas SA, Marck BT, Palmiter RD, et al. Restoration of norepinephrine and reversal of phenotypes in mice lacking dopamine beta-hydroxylase. J. Neurochem. 1998;70:2468-76.
29. Cho MC, Rao M, Koch WJ, et al. Enhanced contractility and decreased β-adrenergic receptor kinase 1 in mice lacking endogenous norepinephrine and epinephrine. Circulation 1999;99:2702-7.
30. Rapacciuolo A, Esposito G, Caron K, et al. Important role of endogenous norepinephrine and epinephrine in the development of in vivo pressure overload cardiac hypertrophy. J Am Coll Card 2001;In Press.
31. Lopez-Ilasaca M, Gutkind JS, Wetzker R. Phosphoinositide 3-kinase gamma is a mediator of Gbetagamma-dependent Jun kinase activation. J Biol Chem 1998;273:2505-8.
32. Lopez-Ilasaca M, Crespo P, Pellici PG, et al. Linkage of G protein-coupled receptors to the MAPK signaling pathway through PI 3-kinase gamma. Science 1997;275:394-7.
33. Rameh LE, Cantley LC. The role of phosphoinositide 3-kinase lipid products in cell function. J Biol Chem 1999;274:8347-50.
34. Vanhaesebroeck B, Leevers SJ, Panayotou G, et al. Phosphoinositide 3-kinases: a conserved family of signal transducers. Trends Biochem Sci 1997;22:267-72.
35. Naga Prasad SV, Esposito G, Mao L, et al. Gβγ dependent phosphoinositide-3-kinase activation in hearts with *in vivo* pressure overload hypertrophy. J. Biol. Chem. 2000;275:4693-8.
36. Naga Prasad SV, Barak LS, Rapacciuolo A, et al. Agonist-dependent recruitment of phosphoinositide 3-kinase to the membrane by {b}-adrenergic receptor kinase 1: A role in receptor sequestration. J Biol Chem 2001;276:18953-59.

12. LITTLE MICE WITH BIG HEARTS: FINDING THE MOLECULAR BASIS FOR DILATED CARDIOMYOPATHY

L.J. De Windt and M.A. Sussman

Introduction

Dilated cardiomyopathy (DCM) is characterized by increased ventricular chamber size and dimension coupled with loss of contractile function, myofibril organization and β-adrenergic responsiveness.[1] Typical microscopic features include high incidence of myocyte hypertrophy and apoptosis, [2-4] varying degrees of fibrosis and the presence of lymphocytes in the interstitial space [5]. Clinically, the disease displays heart failure symptoms consistent with New York Heart Association (NYHA) functional class III or IV at the time of diagnosis [6] and remains the principal indication for heart transplantation in both adults and pediatric cardiology [6-8]. The clinical treatment of human DCM is a substantial economic burden estimated to cost $10 to $40 billion annually in the United States alone [9]. First and foremost in terms of pathologic consequences, DCM is distinguished by a profound loss of myocardial contractile function. The degeneration of structure and function in cardiomyocytes observed in dilation is less amenable to *in vitro* analysis compared to hypertrophic changes associated with enhanced contractile function and enlarged cardiomyocyte size that are readily apparent by analyses of cultured cells. Cardiomyocyte remodeling in dilation results from a combination of physical stresses and signal transduction leading to profound structural changes of the entire heart. This combinatorial environment is best reproduced *in vivo*, as evidenced by the emergence of transgenic and gene targeted animal models that display the characteristic features of DCM. These models are useful to study the temporal progression of pathogenic alterations that occur at the molecular, cellular and organ levels. Many genetic mouse models of cardiomyopathy have emerged over the past decade, many of which display hypertrophic remodeling, whether or not in combination with features of DCM. The scope of this review, however, is to concentrate on those genetic models that display multiple characteristics of human

DCM. These models can be categorized based upon the distinctive manipulation of cardiomyocyte homeostasis that led to their creation. Accordingly, the manipulative loss of various structural and/or functional properties of cardiomyocytes provokes development of the DCM phenotype in mice. Specifically, these categories can be subdivided into (1) the cytoskeletal architecture, (2) the rate of myofibrillogenesis and function of specific sarcomeric proteins, (3) sarcoplasmic reticulum (SR) calcium (Ca^{2+}) handling and (4) the relative activation status of intracellular signaling pathways. These four intrinsic properties of cardiomyocytes and their relationship to the development of DCM have been demonstrated by recently developed genetic DCM models discussed in this review (Table 1).

Table 1. Murine DCM paradigms grouped by structural/functional defects.

Cytoskeleton	Myofibrillogenesis	Calcium handling	Signal Transduction
Mucle LIM protein[1]	Tropomodulin[2]	FKBP12[1]	Cytokines (gp130[1], TNFα[2])
Activated rac1 expression[2]	Myosin binding protein-C[2]	Calsequestrin[2]	Ca2+/Calmodulin activated kinases (calcineurin[2], CamK[2])
Dystrophin[1]			Protein kinases (TAK-1[2], MEK5[2])
			Transcription factor (CREB[2])

[1] gene targeted knockout models
[2] altered or mutant protein expression models

Biomechanical defects due to alterations in cardiac cytoskeletal architecture

Cumulative evidence demonstrates a correlation between loss of cardiomyocyte cytoskeletal integrity and the development of DCM. The initial evidence for this assocoation was derived from identification of frameshift mutations in the *dystrophin* gene caried by patients with later-onset X-linked DCM [10,11]. Additionally, mutations in the genes encoding *cardiac actin*, *desmin* and *lamin A/C*, all direct or indirect cytoskeletal partners of the dystrophin-associated protein complex, have been demonstrated to be the primary defect in a family with hereditary DCM [12-14]. Animal models with mutations leading to loss of function in cytoskeletal components also have a propensity to develop DCM, such as the mutation in the *δ-sarcoglycan* gene leading to DCM in the BIO14.6 and TO-2 hamster [15] and targeted loss of both dystrophin and utrophin in mice [16]. Targeted disruption of the striated muscle-specific LIM-only protein MLP (muscle LIM protein) provided direct experimental evidence that DCM is a disease of the cytoskeleton. The LIM motif, a double zinc-finger structure that participates in protein-protein interactions, is found in proteins involved with processes of cellular determination and differentiation [17]. MLP, a member of this class of proteins, acts as a positive regulator of myogenic differentiation [18] and is involved in targeting interacting proteins to the actin-based cytoskeleton in striated muscle (figure 1) [19]. MLP-deficient mice display disruption of cardiomyocyte cytoarchitecture that transitions into cardiac hypertrophy and failure [20]. Although both skeletal and cardiac muscle lacked rigor in MLP-deficient mice, the defects were most striking in the developing heart, which enlarged dramatically after birth. Histologic analysis revealed myofibrillar disarray, interstitial cell proliferation, and fibrosis analogous with microscopic aspects of human DCM. Finally, echocardiographic and hemodynamic analysis demonstrated severe systolic and diastolic dysfunction with signs of overt LV pump failure in adult mice. A specific and direct interaction between MLP and β-spectrin or α-actinin has been demonstrated [21].

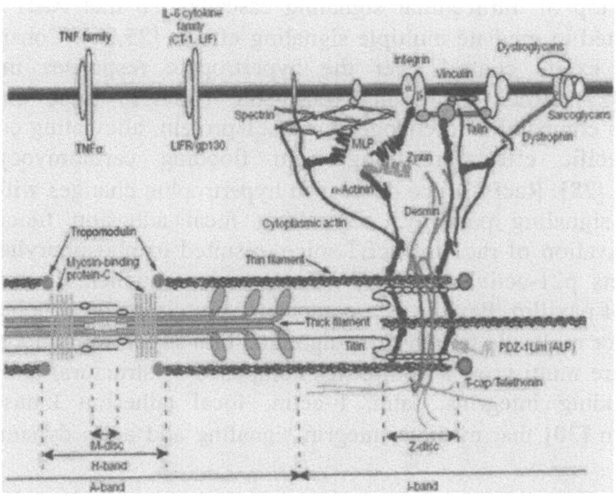

Figure 1 *Cytoskeletal and myofibrilar components in the cardiocyte. Schematic representation of interconnection between cystokeletal and myofibrillar organization. Several proteins associated with DCM are indicated in this figure including MLP, tropomodulin, dystrophin, and myosin binding protein C. Signaling pathwats for TNFα and gp130 shown at the membrane have also been implicated in development of DCM. Modified from Chien, K. Nature 2000; 407:227-32.*

Given that both α-actinin and β-spectrin are costameric components, it is tempting to speculate that MLP stabilizes the association of the contractile apparatus and the sarcolemma by linking the β-spectrin network to the α-actinin crosslinked actin filaments of the myofibril. Accordingly, it was hypothesized that MLP might be involved in the linkage between tension and muscle growth through the organization of the myofibrillar apparatus along the actin skeleton of the cardiomyocyte. MLP associates with Z-line structures through its second LIM domain to establish and maintain proper cardiomyocyte architecture. Deletion of MLP results in cytoarchitectural perturbations that lead to impaired cell and tissue tension upon postnatal elevation of hemodynamic demands. MLP-deficient cardiomyocytes are impaired in their ability to develop sufficient tension to control the developmentally associated hypertrophic response, which translates into DCM at the whole organ level [23].

The importance of regulated cytoskeletal organization and stable cellular adhesion in the heart was demonstated in mice expressing a constitutively active form of the small GTP-binding protein Rac1 (V12rac1) [24]. Rac1 is notorious for inducing cytoskeletal remodeling in various non-muscle cells. Rac1, like Ras, is hierarchically positioned at the top of intracellular signaling cascades, so that Rac1 activation would be anticipated to mediate multiple signaling effects [25,26]. Consistent with this view, Rac1 exerts control over the hypertrophic responses in cultured cardiomyocytes [27]. Rac1 expressing transgenic (RacET) mice express the transgene at levels comparable to endogenous Rac1 protein, alleviating concerns of irrelevant nonspecific effects resulting from flooding cardiomyocytes with transgenic protein [28]. RacET mice developed hypertrophic changes within weeks after birth, and signaling pathways controlling focal adhesion function were dysregulated. Activation of rac1 in racET mice, resulted in phosphorylation of its downstream targets p21-activated kinase (PAK) and src which, in turn, led to phosphorylation of paxillin. Paxillin is a component protein of focal adhesions that serves as an adaptor molecule to transduce signaling into structural remodeling [29]. Focal adhesions are multi-protein complexes composed of structural and signaling components including integrins, talin, F-actin, focal adhesion kinase (FAK), paxillin, and zyxin [30] that mediate integrin signaling and actin dynamics in the cardiomyocyte.

Consequently, the loss of cytoskeletal-based cardiomyocyte adhesion in juvenile RacETs led to a lethal DCM phenotype and heart failure in some individuals. Interestingly, myofiber disruption was not observed, unlike other mouse models of DCM [20,31]. Rather, constitutive activation of Rac1 in RacET mice lead to two important characteristics of DCM: impaired force transmission and severe loss of systolic function. As such, the MLP and the racET mouse models provide experimental paradigms to explore the relationship between cytoskeletal disruption, cytoskeletal adhesion instability and dilation that is the molecular basis for some forms of human DCM [20,22,31].

Abnormalities in cytoskeletal architecture may not only be isolated to specific rare forms of hereditary human DCM, but might also form the basis for pathogenesis of common acquired forms of DCM. For example, a subset of acquired human DCM is associated with enteroviral infection of the myocardium caused by Coxsackie B viruses [32]. Enterovirally encoded protease 2A specifically cleaved human dystrophin *in vitro* and murine dystrophin *in vivo* [33]. Dystrophin cleavage products accumulated in the cytosolic fraction following viral infection of the heart, leading to disruption of cytoskeletal architecture reminiscent of human DCM and the abovementioned murine DCM paradigms.[33]

Mechanical overload, as occurs during hypertension, is a main risk factor for subsequent development of heart failure and DCM. Mechanical stress is known to elicit functional and structural adaptations of the myocardium. Franchini and coworkers used a pressure overload rat model to demonstrate tyrosine phosphorylation of FAK within 3 minutes after onset of the biomechanical stress [34]. FAK activation paralled src activation, and FAK associated with several intracellular signaling modules, indicating a role for FAK as a cytoskeletal scaffold adaptor molecule where signal transduction modules anchor and become involved in hypertrophic remodeling and cytoskeletal adhesion rearrangement [34].

In summary, cumulative experimental evidence indicates disruption of the cardiac cytoskeletal architecture and/or contact sites of the cytoskeleton results in myocyte dysfunction due to impaired force transmission, which is exacerbated by increased hemodynamic workload. Significant loss of cardiomyocyte cytoskeletal structure and corresponding intra- and intercellular contact sites appear to be sufficient to predispose individuals to the cardiomyopathic phenotype observed in both hereditary and acquired forms of human DCM.

Loss of myofibrillogenesis and altered intrasarcomeric components

The majority of human hypertrophic cardiomyopathies (HCM) have been associated with specific mutations in genes encoding sarcomeric component proteins leading to impaired force generation [35,36]. In comparison, impaired structural remodeling coupled with hypertrophic stimulation leads to dilated cardiomyopathies. Inability to mount a structural response in the failing heart may be associated with impaired myofibrillogenesis, as evidenced by the myofiber disarray and loss that is characteristic of human DCM [1]. Thus, hypertrophic and dilated cardiomyopathy share common molecular origins on the signaling level, with the outcome of disease determined by the capacity of the heart to respond by creating muscle mass to increase cardiac output [31,37]. Impairment of myofibril organization as a basis for DCM was studied experimentally in the mouse heart by increasing the expression level of tropomodulin (Tmod). Tmod is a component of the thin filament complex that colocalizes with pointed ends of actin filaments and controls thin filament length in sarcomeres (figure 1). Overexpression of Tmod in cardiomyocyte cultures leads to myofiber degeneration, consistent with a role for Tmod in the maintanance of myofiber length specification [38]. Tmod overexpressing transgenic (TOT) mice display multiple features of end-stage human heart failure and DCM, including a severely decreased systolic function, enlargement of the heart accompanied by thinning of the ventricular walls, myocyte disarray, widespread disruption of myofibril organization [31]. Additionally, increased apoptotic cardiomyocyte dropout is evident in TOT mice (Sussman and Anversa, unpublished observations). Echocardiographic analysis documented severe dilation, leading to elevation of peak systolic wall stress, which in turn initiates a "failed" hypertrophic response as interpreted by the activation of a fetal gene expression profile. Based upon this and subsequent studies of the TOT model, a profile of compensatory signaling mechanisms has emerged that is initiated by impaired myofibril organization. Since TOT mice cannot mount an appropriate hypertrophic response by increasing myofibril density, the reactive hypertrophic stimulation becomes the chronic driving force behind a downward spiral that leads to cardiac remodeling and severe dilation [31,37].

Another model of murine DCM was established by insertional mutagenesis of the myosin binding protein C (MyBP-C) gene using gene targeting approaches [39]. These mice expressed a truncated form of MyBP-C peptides, analogous to that found in human HCM. Human patients with this familial HCM disease and heterozygous mice develop HCM. In contrast, homozygous mice bearing the mutation on both alleles are unable to generate wildtype MYBP-C, leading to postnatal ventricular dysfunction which transitions into a severe form of DCM. Histopathological analysis revealed typical features of human DCM, including prominent cardiomyocyte hypertrophy, myofibrillar disarray, fibrosis and calcification. Similar to TOT mice, poorly defined M-bands were observed in homozygous mutant MyBP-C mice at the ultrastructural level of the sarcomere, suggesting that altered myofibril organization was a causal defect in the dilated heart [39].

The observation that heterozygous mutant MyBP-C mice develop different cardiac pathology than their homozygous counterparts raises the interesting concept that selected parameters of the sarcomere, such as force generation and myofibril organization, may be continuously monitored by the cardiomyocyte. Intrasarcomeric abnormalities such as myofibrillar degeneration or altered contractility could initiate reactive signaling cascades in an attempt to remodel the heart. Depending upon the severity of the primary defect, the heart can respond with development of either hypertrophy in the case of relatively mild alterations (e.g. heterozygous MyBP-C mice) or DCM in the case of severe sarcomeric abnormalities (e.g. homozygous MyBP-C and TOT mice). Concurrent with this intracellular sensing mechanism of the cardiomyocyte, extracellular stimuli such as hemodynamic load or matrix remodeling, together with environmental factors [37] can have a profound impact on outcome and can tip the balance from stages of compensated hypertrophy to decompensation and DCM.

Alterations in SR calcium handling

A number of mouse models have implicated a pivotal role for the altered movement of cytosolic Ca^{2+} during each cardiac cycle in the pathogenesis of DCM. In the process referred to as excitation-contraction coupling, Ca^{2+} stored in the sarcoplasmic reticulum (SR) is released into the cytosol to initiate contraction of the cardiac muscle and is re-sequestered into the SR to achieve relaxation in response to rhythmic voltage-dependent calcium release. The release of SR Ca^{2+} for EC coupling is coordinated by SR Ca^{2+} release channels called ryanodine receptors (RyR) in striated muscle [40], which are biphasically activated such that low [Ca^{2+}] activates the channels (during diastole) and high [Ca^{2+}] inactivates them (during systole). FK506-binding proteins (FKBPs) are regulatory subunits that stabilize RyR channel function [41] and facilitate coupled gating between neighboring channels on the SR [42]. During relaxation, activity of the SR Ca^{2+} ATPase (SERCA2a) determines the rate of Ca^{2+} reuptake, while SERCA activity itself is regulated by the inhibitory protein phospholamban (PLB) [43,44]. Latter system forms an integral part of the β–adrenergic signaling cascade that can augment myocardial inotropy. Agonist stimulation of the β-adrenergic system results in activation of cAMP-dependent protein kinase (PKA), which phosphorylates PLB and its dissociation from SERCA-2a, resulting in proportional increases in the rate of Ca^{2+} reuptake into the SR and increased ventricular lusitropy [45].

Genetic disruption of FKBP12 resulted in severe DCM and ventricular septal defects in mice [46]. Many FKBP12-deficient animals displayed premature embryonic death related to neural tube closure defects and excencephaly, whereas surviving mice exhibited increased LV chamber dimensions and reduced systolic performance [46]. RyR2 is a multi-protein complex consisting of FKBP12.6, protein phosphatases 1a and 2a, A-kinase anchoring protein, as well as PKA that regulates RyR2 function. [47]. Altered RyR2 channel function has been postulated to be involved in human DCM, since hyperphosphorylation of the RyR2 complex has been demonstrated in human heart failure biopsies. Consequently, FKBP12.6 dissociates from its cognate receptor leading to increased Ca^{2+} sensitivity for activation and elevated channel open probability, impairing calcium homeostasis [47].

Another example of altered calcium homeostasis leading to DCM can be observed following targeted overexpression of calsequestrin in the heart. Calsequestrin is a high capacity Ca^{2+} binding protein located in the lumen of the junctional SR where it forms a complex with the RyR, junctin and triadin to cooperatively regulate homeostatic Ca^{2+}-triggered Ca^{2+} release [48]. The increased SR Ca^{2+} storage capacity of the calsequestrin TG cardiomyocytes leads to elevation of SR Ca^{2+} content that was not available for RyR-mediated Ca^{2+} release. A marked reduction of Ca^{2+}-induced Ca^{2+} release and frequency of Ca^{2+} sparks was observed, associated with an attenuation of the Ca^{2+} transient and decreased contractile parameters.

Decreased Ca^{2+} reuptake is a central feature of human DCM. The first genetic evidence in support a causal role for defective SR Ca^{2+} reuptake in DCM pathogenesis was demonstrated by creation of a hybrid mouse line deficient for both MLP (see section 1) and PLB [49]. Ablation of PLB prevented development of cardiomyopathic changes in the MLP mouse. It is important to note that protein levels for SERCA2a, PLB and RyR2 in the cardiomyopathic MLP-deficient were not dramatically different from their wildtype counterparts, [49] unlike the situation observed in human DCM [50]. This observation suggests that SERCA2a activity and regulation, rather than its absolute levels, appears contributory in MLP-deficient mouse cardiomyopathy. Accordingly, ablation of phospholamban activates SERCA2a leading to reduced wall stress and enhanced cardiac relaxation. In comparison, TOT mice also exhibit impaired Ca^{2+} reuptake coincident with marked decreases in SERCA2a protein levels by immunoblot analyses (Sussman, unpublished observations). Ablation of PLB in the TOT background produced no change in the disease progression [51], probably due to the severely decreased SERCA2a levels of the TOT mouse.

Thus, defects in SR Ca^{2+} are commonly found in multiple forms of DCM. However, modulation of SERCA2a function by decreasing phospholamban:SERCA2a stoichiometry will not be beneficial in circumstances where SERCA2a levels are diminished below a critical threshold. Increasing SERCA2a expression together with PLB inhibition might therefore present a more attractive future therapeutic target in the treatment of DCM associated with severe SERCA2a deficiency.

Hypertrophic signaling and transcriptional events

Single cardiomyocytes interpret and respond to autocrine and paracrine signals through a vast complex of intracellular signaling cascades, often leading to a reactive hypertrophic response. Although hypertrophy could be considered initially beneficial to maintain cardiac output and compensate for increased wall stress, chronic hypertrophic stimulation eventually promotes decompensation and heart failure [52,53]. Since intracellular signaling pathways participate in the initiation and maintenance of cardiac enlargement, characterization of reactive signaling is currently the focus of intense investigation [54]. Several approaches to understand the role of intracellular signaling in cardiomyopathy involve selective increases or inhibition of individual signaling molecules using transgenesis and gene targeting. The majority of these molecules participate in signaling to activate hypertrophic responses, but a select number of extracellular and intracellular signals lead to DCM in the mouse. One such pathway was discovered by ventricular-restricted disruption of the gp130 cytokine receptor [55]. Gp130 is a common signal transducer for cytokines like cardiotrophin-1 (CT-1) and members of the interleukin-6 (IL-6) family. CT-1 induces hypertrophy and is a potent anti-apoptotic agent for cultured cardiomyocytes [56]. Mice lacking gp130 receptor suffer from lethal DCM characterized by cardiomyocyte apoptosis, increased chamber dimension and significant loss of LV function following pressure overload-induced stress mediated by transverse aortic constriction that produces compensated hypertrophy in wildtype animals. This suggests that gp130 is part of an essential stress-activated myocyte survival pathway and establishes direct genetic evidence for sufficiency of cardiomyocyte death in the pathogenesis of DCM [55].

Another cytokine that has been implicated in human DCM is tumor necrosis factor alpha (TNFα). Failing human myocardium expresses TNFα abundantly, derived in part from ventricular myocytes [57,58]. Targeted overexpression of TNFα in the murine heart results in severe dilation, lymphocytic interstitial infiltrate, and increased lung weight consistent with congestive heart failure [59-61]. Overexpression level of TNFα appears to determine the actual phenotypic outcome, with moderate expression leading to DCM, while more robust expression induces lethal myocarditis. These studies demonstrate that TNFα expression in the failing human heart may not be an epiphenomenon, but rather an important contributory factor in the pathogenesis of DCM [59-61].

The Ca^{2+}-calmodulin activated serine/threonine phosphatase calcineurin controls cardiac hypertrophy [62,63], possibly by targeting members of the NFAT family of transcription factors. Generation of transgenic animals encoding a constitutively active form of either calcineurin or NFATc4 displayed massive hypertrophy [64]. High expressing calcineurin transgenic mice eventually displayed some features of DCM (enlargement of LV cavity size, poor contractile function) but, unlike human DCM, calcineurin activation was associated with protection against multiple apoptotic insults [65]. Therefore, it is likely that the extreme degree of hypertrophy observed in these mice due to the robust overexpression poses geometrical constraints to the heart, eventually resulting in dilation. Thus, current interpretation of the data is consistent with calcineurin as a participant of early reactive hypertrophic signaling [66] rather than being responsible for late onset deleterious signaling. An interesting aspect was that administration of the calcineurin inhibitor cyclosporin A prevented the onset of cardiomyopathy in mutiple hypertrophic and dilated cardiomyopathic mouse models, including TOT mice [67]. Ca^{2+}/calmodulin activated signaling may play an integral part in the pathogenesis of heart failure and DCM. Recently, it was demonstrated that myocardial Calmodulin-activated protein kinase (CamK) and concomitant activation of its downstream target myocyte enhancing factor (MEF-2) resulted in a massive hypertrophic remodeling in transgenic mice with late onset LV wall thinning and DCM, reminiscent of the cardiomyopathic phenotype of the calcineurin TG mouse [68]. Whether Ca^{2+}-activated signaling represents a common underlying paradigm for all forms of DCM or in response to biomechanical stress awaits genetic confirmation and further experimentation.

Another family of notorious hypertrophic signal transducers are the separate members of the MAPK superfamily (i.e., p38, JNK1/2, Erk1/2 and Erk5), all of which have been implicated in hypertrophic remodeling [69-71]. Recent evidence supports the notion of some functional diversification in the cardiomyopathic remodeling following activation of the single members of the MAPK branches. Transgenic expression of an activated mutant of the transforming growth factor-β activated protein kinase (TAK1) targets and activates members of the p38 MAPK family, producing massive neonatal cardiac hypertrophic remodeling, interstital fibrosis, increased apoptosis, fulminant heart failure and DCM in mice [72]. More recently, expression of a constitutive active mutant of MEK5, the upstream kinase of big MAPK or ERK5 produced a hypertrophic response where cardiomyocytes acquired an elongated morphology accompanied by serial sarcomerogenesis [73]. Juvenile mice showed an extreme form of eccentric hypertrophy and DCM with severe LV dysfunction and increased mortality [73]. These reports contrast to the finding that transgenic activation of the Erk1/2 MAPK branch produced a fundamentally distinct form of hypertrophy characterized by a mild, concentric hypertrophy, persistent enhanced myocardial contractility and resistence against pleiotropic apoptotoic stimuli [74]. Cytokine-based signaling undoubtedly operates, in part, via non-genomic kinase-mediated signal transduction mechanisms. However, mouse models have shown that genomic effects of signal transduction can also lead to DCM. In one example, transgenic mice were created with overexpression of mutant CREB protein ($CREB_{133}$) that exerts dominant-negative effects upon transcriptional activity[75]. CREB is a basic leucine zipper transcription factor with transactivational activity both as a homodimer and as a heterodimer with other members of the CREB/ATF family and AP-1. $CREB_{133}$ mice displayed severe dilation of all four chambers, systolic and diastolic dysfunction and reduced β–adrenergic responsiveness. Congestive heart failure was underscored by

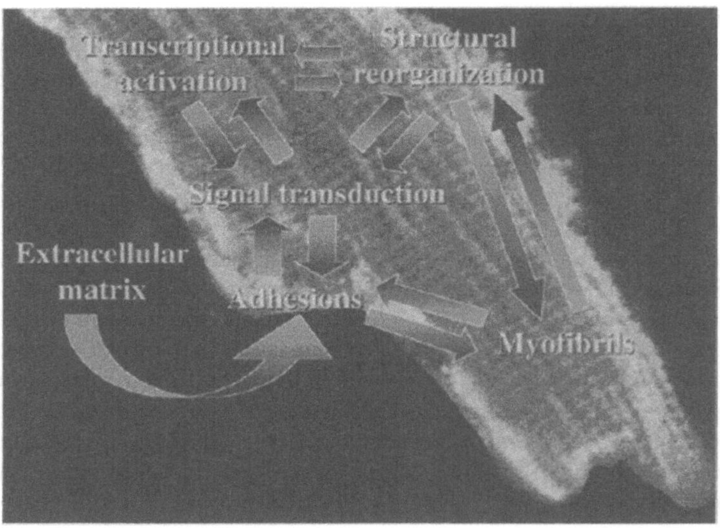

Figure 2: *Interdepence of structural and signaling pathways in the cardiomyocyte Illustration of the bidirectional nature of cardiac signaling pathways and the interdependence of structure, signal transuction, and function that normally exists within cardiomyocytes. Altering any one area of the diagrammatic flow has consequences for all of the other interconnected areas due to tightly linked compensatory mechanism that are inherent to cardiomyocyte function.*

observations of hepatic and pulmonary congestion, anasarca due to peripheral edema, and early lethality [75].

Conclusion

Genetically altered mouse models have proven to be invaluable systems for understanding the molecular basis for dilated cardiomyopathy and heart failure. The field is constantly expanding and novel models for DCM are being reported every year. As such, this review was not intended to be comprehensive for the field, but rather highlight certain examples germane to various subcellular domains of the cardiomyocyte that have been established in the literature. Collectively, research over the last few years has conclusively demonstrated that DCM is a complex disease with multiple etiologies, often involving multiple structural and signaling proteins in the onset and progression of the disease. The impact of altering cytoskeletal structure or signaling molecule activation state is inextricably linked to the web of sensing and compensatory pathways that is an interal part of cardiomyocyte biology (figure 2). As such, it is important to remember that disruption of any area within the cell has the potential for far-reaching consequences in subcellular compartments that may seem superficially unrelated to the primary defect. Murine models have provided significant insight regarding the relationships between structure, signaling, and function within the cardiomyocyte and the impact for cardiac function as evidenced by the paradigms discussed in this review. Future studies will refine our understanding of this dynamic balance in cardiomyocyte cellular regulation, allow testing of interventional approaches designed to interrupt deleterious compensatory signaling mechanisms, and uncover novel participants that lead to the development of big hearts in little mice.

References

1. Dec GW, Fuster V. Idiopathic dilated cardiomyopathy. N Engl J Med. 1994;331:1564-75.
2. Rayment NB, Haven AJ, Madden B, Murday A, Trickey R, Shipley M, Davies MJ, Katz DR. Myocyte loss in chronic heart failure. J Pathol. 1999;188:213-9.
3. Narula J, Haider N, Virmani R, DiSalvo TG, Kolodgie FD, Hajjar RJ, Schmidt U, Semigran MJ, Dec GW, Khaw BA. Apoptosis in myocytes in end-stage heart failure. N Engl J Med. 1996;335:1182-9.
4. Hong BK, Kwon HM, Byun KH, Kim D, Choi EY, Kang TS, Kang SM, Chun KJ, Jang Y, Kim HS, Kim M. Apoptosis in dilated cardiomyopathy. Korean J Intern Med. 2000;15:56-64.
5. Tazelaar HD, Billingham ME. Leukocytic infiltrates in idiopathic dilated cardiomyopathy. A source of confusion with active myocarditis. Am J Surg Pathol. 1986;10:405-12.
6. Sugrue DD, Rodeheffer RJ, Codd MB, Ballard DJ, Fuster V, Gersh BJ. The clinical course of idiopathic dilated cardiomyopathy. A population- based study. Ann Intern Med. 1992;117:117-23.
7. Kaye DM, Anderson ST, Federman J. Electrocardiographic and echocardiographic features of left atrial size after orthotopic cardiac transplantation. Am J Cardiol. 1992;70:1096-9.
8. Mudge GH, Goldstein S, Addonizio LJ, Caplan A, Mancini D, Levine TB, Ritsch ME, Jr., Stevenson LW. 24th Bethesda conference: Cardiac transplantation. Task Force 3: Recipient guidelines/prioritization. J Am Coll Cardiol. 1993;22:21-31.
9. O'Connell JB, Bristow MR. Economic impact of heart failure in the United States: time for a different approach. J Heart Lung Transplant. 1994;13:S107-12.
10. Beggs AH. Dystrophinopathy, the expanding phenotype. Dystrophin abnormalities in X-linked dilated cardiomyopathy. Circulation. 1997;95:2344-7.
11. Towbin JA, Hejtmancik JF, Brink P, Gelb B, Zhu XM, Chamberlain JS, McCabe ER, Swift M. X-linked dilated cardiomyopathy. Molecular genetic evidence of linkage to the Duchenne muscular dystrophy (dystrophin) gene at the Xp21 locus. Circulation. 1993;87:1854-65.
12. Olson TM, Michels VV, Thibodeau SN, Tai YS, Keating MT. Actin mutations in dilated cardiomyopathy, a heritable form of heart failure. Science. 1998;280:750-2.
13. Li D, Tapscoft T, Gonzalez O, Burch PE, Quinones MA, Zoghbi WA, Hill R, Bachinski LL, Mann DL, Roberts R. Desmin mutation responsible for idiopathic dilated cardiomyopathy. Circulation. 1999;100:461-4.
14. Fatkin D, MacRae C, Sasaki T, Wolff MR, Porcu M, Frenneaux M, Atherton J, Vidaillet HJ, Jr., Spudich S, De Girolami U, Seidman JG, Seidman C, Muntoni F, Muehle G, Johnson W, McDonough B. Missense mutations in the rod domain of the lamin A/C gene as causes of dilated cardiomyopathy and conduction-system disease. N Engl J Med. 1999;341:1715-24.
15. Nigro V, Okazaki Y, Belsito A, Piluso G, Matsuda Y, Politano L, Nigro G, Ventura C, Abbondanza C, Molinari AM, Acampora D, Nishimura M, Hayashizaki Y, Puca GA. Identification of the Syrian hamster cardiomyopathy gene. Hum Mol Genet. 1997;6:601-7.
16. Grady RM, Teng H, Nichol MC, Cunningham JC, Wilkinson RS, Sanes JR. Skeletal and cardiac myopathies in mice lacking utrophin and dystrophin: a model for Duchenne muscular dystrophy. Cell. 1997;90:729-38.
17. Feuerstein R, Wang X, Song D, Cooke NE, Liebhaber SA. The LIM/double zinc-finger motif functions as a protein dimerization domain. Proc Natl Acad Sci U S A. 1994;91:10655-9.
18. Arber S, Halder G, Caroni P. Muscle LIM protein, a novel essential regulator of myogenesis, promotes myogenic differentiation. Cell. 1994;79:221-31.
19. Arber S, Caroni P. Specificity of single LIM motifs in targeting and LIM/LIM interactions in situ. Genes Dev. 1996;10:289-300.

20. Arber S, Hunter JJ, Ross J, Jr., Hongo M, Sansig G, Borg J, Perriard JC, Chien KR, Caroni P. MLP-deficient mice exhibit a disruption of cardiac cytoarchitectural organization, dilated cardiomyopathy, and heart failure. Cell. 1997;88:393-403.
21. Flick MJ, Konieczny SF. The muscle regulatory and structural protein MLP is a cytoskeletal binding partner of betaI-spectrin. J Cell Sci. 2000;113:1553-64.
22. Pashmforoush M, Pomies P, Peterson KL, Kubalak S, Ross J, Jr., Hefti A, Aebi U, Beckerle MC, Chien KR. Adult mice deficient in actinin-associated LIM-domain protein reveal a developmental pathway for right ventricular cardiomyopathy. Nat Med. 2001;7:591-7.
23. McKoy G, Protonotarios N, Crosby A, Tsatsopoulou A, Anastasakis A, Coonar A, Norman M, Baboonian C, Jeffery S, McKenna WJ. Identification of a deletion in plakoglobin in arrhythmogenic right ventricular cardiomyopathy with palmoplantar keratoderma and woolly hair (Naxos disease). Lancet. 2000;355:2119-24.
24. Sussman MA, Welch S, Walker A, Klevitsky R, Hewett TE, Price RL, Schaefer E, Yager K. Altered focal adhesion regulation correlates with cardiomyopathy in mice expressing constitutively active rac1. J Clin Invest. 2000;105:875-86.
25. Mackay DJ, Esch F, Furthmayr H, Hall A. Rho- and rac-dependent assembly of focal adhesion complexes and actin filaments in permeabilized fibroblasts: an essential role for ezrin/radixin/moesin proteins. J Cell Biol. 1997;138:927-38.
26. Tapon N, Hall A. Rho, Rac and Cdc42 GTPases regulate the organization of the actin cytoskeleton. Curr Opin Cell Biol. 1997;9:86-92.
27. Pracyk JB, Tanaka K, Hegland DD, Kim KS, Sethi R, Rovira, II, Blazina DR, Lee L, Bruder JT, Kovesdi I, Goldshmidt-Clermont PJ, Irani K, Finkel T. A requirement for the rac1 GTPase in the signal transduction pathway leading to cardiac myocyte hypertrophy. J Clin Invest. 1998;102:929-37.
28. Huang WY, Aramburu J, Douglas PS, Izumo S. Transgenic expression of green fluorescence protein can cause dilated cardiomyopathy. Nat Med. 2000;6:482-3.
29. Brown MC, Perrotta JA, Turner CE. Serine and threonine phosphorylation of the paxillin LIM domains regulates paxillin focal adhesion localization and cell adhesion to fibronectin. Mol Biol Cell. 1998;9:1803-16.
30. Critchley DR. Focal adhesions - the cytoskeletal connection. Curr Opin Cell Biol. 2000;12:133-9.
31. Sussman MA, Welch S, Cambon N, Klevitsky R, Hewett TE, Price R, Witt SA, Kimball TR. Myofibril degeneration caused by tropomodulin overexpression leads to dilated cardiomyopathy in juvenile mice. J Clin Invest. 1998;101:51-61.
32. Baboonian C, Davies MJ, Booth JC, McKenna WJ. Coxsackie B viruses and human heart disease. Curr Top Microbiol Immunol. 1997;223:31-52.
33. Badorff C, Lee GH, Lamphear BJ, Martone ME, Campbell KP, Rhoads RE, Knowlton KU. Enteroviral protease 2A cleaves dystrophin: evidence of cytoskeletal disruption in an acquired cardiomyopathy. Nat Med. 1999;5:320-6.
34. Franchini KG, Torsoni AS, Soares PH, Saad MJ. Early activation of the multicomponent signaling complex associated with focal adhesion kinase induced by pressure overload in the rat heart. Circ Res. 2000;87:558-65.
35. Maass A, Leinwand LA. Animal models of hypertrophic cardiomyopathy. Curr Opin Cardiol. 2000;15:189-96.
36. Bonne G, Carrier L, Richard P, Hainque B, Schwartz K. Familial hypertrophic cardiomyopathy: from mutations to functional defects. Circ Res. 1998;83:580-93.
37. Sussman MA, Welch S, Gude N, Khoury PR, Daniels SR, Kirkpatrick D, Walsh RA, Price RL, Lim HW, Molkentin JD. Pathogenesis of dilated cardiomyopathy: molecular, structural, and population analyses in tropomodulin-overexpressing transgenic mice. Am J Pathol. 1999;155:2101-13.
38. Sussman MA, Baque S, Uhm CS, Daniels MP, Price RL, Simpson D, Terracio L, Kedes L. Altered expression of tropomodulin in cardiomyocytes disrupts the sarcomeric structure of myofibrils. Circ Res. 1998;82:94-105.

39. McConnell BK, Jones KA, Fatkin D, Arroyo LH, Lee RT, Aristizabal O, Turnbull DH, Georgakopoulos D, Kass D, Bond M, Niimura H, Schoen FJ, Conner D, Fischman DA, Seidman CE, Seidman JG, Fischman DH. Dilated cardiomyopathy in homozygous myosin-binding protein-C mutant mice. J Clin Invest. 1999;104:1235-44.
40. Bezprozvanny I, Watras J, Ehrlich BE. Bell-shaped calcium-response curves of Ins(1,4,5)P3- and calcium-gated channels from endoplasmic reticulum of cerebellum. Nature. 1991;351:751-4.
41. Brillantes AB, Ondrias K, Scott A, Kobrinsky E, Ondriasova E, Moschella MC, Jayaraman T, Landers M, Ehrlich BE, Marks AR. Stabilization of calcium release channel (ryanodine receptor) function by FK506-binding protein. Cell. 1994;77:513-23.
42. Marx SO, Ondrias K, Marks AR. Coupled gating between individual skeletal muscle Ca2+ release channels (ryanodine receptors). Science. 1998;281:818-21.
43. Koss KL, Kranias EG. Phospholamban: a prominent regulator of myocardial contractility. Circ Res. 1996;79:1059-63.
44. Tada M, Yabuki M, Toyofuku T. Molecular regulation of phospholamban function and gene expression. Ann N Y Acad Sci. 1998;853:116-29.
45. Luo W, Chu G, Sato Y, Zhou Z, Kadambi VJ, Kranias EG. Transgenic approaches to define the functional role of dual site phospholamban phosphorylation. J Biol Chem. 1998;273:4734-9.
46. Shou W, Aghdasi B, Armstrong DL, Guo Q, Bao S, Charng MJ, Mathews LM, Schneider MD, Hamilton SL, Matzuk MM. Cardiac defects and altered ryanodine receptor function in mice lacking FKBP12. Nature. 1998;391:489-92.
47. Marx SO, Reiken S, Hisamatsu Y, Jayaraman T, Burkhoff D, Rosemblit N, Marks AR. PKA phosphorylation dissociates FKBP12.6 from the calcium release channel (ryanodine receptor): defective regulation in failing hearts. Cell. 2000;101:365-76.
48. Jones LR, Suzuki YJ, Wang W, Kobayashi YM, Ramesh V, Franzini-Armstrong C, Cleemann L, Morad M. Regulation of Ca2+ signaling in transgenic mouse cardiac myocytes overexpressing calsequestrin. J Clin Invest. 1998;101:1385-93.
49. Minamisawa S, Hoshijima M, Chu G, Ward CA, Frank K, Gu Y, Martone ME, Wang Y, Ross J, Jr., Kranias EG, Giles WR, Chien KR. Chronic phospholamban-sarcoplasmic reticulum calcium ATPase interaction is the critical calcium cycling defect in dilated cardiomyopathy. Cell. 1999;99:313-22.
50. Alpert NR, Hasenfuss G, Leavitt BJ, Ittleman FP, Pieske B, Mulieri LA. A mechanistic analysis of reduced mechanical performance in human heart failure. Jpn Heart J. 2000;41:103-15.
51. Delling U, Sussman MA, Molkentin JD. Re-evaluating sarcoplasmic reticulum function in heart failure. Nat Med. 2000;6:942-3.
52. Braunwald E, Bristow MR. Congestive heart failure: fifty years of progress. Circulation. 2000;102:IV14-23.
53. Lorell BH, Carabello BA. Left ventricular hypertrophy: pathogenesis, detection, and prognosis. Circulation. 2000;102:470-9.
54. Molkentin JD, Dorn IG, 2nd. Cytoplasmic signaling pathways that regulate cardiac hypertrophy. Annu Rev Physiol. 2001;63:391-426.
55. Hirota H, Chen J, Betz UA, Rajewsky K, Gu Y, Ross J, Jr., Muller W, Chien KR. Loss of a gp130 cardiac muscle cell survival pathway is a critical event in the onset of heart failure during biomechanical stress. Cell. 1999;97:189-98.
56. Sheng Z, Knowlton K, Chen J, Hoshijima M, Brown JH, Chien KR. Cardiotrophin 1 (CT-1) inhibition of cardiac myocyte apoptosis via a mitogen-activated protein kinase-dependent pathway. Divergence from downstream CT-1 signals for myocardial cell hypertrophy. J Biol Chem. 1997;272:5783-91.
57. Mann DL. Recent insights into the role of tumor necrosis factor in the failing heart. Heart Fail Rev. 2001;6:71-80.
58. Bolger AP, Anker SD. Tumour necrosis factor in chronic heart failure: a peripheral view on pathogenesis, clinical manifestations and therapeutic implications. Drugs. 2000;60:1245-57.

59. Kubota T, Miyagishima M, Frye CS, Alber SM, Bounoutas GS, Kadokami T, Watkins SC, McTiernan CF, Feldman AM. Overexpression of Tumor Necrosis Factor- alpha Activates Both Anti- and Pro-Apoptotic Pathways in the Myocardium. J Mol Cell Cardiol. 2001;33:1331-44.
60. Kubota T, McTiernan CF, Frye CS, Slawson SE, Lemster BH, Koretsky AP, Demetris AJ, Feldman AM. Dilated cardiomyopathy in transgenic mice with cardiac-specific overexpression of tumor necrosis factor-alpha. Circ Res. 1997;81:627-35.
61. Kubota T, McTiernan CF, Frye CS, Demetris AJ, Feldman AM. Cardiac-specific overexpression of tumor necrosis factor-alpha causes lethal myocarditis in transgenic mice. J Card Fail. 1997;3:117-24.
62. Olson EN, Williams RS. Remodeling muscles with calcineurin. Bioessays. 2000;22:510-9.
63. Olson EN, Williams RS. Calcineurin signaling and muscle remodeling. Cell. 2000;101:689-92.
64. Molkentin JD, Lu JR, Antos CL, Markham B, Richardson J, Robbins J, Grant SR, Olson EN. A calcineurin-dependent transcriptional pathway for cardiac hypertrophy. Cell. 1998;93:215-28.
65. De Windt LJ, Lim HW, Taigen T, Wencker D, Condorelli G, Dorn GW, 2nd, Kitsis RN, Molkentin JD. Calcineurin-mediated hypertrophy protects cardiomyocytes from apoptosis in vitro and in vivo: An apoptosis-independent model of dilated heart failure. Circ Res. 2000;86:255-63.
66. De Windt LJ, Lim HW, Haq S, Force T, Molkentin JD. Calcineurin promotes protein kinase C and c-Jun NH2-terminal kinase activation in the heart. Cross-talk between cardiac hypertrophic signaling pathways. J Biol Chem. 2000;275:13571-9.
67. Sussman MA, Lim HW, Gude N, Taigen T, Olson EN, Robbins J, Colbert MC, Gualberto A, Wieczorek DF, Molkentin JD. Prevention of cardiac hypertrophy in mice by calcineurin inhibition. Science. 1998;281:1690-3.
68. Passier R, Zeng H, Frey N, Naya FJ, Nicol RL, McKinsey TA, Overbeek P, Richardson JA, Grant SR, Olson EN. CaM kinase signaling induces cardiac hypertrophy and activates the MEF2 transcription factor in vivo. J Clin Invest. 2000;105:1395-406.
69. Sugden PH, Clerk A. Cellular mechanisms of cardiac hypertrophy. J Mol Med. 1998;76:725-46.
70. Clerk A, Sugden PH. Activation of protein kinase cascades in the heart by hypertrophic G protein-coupled receptor agonists. Am J Cardiol. 1999;83:64H-69H.
71. Clerk A, Sugden PH. Small guanine nucleotide-binding proteins and myocardial hypertrophy. Circ Res. 2000;86:1019-23.
72. Zhang D, Gaussin V, Taffet GE, Belaguli NS, Yamada M, Schwartz RJ, Michael LH, Overbeek PA, Schneider MD. TAK1 is activated in the myocardium after pressure overload and is sufficient to provoke heart failure in transgenic mice. Nat Med. 2000;6:556-63.
73. Nicol RL, Frey N, Pearson G, Cobb M, Richardson J, Olson EN. Activated MEK5 induces serial assembly of sarcomeres and eccentric cardiac hypertrophy. Embo J. 2001;20:2757-67.
74. Bueno OF, De Windt LJ, Tymitz KM, Witt SA, Kimball TR, Klevitsky R, Hewett TE, Jones SP, Lefer DJ, Peng CF, Kitsis RN, Molkentin JD. The MEK1-ERK1/2 signaling pathway promotes compensated cardiac hypertrophy in transgenic mice. Embo J. 2000;19:6341-50.
75. Fentzke RC, Korcarz CE, Lang RM, Lin H, Leiden JM. Dilated cardiomyopathy in transgenic mice expressing a dominant- negative CREB transcription factor in the heart. J Clin Invest. 1998;101:2415-26.

13. CARDIAC HYPERTROPHIC SIGNALING THE GOOD, THE BAD AND THE UGLY

O.F.Bueno, E.van Rooij, D.J.Lips, P.A.Doevendans, L.J.De Windt

Introduction

All cells in multi-cellular organisms must be able to sense their surrounding environment and and respond based upon this information. In a similar fashion, cardiac muscle cells are equipped with a specialized protein machinery composed of detection systems (receptors), intermediate proteins within the cell for information transduction (intracellular signal transducers) and nuclear components specialized in changing the genetic profile of the cell (transcription factors). This integrated system is the subject of part of the biological sciences that studies "signal transduction" or shortly "signaling", and topics the molecular mechanisms by which transfer of biological - information at the cellular level is converted. Cardiac signaling systems provide crucial information for cells to decide about differentiation status, death or metabolic control. As such, it is not surprising that many signaling malfunctions underly human diseases. For example, cancer evolves following inactivating mutations in growth-inhibitory pathways, resulting in specialized cells with proliferative advantages over its neighbouring cells[1]. Diabetes results from defects in the insulin-signaling pathway used to control blood glucose levels[2]. Certain forms of achrondoplasia (dwarfism) result from mutations in the receptor tyrosine kinase for fibroblast growth factor,[3] while in agammaglobulinaemia (failure to produce immunoglobulins in the blood), a mutation in the B-cell tyrosine kinase Btk results in a failure of this enzyme to respond to activation of the enzyme phosphatidylinositol-3-OH kinase (PI-3K)[4].

Given the fact that adult cardiomyocytes are resistant to cell cycle reentry, individual cardiomyocytes undergo enlargement following stimulation by a wide array of neurohumoral factors or when faced with increased ventricular wall tension. This process, referred to as hypertrophic growth, forms the basis of sustained pathological hypertrophy of the whole organ and is the leading predictor of development of congestive heart failure, now the leading cause of mortality and morbidity in developed countries worldwide. Signal transduction also plays a role in transducing death signals that lead

to apoptotic cardiomyocyte death and tissue remodeling following thrombotic occlusion of coronary arteries. In a rapid pace, the extracellular ligands, the intracellular information transducers and the transcriptional effectors are being unraveled to provide more insight into the molecular mechanisms behind acquired or congenital heart disease. Given this, it is the hope of cardiovascular researchers that a detailed understanding of hypertrophic signaling pathways will eventually lead to an ability to intervene in heart disease. Developments in other biological fields can stem us optimistic: a drug called STI571, which inhibits the tyrosine kinase Abl, is now successfully used to treat various forms of leukemia in which the enzyme has been erroneously activated [5] while PD184352, a drug that specifically inhibits the extracellular-signal activated kinase (ERK) members of the mitogen activated protein kinase (MAPK) superfamily, was demonstrated to be highly effective in preventing colon tumor development [6].

The present chapter first covers the key themes and components in cellular signaling, illustrated with some well-understood signaling pathways. Subsequently, the molecular mode of action of some important signaling pathways, which govern the cardiac hypertrophic response are delineated. In the remainder of the chapter a novel biological paradigm is introduced, one which attempts to correlate different cardiomyopathic outcomes to a differential activation pattern of isolated hypertrophic signaling cascades.

Signal transduction components

Receptors

Transmembrane signaling is initiated at the receptor level once endogenous ligands bind to their respective receptor protein. Circulating extracellular molecules such as hormones, growth factors, or neurotransmitters bind with high affinity to their receptor producing conformational changes on them that initiate specific responses that allow cells to respond/adapt to changes in the extracellular environment. So far, four basic mechanisms of transmembrane signaling have been described: 1) lipid-soluble ligand that crosses the plasma membrane and acts on intracellular receptors (e.g. thyroid hormone, steroid hormones, nitric oxide, etc), 2) transmembrane receptors that have intracellular enzymatic activity allosterically regulated by the binding of a ligand to an extracellular domain on the receptor (e.g. insulin, EGF, ANF, TGFß, etc), 3) ligand-gated ion-channels whose permeability to diverse ions is regulated by the binding of the ligand (e.g. acetycholine, GABA, etc), and 4) transmembrane receptors which are coupled to a GTP-binding signal transducer protein (G proten) (e.g. epinephrine, angiotensin, enndothelin, etc). The majority of signaling pathways initiate at the cellular outer surface with receptor proteins that are specialized in sensing changes in the extracellular environment. Receptors are most commonly found at the cell membrane, where they bind to extracellular molecules (e.g. hormones or growth factors) that cannot penetrate the hydrophobic membrane (figure 1). These receptors are large molecules that span across the plasma membrane and utilize peptide domains which allow it to span the membrane once or even multiple times. Binding of a signal molecule (or ligand) can result in a change of the three dimensional shape

(conformational change) and, in some cases, engagement of two receptors to form dimers (homo/heterodimerization).
Another aspect of ligand-induced activation is stimulation of integral activation of the intracellular part of the receptor. The most common – and best-studied – receptor involves kinase activity. Two principal classes of receptor kinases exist: those that transfer phosphate to serine and threonine amino acid residues of target proteins (e.g. members of the activin/transforming growth factor (TGF) β receptor family) or those that phosphorylate tyrosine residues (e.g. members of the growth factor receptor family). Other families of transmembrane receptors display different activities: intracellular receptor domains may interact with and induce conformation changes of interacting molecules such as G-protein coupled receptors, while receptors of the integrin family bind to the extracellular matrix and as such have no defined ligands but are responsible for cell-cell connections and structural architecture. Distinct classes of receptors reside inside cells and await ligands that can penetrate the cellular membrane such as steroid hormones or nitric oxide (NO).

Intracellular mediators

Once inside the cardiac muscle cell, the signal is amplified by distinct classes of enzymes with specific activities that allow them to either transfer a phosphate group to target proteins (kinase) or remove phosphates from residues (phosphatase, figure 1). The human genome predicts an overwhelming complexity of combinations for intracellular signal transduction, since some estimates predict that of the ~50,000 human genes, ~1,100 are kinases and ~300 are phosphatases. Current conceptualization therefore suggests that these numbers allow for both specification and a certain degree in functional redundancy between these intracellular mediators. Often, phosphorylation by a kinase leads to activation of kinase activity in the target protein, allowing for perpetualization of the signal. In other instances, phosphorylation can affect the affinity with other molecules, allowing for complex formation of a whole subset of specialized enzymes in close vicinity to each other. Some proteins in these complexes may only function to bring together other signaling molecules (adaptor or docking proteins). For example, binding of a growth factor to its corresponding receptor leads to self-phosphorylation of the receptor. The SH2 domain of an adaptor protein, Grb2, attaches to phosphorylated tyrosine residues of the receptor and links an enzyme called Sos to the activated receptor. Sos itself is a major regulator of Ras activity and the crucial outcome of these phosphorylation steps is a complex formation event so that Sos has increased access to Ras. Activation of Ras on its turn culminates in phosphorylation (activation) of protein kinase cascades belonging to the family of mitogen-activated protein kinases (MAPK), the activity of which directly influences gene expression (figure 1).

Nuclear activities

The final outcome of these signaling events often constitutes a change in gene expression profile. This requires the transfer of a final target protein or enzyme to physically move from the cytosol into the cellular nucleus. Specific recognition motifs (which may be functionally altered by (de)phosphorylation events) within amino acid sequence of proteins govern the import and export of proteins to and from the nucleus (figure 1). For example, the Ras pathway impinges upon the ERK member of the MAPK superfamily. ERK is able to translocate into the nucleus upon activation by upstream kinases. Once inside the nucleus, ERK phosphorylates and activates multiple target proteins that are involved in transcriptional activity of target genes. The result of this may be an altered mRNA formation, thereby altering the protein composition of the cell, which lead to altered cell function.

Cardiac hypertrophic signaling pathways

G proteins

A well-characterized signaling paradigm in hypertrophic signaling pathways involves the recruitment of GTP-binding proteins that follows the binding of a ligand to a G protein-coupled receptor. Signals derived from G protein activation can result in the generation of either stimulatory or inhibitory signaling. All heterotrimeric G proteins constitute of separate G_α and $G_{\beta\gamma}$ subunits. Activation of G proteins involve the catalysis of GTP to GDP on the G_α which results in the dissociation of the G_α from the $G_{\beta\gamma}$ subunits. Activation of G protein paradigms involve the catalyzation of GTP to GDP exchange on G_α subunits by e.g. occupational activation of transmembrane receptors,

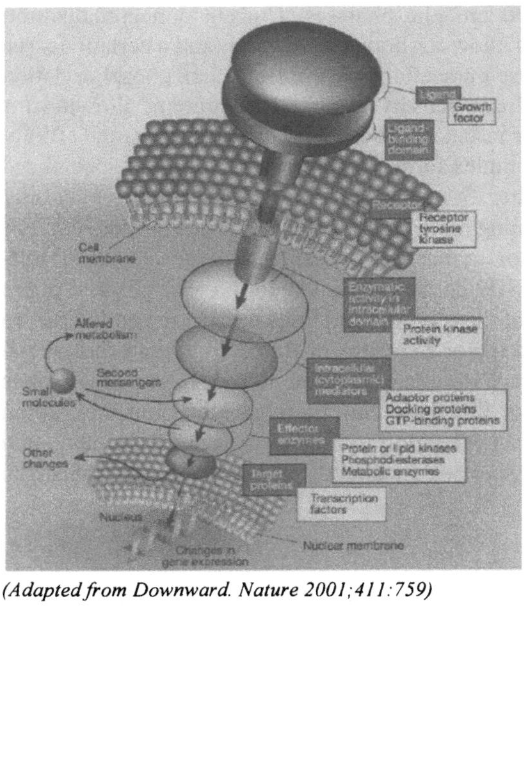

(Adapted from Downward. Nature 2001;411:759)

Figure 1 *Principle of signaling pathway. The grey boxes indicate general components of signaling pathways; the white boxes indicate specific examples. On the outside of the cellular membrane, receptors bindt to active ligands such as hormones, cytokines or growth factors. As a result, enzymatic activity of the intracellular domain of transmembrane receptors is altered. This can affect the association of the receptor with intracellular mediators, or the localization of intracellular mediators. Some intracellular effectors can move directly to the nucleus and control gene expression, or they can induce other proteins to do so. Other targets include small molecules, either generating further signaling mediators (second messengers) or controlling the metabolic state of the cell. Real signling pathways may ypass entire classes of these molecules, or may have several components in one or more class, working together in series or parallel.*

which results in dissociation of the G_{α} and $G_{\beta\gamma}$ subunits. G_{α} subunits subsequently target the activation status of several adenylyl cyclase (AC) and phospholipase C (PLC) isoforms. Free $G_{\beta\gamma}$ subunits have been linked to the activation of MAPK, PI-3K and the small G protein Ras. In the heart, three major classes of G_{α} subunits exist ($G_{\alpha,q}$, $G_{\alpha,s}$ and $G_{\alpha,i}$) and can be categorized based upon the distinct functional classes of cardiovascular receptors that couple to them. The involvement of these physiologically distinct Gα proteins will be discussed in relation to their putative role in mediating cardiomyopathy.

The β-adrenergic receptors (βAR) that are responsible for direct changes in heart rate and contractility directly couple to $G_{\alpha,s}$ in response to epinephrine or norepinephrine occupation [7]. To confirm whether chronic $G_{\alpha,s}$ stimulation enhances contractile reserve and whether it plays a role in hypertrophy and heart failure (a condition known to suffer from increased systemic catecholamine circulation), transgenic mice were created with approximately 3 fold $G_{\alpha,s}$ overexpression relative to base-line and ~90% increased $G_{\alpha,s}$ activity, but without the expected effects on AC activity [8]. $G_{\alpha,s}$ transgenic mice displayed an increased proportion of βARs coupled to $G_{\alpha,s}$, resulting in normal baseline function but enhanced isoproteronol responsiveness. However, transgenic mice did develop cardiomyocyte atrophy and increased myocardial fibrosis, and increased apoptotic index. Latter findings are consistent with the deleterious phenotype of β1AR and β2AR overexpressing mice and human heart failure patients with increased catecholamine stimulation [9-11]. It is unclear, however, whether these unfavourable effects are due to AC stimulation or the simultaneous activation of other pathways (figure 2). A more definitive role for Gα subunit stimulation and cardiac hypertrophy and decompensation has been found for $G_{\alpha,q}$. Agonist stimulation of receptors that are coupled to PLC through $G_{\alpha,q}$ with phenylephrine, angiotensin II, endothelin-1 and prostaglandin F2α has been demonstrated to result in increased cell size in cardiomyocyte cultures by different approaches. Overexpression *in vitro* of activated $G_{\alpha,q}$ or receptors that directly couple to $G_{\alpha,q}$ have also been demonstrated to be sufficient to drive a significant hypertrophic phenotype. More definitive evidence for a role of $G_{\alpha,q}$ coupled signaling in hypertrophic remodeling came from genetically engineered mice overexpressing $G_{\alpha,q}$ in the cardiac compartment. These mice exhibited a concentric hypertrophic response reminiscent of pressure overload hypertrophy accompanied by sinus bradicardia at rest, indicating a potentially important role for this particular Gα subunit in the genesis of heart failure (figure 2)[12]. $G_{\alpha,i}$ signaling, in contrast, has been linked with blocking βAR-mediated augmentation of calcium cycling and contractility [13]. $G_{\alpha,i}$ has been demonstrated to inhibit the activity of adenylyl cyclase (AC) which occurs following βAR stimulation [13]. Since some studies using biopsies from human heart failure patients have indicated an upregulation of $G_{\alpha,i}$ content, a role for $G_{\alpha,i}$ in the processes that may contribute to decompensation and heart failure has been postulated. Multiple studies have now confirmed increased $G_{\alpha,i}$ content and reduced AC activity in human and experimental models of heart failure. In line with this assumption, it was recently found that mice with conditional activation of a $G_{\alpha,i}$-coupled receptor exhibit bradycardia and signs of failure (figure 2)[14].

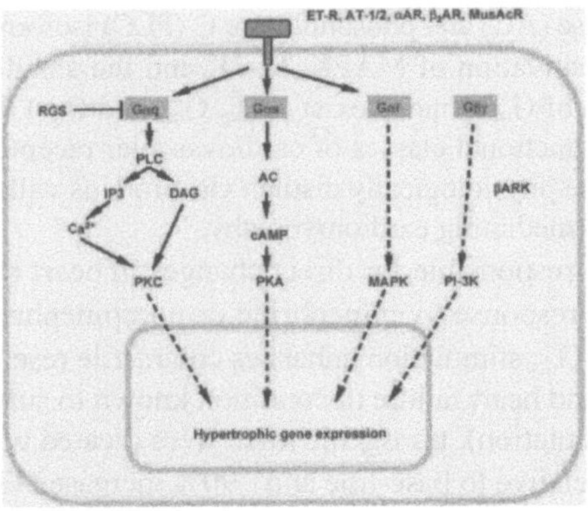

Figure 2 *Cardiac G protein signaling. A variety of seven-transmembrane spanning receptors couple to G proteins. As a result, GDP-GTP exchange takes place and dissociation of the Gα and Gβγ subunit. Each subunit (different Gα isoforms and Gβγ) is able to then trigger the activation of downstream effector proteins with distinct results on cardiomyocyte homeostasis. ET-R, endo-thelin-1 receptor; AT-1/2, angiotensin 1/2 receptor; AR, adrenergic receptor; muscAcR, muscarinic acetylcholine receptor; RGS, regulator of G protein signaling; PLC, phospholipase C; AC, adenylate cyclase; IP3, inositol 3-phosphate; DAG, diacylglycerol; PKC, protein kinase C; cAMP, cyclic AMP; MAPK, mitogen activated protein kinase; PI-3K, phosphatidylinositol 3-kinase*

Small guanine nucleotide-binding proteins

The small guanine nucleotide-binding protein (small G-protein) superfamily comprises a multitude of proteins with relative molecular mass of ~ 21 kDa. In their inactive state, they are bound to GDP and are activated by exchange of GDP for GTP. This process is enhanced and regulated by guanine nucleotide exchange factors (GEFs). The innate GTPase activity of small G proteins hydrolyzes bound GTP to GDP, returning them to their inactive state. The GTPase activity is enhanced by GTPase-activating proteins (GAPs; figure 3). Five subfamilies of small G proteins have been characterized (Ras, Rho, ADP ribosylation factors or ARFs, Rab and Ran), each consisting of multiple members. All are lipid modified and associated with membranous structures within the cell. To date, only the Ras and Rho subfamilies have been studied in the heart, primarily in the context of hypertrophic signaling. The Ras family included the 3 classical Ras isoforms (Harvey [HRas], Kirsten [KRas] and NRas), Rap and Ral. The Rho family includes RhoA, cdc42 and Rac1. In the studies investigating their contribution in hypertrophic signaling, mutants have often been employed, including constitutively activating mutations (Gly-12 to Val)Ras[V12Ras], V12Rac-1 and V14RhoA and dominant inhibitory (dominant negative) mutants such as (Ser-17 to Asn)Ras [N17Ras], N17Rac1 and N19RhoA (reviewed in [15]).

The implication of Ras as a hypertrophic signal mediator probably consistutes one of the earliest studied examples in cardiac signaling. The classical Ras isoforms are rapidly activated in myocytes exposed to ET-1, AngII or phenylephrine (PE), which activate the GPCR systems, with upstream activators $G_{q/11}$PCRs, the $G_{q/11}$heterotrimeric proteins, phospholipase Cβ and protein kinase C (PKC) members. As with receptor tyrosine kinase stimulation of Ras, adaptor proteins (Grb2, Shc) may be involved, which activate the Ras guanine nucleotide exchange factor (GEF) Sos.

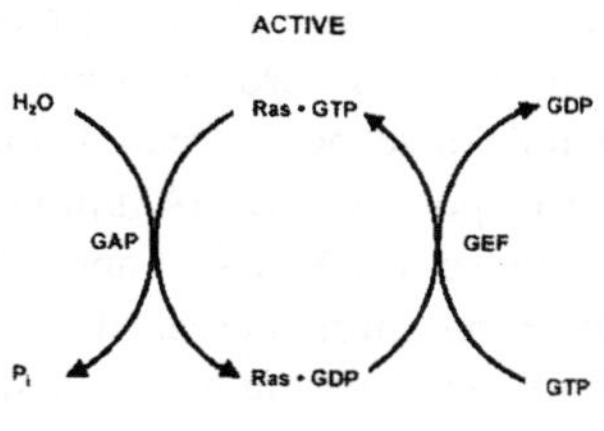

Figure 3 *Ras GTPase cycle. After stimulation of membrane receptors, inactive Ras.GDP is activated by the exchange of the GDP moiety for GTP. This reaction is stimulated by GEFs. Binding of GTP induces a conformational change in Ras, allowing it to couple to its downstream effectors. The innate GTPase activity of Ras hydrolyzes the bound GTP, returning the protein to its active GDP-ligated state, a reaction that is enhanced by GAPs. Although the cycle is shown for Ras, it holds true for all other small G proteins, such as RhoA and Rac1 family members.*

Multiple *in vitro* studies have demonstrated that during hypertrophic signaling, Ras binds and activates several signaling molecules, including c-Raf, PI-3K and Ral GDS (a GEF for the Ras-like protein Ral). The best-studied downstream Ras targets include the ERK and c-Jun NH_2 terminal protein kinase (JNK) members of the MAPK superfamily (figure 4). Transgenic mice have been created expressing a V12Ras transgene in a postnatal, ventriculo-specific manner using the MLC-2v promoter [16-18]. Homozygous transgenic animals displayed hallmarks of LV hypertrophy, including increased indexed LV weights, increased myocyte cross-sectional area and abundance of a fetal gene expression pattern. The role of Ras in the hypertrophic response was not completely unambiguous because there was no RV hypertrophy and heterozygous transgenic animals displayed no phenotype at all despite significant transgene expression. Subsequent studies using transgenic offspring derived from V12Ras animals that were echocardiographically selected with demonstratable hypertrophy, suggest substantial interindividual genetic variability and susceptibility to the phenotype. These animals demonstrated selective diastolic dysfunction and concentric hypertrophy without LV cavity enlargement. Interestingly, hypertrophic V12Ras hearts demonstrated mild JNK MAPK activation, in the absence of ERK activation. Current conceptualization is that Ras promotes cardiac hypertrophy in this genetic context and reproduces features of hypertrophic rather than dilated cardiomyopathy [15].

The involvement of the Rho subfamily members in hypertrophic signaling is more complex. RhoA activates two groups of protein kinases: (1) protein kinase N/PKC-related kinases and (2) Rho kinases. Recently, protein kinase N (PKN) was implicated in transducing signals to upstream kinase members of the MAPK superfamily, leading to activation of the p38 terminal MAPK branch and, ultimately, the transcription factor myocyte-enhancing factor-2 (MEF-2) (figure 4)[19]. It is likely that in cardiomyocytes a similar pathway is active and responsible for the hypertrophic effects of RhoA. In other cell types, RhoA and Rac1 activation is associated with cytoskeletal remodeling and changes in cell morphology. The Rho kinases include Rho kinase itself (ROKα or ROCK2) and p160ROCK (ROKβ or ROCK1). ROKα stimulates LIM kinase to phosphorylate and inactivate cofilin. Since cofilin activation leads to actin depolymerization, the net effect of this pathway may be to promote formation of actin fibrils (figure 4). Rho kinase also promotes phosphorylation of MLCs and inhibition of

MLC phosphatase, which would constitute an alternative pathway for cytoskeletal organization in cardiac myocytes. Although multiple in vitro studies have demonstrated a clear hypertrophy reducing effect of Rho inhibition in cultured cardiomyocytes, transgenic animals expressing either wildtype or V14RhoA have been created, but surprisingly did not display increased LV mass [20]. Heterozygous V14RhoA animals died prematurely with signs of ventricular dilation (no increase in LV mass) and LV dysfunction, accompanied by changes in hypertrophic gene expression, increased atrial mass, and conduction abnormalities. Given the role of RhoA in regulating cytoskeletal organization in other cell types, the phenotype of these animals is so dissimilar from LV hypertrophy that firm conclusions regarding a role for RhoA in hypertrophic signaling cannot be drawn based upon the available literature.

The phenotype of Rac1 transgenic animals is more in accordance with its expected dichotomous role in cardiac remodeling. Rac1 is known to activate p21 activated kinase (PAK), which promotes activation of ERK and JNK MAPK on the one hand, and cell-cell interactions and cellular architecture by modulating the actomyosin cytoskeleton. V12Rac1 animals displayed a complex phenotype consisting of a lethal neonatal dilated cardiomyopathy and a resolving transient cardiac hypertrophy in juvenile animals [20]. The authors observed phosphorylation of Rac1's immediate downstream targets PAK and Src, which in their turn lead to phosphorylation of Paxillin. Paxillin is a component protein of focal adhesions that serves as an adaptor molecule to transduce signaling into structural remodeling. Focal adhesions are multi-protein complexes composed of structural and signaling components including integrins, talin, F-actin, focal adhesion kinase (FAK), paxillin, and zyxin, which are thought to mediate integrin signaling and actin dynamics in the cardiomyocyte. Consistent with its hypothesized role in morphological remodeling of the myocardium, V12Rac1 mice developed hypertrophic changes within weeks after birth, and signaling pathways controlling focal adhesion function were dysregulated. In addition, focal adhesion structures were altered, consistent with a role for Rac1 in controling the actin skeleton. Consequently, the loss of cytoskeletal-based cardiomyocyte adhesion in juvenile V12Rac1 mice led to lethal DCM and heart failure in some individuals [21].

Thus, overwhelming data from in vitro studies strongly implicate both the Ras and Rho members of the small G protein superfamily in controlling important aspects of development of hypertrophy, yet the signaling pathways remain to be characterized in detail. Furthermore, transgenic models expressing constitutively activated mutants of Ras, RhoA and Rac1 in the cardiac compartment largely support this view. In addition, the current literature also suggests that the pathological changes brought about by these small G proteins may not necessarily be related to myocyte growth.

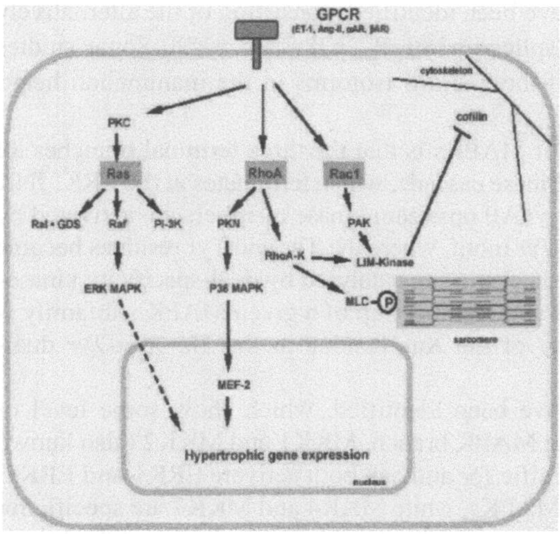

***Figure 4** Regulation of cardiac myocyte responses by Ras, RhoA and Rac1. GPCR's such as those for endothelin-1, angiotensin-II or adrenergic stimuli activate Ras, RhoA or Rac1. Activation of Ras is most probably mediated by PKC and signals downstream into c-Raf which is the first component of the ERK MAPK pathways, or signals through the PI-3K pathway. RhoA and Rac1 promote hypertrophic gene expression through e.g. p38 MAPK pathways, but also regulate cellular architecture by modulating the actin/myosin cytoskeleton. Abbreviations: GPCR, G protein coupled receptor; ET-1, endothelin-1; Ang-II, angiotensin II; AR, adrenergic receptor; PKC, protein kinase C; MAPK, mitogen activated protein kinase; PKN, protein kinase N; RhoA-K, rhoA kinase; MLC, myosin light chain; PAK, p21-activated kinase.*

Mitogen-activated protein kinase cascades

Terminology, characteristics and involvement in hypertrophic signaling

By far the most-extensively studied signal transduction paradigm includes the individual members of the superfamily of MAPK. This superfamily can be subdivided into 3 distinct terminal branches based upon sequence-homology and each branch contains multiple isoforms. These three branches include: (1) the extracellular-signal activated protein kinases (ERK), c-Jun NH_2-terminal kinases (JNK), and (3) the p38-MAPKs. Once activated, the ERK, JNK and p38 MAPKs phosphorylate a wide array of transcription factors that are thought to be involved in reprogramming the cardiac gene expression profile (table I). An ever-increasing number of individual members of the separate MAPK branches are being uncovered, displaying varying degrees of homology to one another.

Several genes of the ERK isoforms have been cloned and encode *erk1*, *erk2* and, *erk3*. More recently another MAPK was identified and designated ERK5 on the basis of sequence homology. The best-described ERK isoforms are ERK1 and ERK2 and encode proteins of 42 and 44 kDa, respectively (and they are also known as p42 and p44). ERK1 and ERK2 are kinases that preferentially phosphorylate the consensus sequence Pro-Xaa-Ser/Thre-Pro in substrate proteins.

At least 10 different JNK isoforms, all derived from differential splicing of three separate genes (*jnk1, jnk2* and *jnk3*) have been identified and the predicted molecular mass of these kinases is ~ 46 or ~ 54 kDa, depending on the presence or absence of a C-terminal extension. Thus, JNK1, -2, and –3 all consist of a p46 and p54 member and these isoforms are indistinguishable on SDS-PAGE gels based upon their migration speed. Of note, the most abundant JNK member in the heart remains to be determined, although a JNK1 antibody immunoprecipitates the vast majority of 46 kDa and a large

part of the 54 kDa proteins.
To date, six p38-MAPK members have been identified, consisting of the alternatively spliced p38α_1/α_2, the alternatively spliced p38β_1/β_2, p38γ and p38δ. Some studies suggest a relative low abundance of the γ and δ isoforms in the mammalian heart, compared to the α and β isoforms.
The most compelling characteristic of MAPKs is that the three terminal branches all consist of a three-membered protein kinase cascade, which terminates at the ERK, JNK or p38 MAPKs members, respectively. All upstream kinase members are activated by a dual-phosphorylation of a *Thr-Xaa-Tyr* motif, where the Thr and Tyr residues become phosphorylated. These phosphorylation events are catalysed by dual-specificity kinases designated MAPK kinases or MKKs and membership of a given MAPK subfamily is assigned on the basis of the identity of the Xaa residue in the *Thr-Xaa-Tyr* dual-phosphorylation motif.
To date, several MKK members have been identified, which show some level of specificity towards the terminal target MAPK branch. MKK1 and MKK2 (also known as MEK1 and MEK2) are highly specific for and can both activate ERK1 and ERK2. MKK3 and MKK6 activate the p38-MAPKs, while MKK4 and MKK7 are specific for the JNKs, although MKK4 can also stimulate p38s. The kinases upstream of the MKKs (MAPK Kinase Kinase or MKK kinase or MKKK) for the JNK and p38 branches have not been fully characterized. In the ERK cascade, c-Raf fulfills this function and it is conceivable that small G proteins such as Ras, cdc42 and Rac1 may activate p21-activated kinase (PAK) and other kinases that activate the specific MKK members for the JNK and p38 MAPK cascades (figure 5, reviewed in [22-23])
In a similar fashion, MAPK members require inactivation following their phosphorylation-dependent activation. This event is carried out by a family of specialized dual-specificity phosphatases termed MAPK-phosphatases (MKPs). MKPs recognize the complex dual-specificity motifs in the single MAPK factors and their coordinated removal of phosphate groups on their MAPK targets allow for continuing cycling of signal transduction and alterations in gene expression. Multiple members of the MKP family exist, and surprisingly little is known about their existence and function in the heart. This is very surprising in view of the fact that they may play a significant role in the genesis of hypertrophic remodeling (figure 5)[24].
A massive number of studies have demonstrated that each of the MAPK terminal branches are (transiently) activated following exposure to either hypertrophic agonists or mechanical load both in cultured myocyte models and isolated perfused hearts. Since a number of excellent reviews exist on these topics, the involvement of MAPKs in hypertrophic signaling will be exemplified by discussing recent findings relating to these kinases. These topics will include (1) the application of genetic strategies as opposed to pharmacological devices to demonstrate their sufficiency and requirement to induce a hypertrophic response in cultured cardiomyocytes; (2) their temporal activation pattern following hypertrophic stimuli *in vivo*, and (3) their activation pattern in human hearts.

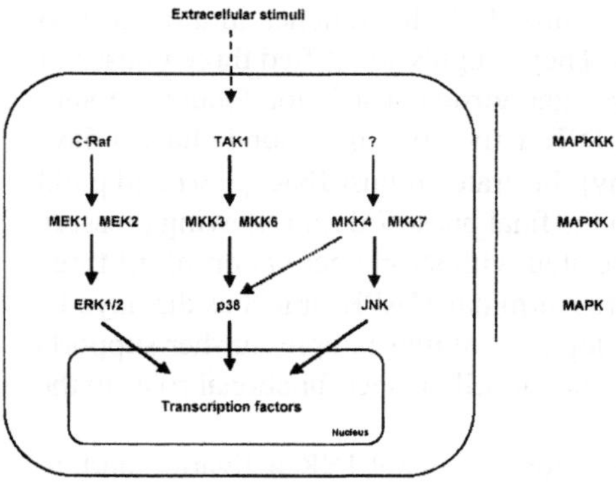

Figure 5 *Principle of mitogen activated protein kinase signaling (MAPK) in cardiac myocyte. MAPK cascades are composed of a signaling cascade paradigm consisting of three terminal kinases, designated p38, ERK and JNK, that are activated by dual phosphorylation by members of the MAPK kinases (MAPKK) group of protein cascades. The MAPKK are activated, in turn, by a group of MAPKK kinases (MAPKKK), the identity of which is poorly defined in cardiac myocytes. The sequential actions of a MAPKKK, a MAPKK and a MAPK creates a MAPK signaling module. abbreviations: TAK1, transforming growth factor activated kinase-1.*

MAPK involvement in hypertrophy

Many of the studies using pharmacological agents to study the role of the single MAPK factors in any cell type are hampered by a lack of specificity or even complete absence (as is the case for the JNK pathway). To this end, researchers have reverted to using viral mediated gene transfer of either activated forms or dominant inhibitory forms of the single MAPK branches to circumvent their low transfection efficiency. Accordingly, our understanding of the MAPK superfamily in terms of cardiopathology has greatly advanced by these strategies. For example, adenoviral-mediated gene transfer of either constitutively activated forms of MKK6 or MKK3 (upstream activators of p38) have demonstrated divergent functional roles for these pathways, with MKK6 – p38α leading to cardiomyocyte hypertrophy and MKK3 – p38β leading to apoptotic cell death [25]. Latter observations are impossible to achieve using pharmacological inhibition since the available inhibitors fail to distinguish between the separate p38 isoforms. Adenoviral expression of activated or inhibitory forms of MKK7 were recently demonstrated to significantly impact the trophic ability of cultured cardiomyocytes [26]. Similarly, expression of an activated mutant of MEK1 or a dominant negative form were shown to significantly enhance hypertrophy or block pleiotropic agonist-induced enlargement of neonatal rat cardiomyocyte cultures, respectively [26,27]. We have recently demonstrated that adenoviral expression of a member of the the dual-specificity phosphatase family (MKP-1) proved to be an extremely efficient strategy to prevent cardiomyocyte hypertrophy in response to all neurohumoral agonists tested so far [29]. MKP-1 prevents phosphorylation of p38, JNK and ERK1/2 simultaneously. Latter observations established the requirement of either MAPK factor to promote myocyte enlargement.

In a recent study, Esposito and coworkers [30] presented evidence for a differential temporal activation pattern of the separate terminal MAPK branches in a controlled pressure overload hypertrophy mouse model. They roughly identified three phases of MAPK induction: the first phase only 7 hours after aortic constriction (acute pressure overload) was accompanied by only a robust JNK activation; the second phase 3 days following pressure overload (early hypertrophy) illustrated robust JNK, p38α and p38β activation, with mild ERK1/2 involvement. In the final phase 7 days following pressure overload (established hypertrophy) was associated with strong activation of all three MAPK branches. The observation that the three terminal MAPK branches displayed a differential activation pattern during the development of hypertrophy further supports the notion that the individual MAPK branches may fulfill distinct functional roles in the process [30].

Haq and coworkers recently provided evidence for prominent JNK activation and, to a lesser extent, the p38 and ERK1/2 pathways, in biopsies from failing human hearts of dilated cardiomyopathy (DCM) and coronary artery disease (CAD)[31]. Earlier studies confirm this activation pattern in CAD patients with respect to the p38 and JNK pathways [32]. Activation of MAPK factors does not appear to be restricted to chronic heart disease but also involved in human tissue in response to ischemia and reperfusion. Patients undergoing coronary artery bypass grafting demonstrated a differential activation pattern before cross-clamping (activation of ERK1/2 and p38-MAPK) and after cornary reperfusion (prominent activation of all three factors). Although these studies were characterized by reasonable group matching in terms of age and sex, studies in human material are inherently hampered by the divergent clinical status of patients, differences in medication, limited availability of control, healthy tissue, and correlative nature of the studies. Nevertheless, important insights are still obtained by these types of studies, which begin to cristalise a view of widely differential activation patterns of the single MAPK factors in distinct human conditions, raising the possibility that their differential regulation may be a causal factor in the transition fom cardiac health via hypertrophy to failure.

Calcium-activated hypertrophic signaling

The myocardium is the only organ that makes such efficient use of calcium for diverse processes. Calcium is an integrated component of cardiac excitation-contraction coupling, and many studies have hold a dysregulated cardiac calcium homeostasis to be an important underlying cause of cardiac growth and failure. Recent studies have emphasized the importance of calcium in the activation of a series of novel hypertrophic signal mediators. A central question for the future will be to decipher how the cardiac myocyte is able to distinguish these calcium fluxes that link extracellular stress to cytosolic and nuclear events in the face of immense calcium fluxes at a beat-to-beat basis. A more in-depth understanding of how calcium/calmodulin activated hypertrophic signaling modules work and are integrated with other signaling motifs may result in the identification of common nodal points, which may become an interesting target for future therapeutic treatment of a variety of heart diseases.

Calcineurin

Ever since its introduction, the significance of the calcineurin-NFAT module has been subject of controversy. The calcineurin-NFAT pathway was one of the first signaling paradigms which provided molecular insight on how extracellular signals travel from the cell membrane into the nucleus. The precise components of the pathway were defined by working backwards from the T- lymphocyte nucleus to the cell membrane. The regulatory regions in the interleukin-2 (IL-2) gene were found to be under control of a CyclosporineA(CsA)-sensitive transcriptional complex designated Nuclear Factor of Activated T-cells (NFAT). NFAT is shuttled between cytoplasmic and nuclear components under influence of the calcium signal, and its nuclear import prevented by the immunosuppressive agents CsA and FK506. Nuclear NFAT import was defined to be regulated by dephosphorylation steps catalyzed by the phosphatase calcineurin, which in its turn was subject to regulation by calcium-calmodulin binding (figure 6). Finally, calcineurin was identified as the cellular target of the immunosuppressive agents CsA and FK506.(reviewed in [33,34])

Calcineurin was first implicated as a hypertrophic signal mediator by the observation that NFAT3 and GATA-4 drove synergistic activation of a brain natriuretic peptide (BNP) promoter-reporter construct in cardiac myocytes. To elucidate the functional significance of the calcineurin-NFAT3 pathway *in vivo*, several lines of transgenic mice were generated containing either truncated activated mutants of NFAT3 or the calcineurin Aα subunit specifically in the heart. Persistent calcineurin activation was sufficient to promote a hypertrophic response in 11 separate transgenic (TG) founder lines, varying from relatively benign forms of concentric hypertrophy to severe forms of dilated cardiomyopathy and early lethality in low and high copy number TG mice, respectively. NFAT3 TG mice also demonstrated a hypertrophic myopathy associated with re-expression of fetal genes. Pharmacological inhibition of calcineurin through CsA injections (25 mg/kg/day) resulted in complete prevention of the pathology. As a testimony to the specificity of the entire signaling cascade, the same dosage of CsA was unable to prevent the morphologic pathology of the NFAT3 TG animals, expressing a calcineurin-independent form of the transcription factor. [35,36]

Since then, much confusion has centered around the significance of this pathway, largely related to studies that used systemic CsA administration in pressure-overload models, with some investigators reporting partial or no rescue of hypertrophy. More recent evidence using genetic, cardiac-selective calcineurin inhibition approaches, however, has consolidated the central involvement of this calcium/calmodulin-activated signaling pathway in both developmental trophy of the myocardium as well as pathological forms of hypertrophy. [37-39]

Calcium/calmodulin-activated protein kinase (CaMK)

Activated calcium-calmodulin-dependent protein kinase (CaMK) isoforms I, II and IV

have been demonstrated to induce hypertrophic gene expression in cultured cardiomyocytes, while the CaMK inhibitor results in block of the Endothelin-1-mediated trophic response. The most compelling evidence in support of a role for CaMK as a hypertrophic signal transducers stems from the observation that overexpression of an activated form of CaMKIV in TG mice resulted in a profound hypertrophic response that transitions into heart failure, quite similar to that seen after calcineurin overexpression. CaMKIV most likely does not represents the most relevant (abundant) isoform of the myocardium, yet the activating mutant most likely mimics the activity of other, more abundant, cardiac isoforms. Indeed, recent evidence supports this notion, since overexpression studies in mice of CaMKIα resulted in a similar pathological remodeling as seen for the CaMKIV isoform.[40]
The nuclear events following CaMK activation are more complex than expected at first glance. A number of elegant studies have elucidated that CaMK stimulates the transcription factor myocyte enhancing factor-2 (MEF-2) family by directly phosphorylating histone deacetylases (HDACs). HDACs are complexed to MEF-2 in quiescent myocytes, thereby repressing its activity. CaMK-dependent phosphorylation results in de-repression of MEF-2 by HDACs, which are expelled from the nucleus and retained in a cytosolic localization by 14-3-3 proteins (figure 6).[41-44] Latter event allows MEF-2 to activate the plethora of muscle-specific genes it is known to control from earlier studies.

Myosin light chain kinase (MLCK)
An integral feature of cardiomyocyte hypertrophy consists of the sarcomerization of this particular cell type to respond appropriately following increased workload. Phosphorylation of myosin light chain 2v (MLC-2v) accompanies this event and is thought to facilitate the incorporation of additional actin/myosin filaments into the sarcomere. Also, MLC-2v phosphorylation increases the calcium sensitivity of the myofilaments, allowing increased contractility. A recent study has demonstrated that the critical phosphorylation event is performed by a calcium-calmodulin activated kinase termed MLC kinase or MLCK. MLCK activity was demonstrated to be the critically component which regulates sarcomeric organization of cultured cardiomyocytes following hypertrophic agonist stimulation. Adenoviral-mediated gene transfer of MLCK resulted in sarcomerization in the absence of a profound cellular trophic effect. Although this in vitro study remains to be consolidated in an in vivo setting, the regulation of MLCK and its profound impact on cardiac architecture does testify to the importance of calcium-calmodulin-dependent signaling in this cell type (Figure 6). [45]

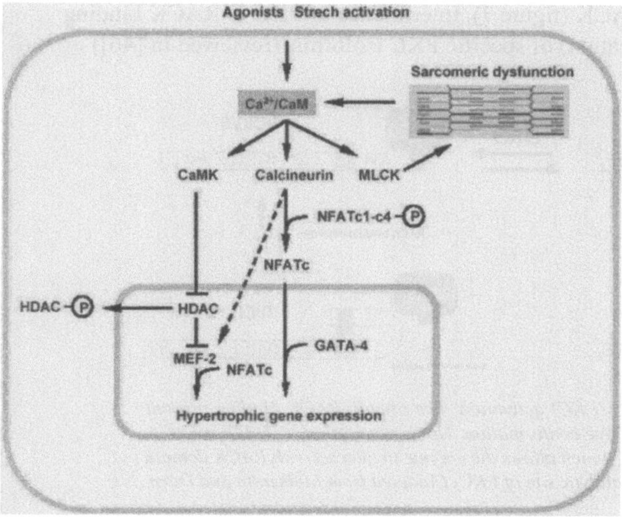

Figure 6 Calcium-activated signaling pathways. Upon agonist stimulation, intrinsic defects or activation of stretch-sensitive receptors, intracellular [Ca^{2+}] rise. The resultant Ca^{2+}/calmodulin complex binds and activates kinases and phosphatases such as Calcineurin, CaMK or MLCK. Calcineurin dephosphorylates NFATc members, which translocate to the nucleus to cooperatively activate gene transcription with the zinc-finger transcription factor GATA-4. Activated CaMK is able to relieve the inhibitory action of HDAC on MEF-2 members. MLCK directly enhances sarcomeric organization by MLC-2v phosphorylation. Abreviations: *CaMK*, Ca^{2+}/calmodulin-activated kinase; *MLCK*, myosin lightc chain kinase, *NFAT*, nuclear factor of activated T-cells; *HDAC*, histone deacytylase.

Protein kinase C

The protein kinase C (PKC) superfamily of calcium and/or lipid activated serine/threonine kinases, consists of at least 12 members and are recruited upon activation of virtually all signal transduction pathways initiated at the cell membrane. All PKC members are encoded by distinct genes, which exhibit different tissue distribution and agonist-activation patterns. PKCs are classified according to their enzymatic properties and divided into the (1) classical PKC isoforms (*cPKCs*: α, βI, βII, γ [alpha, beta and gamma]), which are activated by both Ca^{2+} and diacylglycerol (DAG) or 12-myristate 13-acetate (PMA); (2) the novel PKC isoforms (*nPKCs*: δ, ε, η, θ [delta, epsilon, eta and theta]), activated by only DAG or PMA; (3) the atypical isoforms (*aPKCs*: λ and ξ [lamda and xi]), not activated by either Ca^{2+}, DAG or PMA; and finally (4) the recently described PKC isozymes (PKCμ and PKCν [mu and nu]). One distinctive property of PKC activation involves their translocation to different intracellular localizations upon activation.

The molecular mechanism for PKC activation and subcellular translocation is remarkably complex and uncovered only in the last few years (figure 7). PKC is regulated by two sequential steps, which are equally critical, the first being a phosphorylation event by the recently discovered 3-phosphoinositide-dependent kinase (PDK)-1 and –2, and the second step being binding to DAG. Unphosphorylated PKC is membrane bound and becomes the target for PDK-dependent phosphorylation in its activation loop. The consequence of this event is an autophosphorylation event in the carboxy terminus (C1 and C2), resulting in the release of the isozyme in the cytosol, maintained in an inactive conformation by binding to an autoinhibitory sequence or pseudosubstrate. Membrane recruitment by DAG or phosphatidylserine (PS) provides the energy to expel the inhibitory pseudosubstrate from the substrate-binding cavity, allowing activation of PKC (figure 7). Activation involves binding to proteins termed RACKs (receptor for activated C kinases), with each PKC isoform having its own specific RACK (figure 7). Interference of PKC to RACK binding can inhibit translocation (and activation) of specific PKC isoforms. (reviewed in [46])

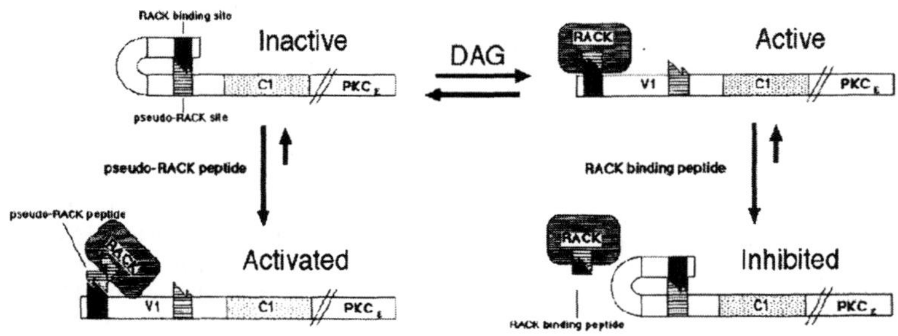

***Figure** 7. Mechanism of protein kinase C (PKC) activation. The pseudo-RACK binding present within PKC retains the kinase in the inactive conformation. Upon stimulation by DAG or Ca^{2+}, PKC undergoes a conformational change, which allows the enzyme to interact with RACK domain containing proteins and exposure of the catalytic site of PKC. (Adapted from Molkentin and Dorn, Annu Rev Physiol 2001;63:391-426]).*

In the last few years, the specific functional roles attributed to the cardiac PKC isoforms is slowy arising, although thorough in vivo correlations are still lacking. AR stimulation was found to stimulate cultured cardiomyocytes and to induce translocation of PKCβI from cytosol to nucleus; PKCβII from fibrillar structures and PKCδ from cytosol to perinuclear regions. PKCε is selectively translocated to the particulate fraction of ventricular myocardium during acute or chronic pressure overload and after AngII stimulation.[47] Conventional and conditional TG approaches have been employed to study the sufficiency for certain PKC isoforms to contribute to cardiac pathology. Accordingly, PKCβII overexpression (αMyHC promoter driven) resulted in cardiac hypertrophy associated with fibrosis and systolic dysfunction, accompanied by Troponin I phosphorylation as a plausible mechanistic explanation for the dysfunctional hemodynamic phenotype.[48,49] Inducible activation of PKCβ using the tetracycline-sensitive conditional expression system resulted in neonatal lethality and mild hypertrophy in the adult mouse. Remarkably, loss of PKCβ did not attenuate pressure overload hypertrophy, indicating substantial functional compensation for PKCβ by other PKC isoforms during mechanical load (and, thus, relative unimportance of this particular PKC isoform in the development of this type of hypertrophy).[50]

Recently, the more abundant cardiac isoforms are under investigation for their contribution to cardiac hypertrophic signaling and crosstalk with other established signal transduction routes. As such, we have demonstrated that calcineurin activation was associated with consistent and specific activation of the abundant PKC isoforms PKCα, βI and θ in cultured cardiac myocytes, calcineurin TG hearts and pressure overloaded rat hearts. Moreover, CsA inhibition of calcineurin was associated with inactivation of PKCα and θ.[51] This study indicated the potential for synergy of select PKC isoforms in reinforcing the hypertrophic circuitry with other potent hypertrophic signal transducers. PKCε-activating and inhibitory peptides were employed as a TG approach in the mouse. This approach demonstrated that relative activation of 20 % already sufficed to have powerful protective effects on cardiac integrity and function in isolated hearts subjected to ex vivo ischemia and reperfusion. PKCε activation was associated with the development of a relatively benign, concentric form of hypertrophy, while a PKCε-selective inhibition approach resulted in an opposite response with severe DCM and heart failure, suggesting a functional role for PKCε as a trophic factor for developmental myotrophy and protection against myocardial pathology. [52-55] Although a PKCα overexpressor mouse has not directly been created, Muth and coworkers have recently reported on the creation a transgenic model overexpressing the L-type Ca^{2+} channel, which displays a remarkable extent of hypertrophy and failure. [53] Interestingly, PKCα appeared to be substantially activated before the onset of hypertrophy, suggesting a potential important role for this isoform in Ca^{2+} activated hypertrophic signaling. Similar approaches will most likely allow the dissection of the functional characteristics of every relevant cardiac PKC isoform for cardiac physiology and pathology.

Cytokine and gp130-mediated signaling

Recent advances in our understanding of neurohumoral factors on cardiac homeostasis have elucidated the molecular mode of action by which members of the interleukin-6 (IL-6) family of cytokines exert their function on cardiomyocyte hypertrophy. A novel member is Cardiotrophin-1 which interacts with the dimerized membrane receptors gp130 and low affinity leukemia inhibitory factor receptor (LIFR). IL-6 cytokine family members binding to the extraceullar part of these heterodimerized receptors initiate a cascade of phosphorylation starting with Janus kinase (JAK), which in turn phosphorylates gp130. This generates a docking site for SH2-domain containing proteins, resulting in recruitment of the family of signal transducer and activator of transcription (STAT) and their nuclear translocation. Activation of gp130 also leads to PI-3K activation by a mechanism involving Jak1-mediated phosphorylation. Accordingly, cardiotrophin-1, IL-1β and IL-6 are potent hypertrophic stimulators in cultured cardiomyocytes. Also, activation of the gp130-LIFR pathway causes cardiac hypertrophy in mice and tissue-specific targeted disruption of *gp130* revealed its positive role in cardiac homeostasis and survival following mechanical stress.[56] Independent STAT3 or PI-3K activation in transgenic mice also induces cardiac hypertrophy,[57,58] establishing a trophic role for the IL-6 family of ligands and all of their their downstream effectors on the ventricular myocardium. (reviewed in[47])

Cardiac hypertrophic signaling: the Good, the Bad, and the Ugly

Genetically modified animals

The general tendency in contemporary literature has been to view cardiac hypertrophy as a single entity with grossly similar outward morphological and molecular manifestations. This view grossly ignores the clinical evidence that divergent primary events can impact importantly on the clinical prognosis. A simplified classification distinguishes several types of clinical hypertrophic manifestations (although even this classification comes short to respecting the wide variety in human hypertrophic syndromes).

For example, (1) *physiological (or benign) hypertrophy* is characterized by preserved or even enhanced contractile performance and hypertrophic remodeling which results from endurance training. In contrast, (2) chronic pressure overload (with its clinical correlates hypertension and aortic stenosis) induces *concentric hypertrophy* (grossly increased wall thickness and little chamber enlargement), and is characterized by single cardiomyocytes displaying an increase in cross-sectional area relative to cell length and sarcomeres added in parallel. Latter form of hypertrophy displays a transient preservation of contractile behaviour, which transitions over time to systolic and diastolic abnormalities. (3) Diastolic heart failure, a relatively "new" clinical entity, which correlates to a large extent with hypertension, old age and female sex, can be viewed upon as the extreme form of concentric remodeling leading to such serious diastolic filling and relaxation behaviour abnormalities that it may be categorized as a single entity. A similar hemodynamic pattern can be observed in *restrictive cardiomyopathy*. Volume overload (4) (during valvular insufficiency or in peripartum cardiomyopathy) results in *eccentric hypertrophy* (i.e. a disproportionaly large increase in chamber volume and little increase in wall thickness), where sarcomeres are added in series, leading to increases in cell length rather than cell width. CAD, hypertension or rare forms of inherited diseases have a high prevalence to develop into (5) dilated cardiomyopathy, the extreme form of eccentric hypertrophy with grossly increased LV cavity, thin walls and severe loss of systolic contractile function.

The presence of functional and genetic correlates to morphologically distinct forms of hypertrophy, suggest that these outcomes are encoded by distinct underlying molecular mechanisms. One such mechanism may well be the relative activation pattern of distinct signaling cascades and subsequent distinct genetic activation patterns. However, hypertrophic syndromes have yet to be classified on the basis of their distinct signaling mechanisms or gene expression pattern. More recently, though, the combined effort of multiple groups, which selectively activated or inhibited separate signaling modules in genetically modified animals has allowed us to attempt to address the first question. One illustrative example comes from the fundamentally distinct forms of cardiomyopathy following activation of single MAPK factors in transgenic models.

Even though MAPKs have been the focuss of particular intense attention (given their importance in controlling proliferation in dividing cells), the vast majority of studies have either used cultured myocytes or were observational studies using animal models, either or not in combination with pharmacological agents.

To assess the in vivo function of myocardial ERK1/2 activation, we have created a series of transgenic lines expressing a constitutively activated mutant of the ERK1/2-selective activator MEK1 in a cardioselective manner. MEK1 TG hearts only demonstrated selective ERK1/2 activation (not of JNK or p38), and, more interstingly, displayed a physiological form of hypertrophy (quite similar to trained athletes hearts) with increased contractility parameters throughout their complete life span. In addition,

MEK1 transgenic hearts were largely resistant to pleiotropic apoptotic stimuli, a finding in accordance with the functional characteristics of ERK1/2 in other cell types. [27]
In sharp contrast, a recently discovered member of the MAPK superfamily is ERK5, also termed big MAPK. ERK5 has homology with ERK1/2 in its N-terminus, but is quite dissimilar in its C-terminus and employs distinct upstream kinases for its activation (i.e. MEK5). TG animals expressing a constitutively active form of MEK5 displayed a pure eccentric form of hypertrophy, which was accompanied by in series assembly of sarcomeres, severe systolic dysfunction and early lethality.[59]
Attempts to create selective p38 activating transgenic animals (αMyHC-driven MKK6) have consistingly failed due to early mortality of multiple founder lines. Autopsy of these animals displayed a completely distinct form of cardiomyopathy, characterized by massive atrial enlargement (containing organized blood cloths), mild ventricular enlargement, smaller myocyte diameter and disarray, and transmural fibrotic lesions, mimicking hypertensive heart disease (De Windt and Molkentin, unpublished observations).
JNK activating TG animals have been created and await publication of their phenotype, and are characterized by mild concentric remodeling, transient increased contractile parameters which decompensate later in life and mild fibrosis (Bueno and Molkentin, unpublished observations). Collectively, these combined studies suggest that the single MAPK factors may fulfill as of yet unrecognized fundamentally distinct roles in the morphological remodeling of the myocardium in vivo.
One potential caveat in our current classification may constitute the growing concern regarding the use of transgenesis and, more specifically, the use of the well-characterized α-MyHC promoter to achieve gain-of-function analyses of intracellular signal transduction mediators. This promoter is currently the most widely used construct to drive TG expression to the myocardium starting at birth as a result of homeostatic MHC isoform switching. This concern was fuelled further by the observation that green fluorescent protein expression in TG animals produced a remarkably malignant cadiomyopathy in high expressing founder animals.[60] It will be hard to exclude the possibility that introducing transgene expression to a level of MHC expression may exaggerate any innate response of that signaling molecule or may cause "spill over" and subsequent activation of intracellular cascades. The assumption that cardiomyopathy is inherent on using this promoter is a grossly incorrect conclusion. In fact, only ~30 % of α-MyHC-driven TG models display some sort of hypertrophic response (Jeff Robbins, Personal Communications). Moreover, a rapidly growing number of models expressing hypothesized anti-hypertrophic factors were indeed resistant towards pleiotropic hypertrophic stimuli. Thus, the overall asumption should be that transgenesis employing the α-MyHC promoter construct (or any construct for that matter) produces a phenotype that closely mimicks the endogenous (hypothesized) function of that particular gene. Gene targeting approaches may not avoid these issues, since loss of gene expression may disrupt the intracellular equilibrium in a similar, albeit opposite, fashion and includes the confounding issue of genetic redundancy.[61] We were aware of these caveats when creating the current classification. Other, more reliable,

conditional and tissue-specific methods to study gain-of-function or loss-of-function designed specifically for cardiovascular biology are clearly needed to avoid these issues. Table I presents the phenotypes of genetically modified animals created to date with altered cardiac-specific activation of hypertrophic signaling modules. The cardiac and whole-body phenotypes were categorized according to histopathological observations (extent of hypertrophy, class of hypertrophy, fibrotic lesions), cellular characterizations (apoptosis), organ function (LV hemodynamics and arrhythmogenesis) and increased mortality of the animals. Accordingly, the phenotypes (and their corresponding animal models) were classified into three groups: "*the Good*", displaying a beneficial form of hypertrophy; "*the Bad*", a large group consisting of variable phenotypes with clear pathological features; and finally "*the Ugly*", phenotypes with left and right-sided cardiac failure that led to pronounced lethality.

A more complex view emerges than initiatally expected, with only very few characteristics directly correlating with the genesis of severe heart failure and DCM. For example, increased apoptotic index (and fibrosis), a feature hypothesized to be a crucial factor in the transition from compensated hypertrophy to decompensation, was variably observed in animals with substantial forms of pathological hypertrophy (the Bad-classification), but, surprisingly, virtually absent in models with clear heart failure (the Ugly-classification).

Table I: Distinct hypertrophic responses in genetically modified mouse models with altered intracellular signaling activation

	Hypertrophy		Apoptosis	Fibrosis	LV hemodynamics	Arrhythmogenic	Lethality
Class (ref)	Extent	Classification					
Good							
ERK1/2-MAPK	++	Concentric	---	-	++	+/-	-
Gp130	+	Concentric	-	-	+	N.D.	-
Stat3	+	Concentric	-	-	+	N.D.	-
IGF	+	Concentric	---	-	++	N.D.	-
PI-3K	+	Concentric	--	-	+	N.D.	-
PKCε	+	Concentric	-	-	+/-	N.D.	-
Bad							
Calcineurin	++	Conc/eccen	--	++	--	+	++
CaMK	++	Conc/eccen	N.D.	++	--	++	++
PKCα	++	Conc/eccen	N.D.	+	--	+/-	+
PKCβII	+	Conc/eccen	N.D.	+	-	+/-	+
JNK1/2-MAPK	++	Concentric	N.D.	+	+/-	+/-	+
P38-MAPK	++	Concentric	+++	+++	--	+/-	++
Gαs	++	Concentric	+	N.D.		+/-	+
Gqα	++	Concentric	++	N.D.	--	+/-	+
Ras	++	Concentric	N.D.	+	-	+/-	+
RhoA	++	Concentric	N.D.	N.D.	-	+/-	+
Ugly							
Rac1	++	Eccentric	-	-	--	N.D.	++
ERK5	++	Eccentric	-	-	--	N.D.	++
TNFα	++	Eccentric	+	+	--	N.D.	++

+, indicates prevalence; - absence/resistance against feature; N.D., not determined; conc, concentric; eccen, eccentric. For explanation of abbreviations in terminology of signaling pathways see text.

Other features such as arrhythmogenesis and LV hemodynamic function neither clearly correlated with the severity of heart failure, even though they were clearly absent in models with beneficial forms of hypertrophy (the Good-classification). One remarkable observation is the presence of cytoskeletal reorganization in Rac1 transgenic animals which led to severe heart failure (the Ugly-classification) in the absence of detectable fibrosis, apoptosis or arrhythmogenicity. Even though cytoskeletal abnormalities were not studied in MEK5 or TNFα transgenic animals, this feature may well play an important characteristic in the genesis of the cardiomyopathic phenotype of these animals, considering the gross extent of eccentric remodeling. Thus, cytoskeletal reorganization emerges as the sole consistent, as of yet underestimated, factor that can independently lead to extreme forms of heart failure, even in the absence of other hypothesized risk factors. Nevertheless, the present functional classification clearly distinguishes a wide variety of future targets for the treatment of common acquired and inherited hypertrophic human heart disease.

Cardiovascular genomics

To date, only one study performed a comparative gene expression profiling study between several mouse models of cardiomyopathy. This analysis did not result in the identification of novel, putative genetic targets, the expression of which correlates with the severity of the heart failure phenotype. In contrast, divergent genetic responses were observed which correlated with the obtained phenotypic characteristics due to the specific activation of hypertrophic signaling molecules.[62]

Another approach would be to find transcriptional nodal points which are activated in maladaptive forms of hypertrophy following activation of signaling modules in the Bad and/or Ugly classification. One such transcriptional activator is the class of myocyte enhancing factor-2 (MEF-2), which appears to be activated by multiple, parallel intracellular signaling factors.[63] Considerably more information has recently been obtained regarding the activation status of MEF-2 due to the creation of MEF-2-dependent Lac-Z reporter transgenic animals, which directly provide a read-out of MEF-2 activation status in vivo. Accordingly, hybrid CamKIV/MEF-2-reporter TG mice demonstrated robust activation of the reporter in a ventricular-specific fashion, indicating a specific role for CamK isozymes in MEF-2 activation.[40] Recent evidence also implicated MEF-2 as downstream transducers of calcineurin signaling in cardiac muscle, skeletal muscle, neurons and T-cells. Although the direct mode of action of calcineurin upon MEF-2 is still unclear, the phosphorylation status of MEF2A was shown to be CsA sensitive and affected its DNA binding efficiency.[64] The sole transcriptional target regulator of the MEK5–ERK5 pathway are members of the MEF-2 superfamily. Although p38 MAPK has multiple downstream effectors, RhoA activation has been demonstrated to activate MEF-2 specifically through p38 MAPK. [65] Finally, TGF has been recently demonstrated to activate transforming growth factor activated kinase (TAK1), a MAPKKK of the p38 MAPK pathway and TAK1 transgenic mice develop a severe form of heart failure. Furthermore, downstream effectors of TGF, transcription factors of the SMAD family, are known to cooperatively synergize with members of the MEF-2 family to activate target genes.[66] It is therefore feasible to suggest that at least part of the hypertrophic effects of TGF are due to MEF-2 activation. Given that many malignant hypertrophic signaling pathways converge in the nucleus at one common downstream endpoint, a promising approach for future anti-hypertrophic therapeutic approaches would be to target common transcriptional nodal points, such as MEF-2.

References

1. Hanahan D, Weinberg RA. The hallmarks of cancer. Cell. 2000;100:57-70.
2. Withers DJ, White M. Perspective: The insulin signaling system--a common link in the pathogenesis of type 2 diabetes. Endocrinology. 2000;141:1917-21.
3. Bellus GA, McIntosh I, Smith EA, Aylsworth AS, Kaitila I, Horton WA, Greenhaw GA, Hecht JT, Francomano CA. A recurrent mutation in the tyrosine kinase domain of fibroblast growth factor receptor 3 causes hypochondroplasia. Nat Genet. 1995;10:357-9.
4. Saffran DC, Parolini O, Fitch-Hilgenberg ME, Rawlings DJ, Afar DE, Witte ON, Conley ME. Brief report: a point mutation in the SH2 domain of Bruton's tyrosine kinase in atypical X-linked agammaglobulinemia. N Engl J Med. 1994;330:1488-91.
5. Mauro MJ, Druker BJ. STI571: a gene product-targeted therapy for leukemia. Curr Oncol Rep. 2001;3:223-7.
6. Sebolt-Leopold JS, Dudley DT, Herrera R, Van Becelaere K, Wiland A, Gowan RC, Tecle H, Barrett SD, Bridges A, Przybranowski S, Leopold WR, Saltiel AR. Blockade of the MAP kinase pathway suppresses growth of colon tumors in vivo. Nat Med. 1999;5:810-6.
7. Lefkowitz RJ, Rockman HA, Koch WJ. Catecholamines, cardiac beta-adrenergic receptors, and heart failure. Circulation. 2000;101:1634-7.
8. Gaudin C, Ishikawa Y, Wight DC, Mahdavi V, Nadal-Ginard B, Wagner TE, Vatner DE, Homcy CJ. Overexpression of Gs alpha protein in the hearts of transgenic mice. J Clin Invest. 1995;95:1676-83.
9. Liggett SB, Tepe NM, Lorenz JN, Canning AM, Jantz TD, Mitarai S, Yatani A, Dorn GW, 2nd. Early and delayed consequences of beta(2)-adrenergic receptor overexpression in mouse hearts: critical role for expression level. Circulation. 2000;101:1707-14.
10. Dorn GW, 2nd, Tepe NM, Wu G, Yatani A, Liggett SB. Mechanisms of impaired beta-adrenergic receptor signaling in G(alphaq)- mediated cardiac hypertrophy and ventricular dysfunction. Mol Pharmacol. 2000;57:278-87.
11. Engelhardt S, Hein L, Wiesmann F, Lohse MJ. Progressive hypertrophy and heart failure in beta1-adrenergic receptor transgenic mice. Proc Natl Acad Sci U S A. 1999;96:7059-64.
12. Adams JW, Sakata Y, Davis MG, Sah VP, Wang Y, Liggett SB, Chien KR, Brown JH, Dorn GW, 2nd. Enhanced Galphaq signaling: a common pathway mediates cardiac hypertrophy and apoptotic heart failure. Proc Natl Acad Sci U S A. 1998;95:10140-5.
13. Neumann J, Schmitz W, Scholz H, von Meyerinck L, Doring V, Kalmar P. Increase in myocardial Gi-proteins in heart failure. Lancet. 1988;2:936-7.
14. Redfern CH, Degtyarev MY, Kwa AT, Salomonis N, Cotte N, Nanevicz T, Fidelman N, Desai K, Vranizan K, Lee EK, Coward P, Shah N, Warrington JA, Fishman GI, Bernstein D, Baker AJ, Conklin BR. Conditional expression of a Gi-coupled receptor causes ventricular conduction delay and a lethal cardiomyopathy. Proc Natl Acad Sci U S A. 2000;97:4826-31.
15. Clerk A, Sugden PH. Small guanine nucleotide-binding proteins and myocardial hypertrophy. Circ Res. 2000;86:1019-23.
16. Hunter JJ, Tanaka N, Rockman HA, Ross J, Jr., Chien KR. Ventricular expression of a MLC-2v-ras fusion gene induces cardiac hypertrophy and selective diastolic dysfunction in transgenic mice. J Biol Chem. 1995;270:23173-8.
17. Tanaka N, Dalton N, Mao L, Rockman HA, Peterson KL, Gottshall KR, Hunter JJ, Chien KR, Ross J, Jr. Transthoracic echocardiography in models of cardiac disease in the mouse. Circulation. 1996;94:1109-17.
18. Gottshall KR, Hunter JJ, Tanaka N, Dalton N, Becker KD, Ross J, Jr., Chien KR. Ras-dependent pathways induce obstructive hypertrophy in echo-selected transgenic mice. Proc Natl Acad Sci U S A. 1997;94:4710-5.

19. Marinissen MJ, Chiariello M, Gutkind JS. Regulation of gene expression by the small GTPase Rho through the ERK6 (p38 gamma) MAP kinase pathway. Genes Dev. 2001;15:535-53.
20. Sah VP, Minamisawa S, Tam SP, Wu TH, Dorn GW, 2nd, Ross J, Jr., Chien KR, Brown JH. Cardiac-specific overexpression of RhoA results in sinus and atrioventricular nodal dysfunction and contractile failure. J Clin Invest. 1999;103:1627-34.
21. Sussman MA, Welch S, Walker A, Klevitsky R, Hewett TE, Price RL, Schaefer E, Yager K. Altered focal adhesion regulation correlates with cardiomyopathy in mice expressing constitutively active rac1. J Clin Invest. 2000;105:875-86.
22. Sugden PH, Clerk A. "Stress-responsive" mitogen-activated protein kinases (c-Jun N-terminal kinases and p38 mitogen-activated protein kinases) in the myocardium. Circ Res. 1998;83:345-52.
23. Sugden PH, Clerk A. Regulation of the ERK subgroup of MAP kinase cascades through G protein- coupled receptors. Cell Signal. 1997;9:337-51.
24. Haneda M, Sugimoto T, Kikkawa R. Mitogen-activated protein kinase phosphatase: a negative regulator of the mitogen-activated protein kinase cascade. Eur J Pharmacol. 1999;365:1-7.
25. Wang Y, Huang S, Sah VP, Ross J, Jr., Brown JH, Han J, Chien KR. Cardiac muscle cell hypertrophy and apoptosis induced by distinct members of the p38 mitogen-activated protein kinase family. J Biol Chem. 1998;273:2161-8.
26. Wang Y, Su B, Sah VP, Brown JH, Han J, Chien KR. Cardiac hypertrophy induced by mitogen-activated protein kinase kinase 7, a specific activator for c-Jun NH2-terminal kinase in ventricular muscle cells. J Biol Chem. 1998;273:5423-6.
27. Bueno OF, De Windt LJ, Tymitz KM, Witt SA, Kimball TR, Klevitsky R, Hewett TE, Jones SP, Lefer DJ, Peng CF, Kitsis RN, Molkentin JD. The MEK1-ERK1/2 signaling pathway promotes compensated cardiac hypertrophy in transgenic mice. Embo J. 2000;19:6341-50.
28. Xiao L, Pimental DR, Amin JK, Singh K, Sawyer DB, Colucci WS. MEK1/2-ERK1/2 mediates alpha1-adrenergic receptor-stimulated hypertrophy in adult rat ventricular myocytes. J Mol Cell Cardiol. 2001;33:779-87.
29. Bueno OF, De Windt LJ, Lim HW, Tymitz KM, Witt SA, Kimball TR, Molkentin JD. The dual-specificity phosphatase MKP-1 limits the cardiac hypertrophic response in vitro and in vivo. Circ Res. 2001;88:88-96.
30. Esposito G, Prasad SV, Rapacciuolo A, Mao L, Koch WJ, Rockman HA. Cardiac overexpression of a G(q) inhibitor blocks induction of extracellular signal-regulated kinase and c-Jun NH(2)-terminal kinase activity in in vivo pressure overload. Circulation. 2001;103:1453-8.
31. Haq S, Choukroun G, Lim H, Tymitz KM, del Monte F, Gwathmey J, Grazette L, Michael A, Hajjar R, Force T, Molkentin JD. Differential activation of signal transduction pathways in human hearts with hypertrophy versus advanced heart failure. Circulation. 2001;103:670-7.
32. Cook SA, Sugden PH, Clerk A. Activation of c-Jun N-terminal kinases and p38-mitogen-activated protein kinases in human heart failure secondary to ischaemic heart disease. J Mol Cell Cardiol. 1999;31:1429-34.
33. Crabtree GR. Generic signals and specific outcomes: signaling through Ca2+, calcineurin, and NF-AT. Cell. 1999;96:611-4.
34. Crabtree GR. Calcium, calcineurin, and the control of transcription. J Biol Chem. 2001;276:2313-6.
35. Molkentin JD, Lu JR, Antos CL, Markham B, Richardson J, Robbins J, Grant SR, Olson EN. A calcineurin-dependent transcriptional pathway for cardiac hypertrophy. Cell. 1998;93:215-28.
36. Sussman MA, Lim HW, Gude N, Taigen T, Olson EN, Robbins J, Colbert MC, Gualberto A, Wieczorek DF, Molkentin JD. Prevention of cardiac hypertrophy in mice by calcineurin inhibition. Science. 1998;281:1690-3.
37. Molkentin JD. Calcineurin and beyond: cardiac hypertrophic signaling. Circ Res. 2000;87:731-8.
38. De Windt LJ, Lim HW, Bueno OF, Liang Q, Delling U, Braz JC, Glascock BJ, Kimball TF, del

Monte F, Hajjar RJ, Molkentin JD. Targeted inhibition of calcineurin attenuates cardiac hypertrophy invivo. Proc Natl Acad Sci U S A. 2001;98:3322-3327.

39. Rothermel BA, McKinsey TA, Vega RB, Nicol RL, Mammen P, Yang J, Antos CL, Shelton JM, Bassel-Duby R, Olson EN, Williams RS. Myocyte-enriched calcineurin-interacting protein, MCIP1, inhibits cardiac hypertrophy in vivo. Proc Natl Acad Sci U S A. 2001;98:3328-33.
40. Passier R, Zeng H, Frey N, Naya FJ, Nicol RL, McKinsey TA, Overbeek P, Richardson JA, Grant SR, Olson EN. CaM kinase signaling induces cardiac hypertrophy and activates the MEF2 transcription factor in vivo. J Clin Invest. 2000;105:1395-406.
41. McKinsey TA, Zhang CL, Olson EN. Activation of the myocyte enhancer factor-2 transcription factor by calcium/calmodulin-dependent protein kinase-stimulated binding of 14-3- 3 to histone deacetylase 5. Proc Natl Acad Sci U S A. 2000;97:14400-5.
42. McKinsey TA, Zhang CL, Olson EN. Identification of a signal-responsive nuclear export sequence in class ii histone deacetylases. Mol Cell Biol. 2001;21:6312-21.
43. Lu J, McKinsey TA, Nicol RL, Olson EN. Signal-dependent activation of the MEF2 transcription factor by dissociation from histone deacetylases. Proc Natl Acad Sci U S A. 2000;97:4070-5.
44. McKinsey TA, Zhang CL, Lu J, Olson EN. Signal-dependent nuclear export of a histone deacetylase regulates muscle differentiation. Nature. 2000;408:106-11.
45. Aoki H, Sadoshima J, Izumo S. Myosin light chain kinase mediates sarcomere organization during cardiac hypertrophy in vitro. Nat Med. 2000;6:183-8.
46. Mochly-Rosen D, Gordon AS. Anchoring proteins for protein kinase C: a means for isozyme selectivity. Faseb J. 1998;12:35-42.
47. Molkentin JD, Dorn IG, 2nd. Cytoplasmic signaling pathways that regulate cardiac hypertrophy. Annu Rev Physiol. 2001;63:391-426.
48. Bowman JC, Steinberg SF, Jiang T, Geenen DL, Fishman GI, Buttrick PM. Expression of protein kinase C beta in the heart causes hypertrophy in adult mice and sudden death in neonates. J Clin Invest. 1997;100:2189-95.
49. Wakasaki H, Koya D, Schoen FJ, Jirousek MR, Ways DK, Hoit BD, Walsh RA, King GL. Targeted overexpression of protein kinase C beta2 isoform in myocardium causes cardiomyopathy. Proc Natl Acad Sci U S A. 1997;94:9320-5.
50. Roman BB, Geenen DL, Leitges M, Buttrick PM. PKC-beta is not necessary for cardiac hypertrophy. Am J Physiol Heart Circ Physiol. 2001;280:H2264-70.
51. De Windt LJ, Lim HW, Haq S, Force T, Molkentin JD. Calcineurin promotes protein kinase C and c-Jun NH2-terminal kinase activation in the heart. Cross-talk between cardiac hypertrophic signaling pathways. J Biol Chem. 2000;275:13571-9.
52. Wu G, Toyokawa T, Hahn H, Dorn GW, 2nd. Epsilon protein kinase C in pathological myocardial hypertrophy. Analysis by combined transgenic expression of translocation modifiers and Galphaq. J Biol Chem. 2000;275:29927-30.
53. Muth JN, Bodi I, Lewis W, Varadi G, Schwartz A. A Ca(2+)-dependent transgenic model of cardiac hypertrophy: A role for protein kinase Calpha. Circulation. 2001;103:140-7.
54. Takeishi Y, Ping P, Bolli R, Kirkpatrick DL, Hoit BD, Walsh RA. Transgenic overexpression of constitutively active protein kinase C epsilon causes concentric cardiac hypertrophy. Circ Res. 2000;86:1218-23.
55. Pass JM, Zheng Y, Wead WB, Zhang J, Li RC, Bolli R, Ping P. PKCepsilon activation induces dichotomous cardiac phenotypes and modulates PKCepsilon-RACK interactions and RACK expression. Am J Physiol Heart Circ Physiol. 2001;280:H946-55.
56. Hirota H, Chen J, Betz UA, Rajewsky K, Gu Y, Ross J, Jr., Muller W, Chien KR. Loss of a gp130 cardiac muscle cell survival pathway is a critical event in the onset of heart failure during biomechanical stress. Cell. 1999;97:189-98.

57. Shioi T, Kang PM, Douglas PS, Hampe J, Yballe CM, Lawitts J, Cantley LC, Izumo S. The conserved phosphoinositide 3-kinase pathway determines heart size in mice. Embo J. 2000;19:2537-48.
58. Li Q, Li B, Wang X, Leri A, Jana KP, Liu Y, Kajstura J, Baserga R, Anversa P. Overexpression of insulin-like growth factor-1 in mice protects from myocyte death after infarction, attenuating ventricular dilation, wall stress, and cardiac hypertrophy. J Clin Invest. 1997;100:1991-9.
59. Nicol RL, Frey N, Pearson G, Cobb M, Richardson J, Olson EN. Activated MEK5 induces serial assembly of sarcomeres and eccentric cardiac hypertrophy. Embo J. 2001;20:2757-67.
60. Huang WY, Aramburu J, Douglas PS, Izumo S. Transgenic expression of green fluorescence protein can cause dilated cardiomyopathy. Nat Med. 2000;6:482-3.
61. Gaussin V, Schneider MD. Surviving infarction one gene at a time: decreased remodeling and mortality in engineered mice lacking the angiotensin II type 1A receptor. Circulation. 1999;100:2043-4.
62. Aronow BJ, Toyokawa T, Canning A, Haghighi K, Delling U, Kranias E, Molkentin JD, Dorn GW, 2nd. Divergent transcriptional responses to independent genetic causes of cardiac hypertrophy. Physiol Genomics. 2001;6:19-28.
63. Naya FS, Olson E. MEF2: a transcriptional target for signaling pathways controlling skeletal muscle growth and differentiation. Curr Opin Cell Biol. 1999;11:683-8.
64. Youn HD, Chatila TA, Liu JO. Integration of calcineurin and MEF2 signals by the coactivator p300 during T-cell apoptosis. Embo J. 2000;19:4323-31.
65. Takano H, Komuro I, Oka T, Shiojima I, Hiroi Y, Mizuno T, Yazaki Y. The Rho family G proteins play a critical role in muscle differentiation. Mol Cell Biol. 1998;18:1580-9.
66. Quinn ZA, Yang CC, Wrana JL, McDermott JC. Smad proteins function as co-modulators for MEF2 transcriptional regulatory proteins. Nucleic Acids Res. 2001;29:732-42.

ESF workshop Maastricht 2001: Session 5

Molecular remodelling in arrhythmias

14. ION CHANNEL REGULATION: FROM ARRHYTHMIAS TO GENES TO CHANNELS (TO CURES?)

J.K. Donahue

Introduction

Cardiac arrest is one of the most dramatic and devastating conditions in all of cardiology. By its very nature, cardiac arrest is sudden and unexpected, leaving friends and family of a victim in a struggle to make sense of the unfortunate situation. Causes of cardiac arrest are myriad, including obstructive or hypertrophic heart disease, pulmonary embolism, aortic dissection or rupture, and cardiac arrhythmias. In adults in the developed world, the most common cause is ventricular tachycardia or fibrillation, often in association with ischemic heart disease [1].

Over the last several decades, a tremendous effort has taken place to understand the nature of cardiac arrhythmias on a pathophysiological and genetic basis. In this review, we will discuss the underlying pathophysiology of arrhythmias and the current status of gene therapy for the treatment of arrhythmias. The length of this review precludes an indepth discussion of these issues; more detail on individual conditions can be found in other chapters in this section and in review articles indicated in the text of this review.

Normal Cardiac Electrophysiology

The cellular basis for all cardiac electrical activity is the action potential (AP). The AP is conventionally divided into 5 sections (phases 0-4, figure 1A). Each phase is defined by the cellular membrane potential and the activity of ion channels that affect that potential. Phase 4 is the resting baseline. The dominant ionic current during this phase is a potassium current, I_{K1}. Phase 0 is the initial depolarization brought on by opening of the sodium channels. Phase 1 is a brief, partial repolarization in the potential from the peak achieved at the end of phase 0. Activation of potassium and chloride currents (I_{to1} and I_{to2}, respectively) and inactivation of the sodium channels are responsible for this portion of the AP. Phase 2 is the plateau period, where a calcium current ($I_{Ca,L}$) maintains the positive depolarization, and potassium currents (I_{Kr} and I_{Ks}) try to force repolarization.

Ultimately, the repolarization occurs in phase 3, when the calcium current is inactivated and the membrane potential returns to the baseline of phase 4. The membrane potential during each phase of the AP is an unsteady equilibrium between positive and negative currents. Any slight perturbation of this balance can affect the shape of the AP and potentially cause an arrhythmia.

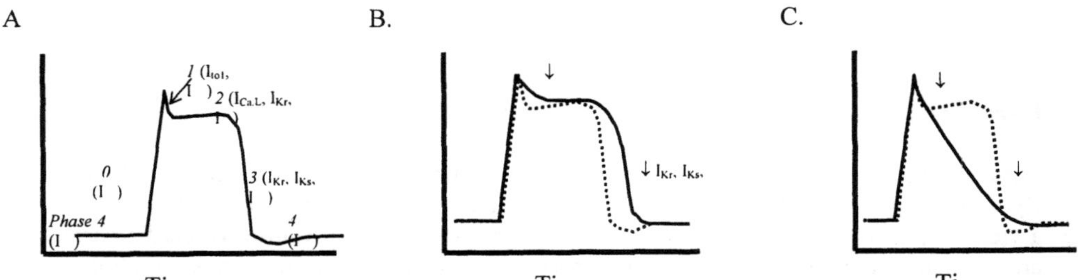

Figure 1. *Schematics of action potential morphologies:* ***A:*** *Normal ventricular action potential; phases 0-4 are noted in italics, and the currents active during each stage are noted in paranthesis.* ***B:*** *Changes in the ventricular action potential in heart failure. A decrease in several potassium currents causes a loss of the notch (phase 1) and a prolongation of the action potential duration.* ***C:*** *Changes in the atrial action potential after prolonged atrial fibrillation. Reduction in the transient outward current, I_{to}, and the l-type calcium current, $I_{Ca,L}$ result in a decreased notch and plateau.*

The action potential is transmitted from one cell to the next by means of gap junctions. Within cardiac tissue, the direction of current flow is anisotropic (e.g. current flow in a longitudinal direction is faster than flow in a transverse direction). This directionality of current flow can be explained by the distribution of gap junctions. Roughly 20% of the gap junction connections are side-to-side; approximately 33% of the connections are end-to-end, and the remaining junctions are end-to side. The average myocyte is connected to approximately 10 neighboring myocytes (for general review, see Beyer et al. [2]).

Pathophysiology of Inherited Arrhythmias

The long QT syndrome was the first arrhythmic disorder characterized at a genetic level. A concentrated effort to identify genetic defects associated with the long QT syndrome led to the discovery of several potassium channel subunits and to a better understanding of the interactions of various components of the action potential. The syndrome has been associated with mutations of the sodium channel α-subunit gene and of the genes for several potassium channel subunits (Table 1). The underlying physiological defect common to all of these genetic mutations is an extension of the plateau phase of the action potential. Either reduction in one of the potassium currents or failed inactivation of the sodium current pushes the balance in the plateau phase toward extended depolarization.

The end result is an increase in propensity toward ventricular arrhythmias from afterdepolarizations or heterogeneous repolarization. A broader review of the genetics and pathophysiology of the long QT syndrome can be found in Priori et al. [3].

Table 1.

Genetics of the Long QT Syndrome			
LQTS variant	Gene	Protein	Reference
1	KvLQT1 (KCNQ1)	I_{Ks} α-subunit	[17]
2	HERG	I_{Kr} α-subunit	[18]
3	SCN5A	INa a-subunit	[19]
4	*	*	[20]
5	MinK (KCNE1)	I_{Ks} β-subunit	[21]
6	MirP1 (KCNE2)	I_{Kr} β-subunit	[22]
7	?	?	?

* not yet identified; maps to locus 4q25-27

The Brugada Syndrome is another inherited arrhythmias that has been characterized on a genetic level. The syndrome consists of idiopathic ventricular fibrillation in patients with right bundle branch block pattern and ST segment elevation in ECG leads V1-3 and with structurally normal hearts [4]. Like one variant of the long QT syndrome, mutations in the cardiac sodium channel gene have been associated with the Brugada syndrome. Unlike the long QT syndrome, where the mutation causes a gain of function in the sodium channel, the sodium channel mutations causing the Brugada syndrome are associated with a loss of sodium channel function. The reduction in the sodium current causes conduction problems that show up on electrophysiology study as prolongation of the His to ventricular conduction time [5]. The changes in sodium current also cause a change in the balance of depolarising and repolarizing forces. This shift is most apparent in the epicardial layer, where the transient outward potassium current is relatively robust. The heterogeneity in repolarization leads to the characteristic ST segment changes seen in the Brugada syndrome, and presumably underlies the ventricular fibrillation risk found in these patients [6].

Pathophysiology of Common Cardiac Arrhythmias

Sudden Death in Congestive Heart Failure

Sudden death in patients with congestive heart failure is a common clinical occurrence. Often, the associated arrhythmia is polymorphic ventricular tachycardia (VT) leading to ventricular fibrillation and death. When observable, the type of VT seen in these patients is similar to that observed in patients with the congenital long QT syndrome. Studies of animal models have documented the similarities between these two diseases on a tissue and cellular level. In both conditions, heterogeneous increases in the action potential duration (APD) have been a consistent finding (figure 1B). In heart failure, the APD prolongation correlates with downregulation of several potassium currents: the transient outward current I_{to}, the inward rectifier current I_{K1}, and the delayed rectifier currents I_{Ks} and I_{Kr}. As discussed above, the plateau of the cardiac action potential is an unstable equilibrium; even small changes of inward or outward currents can markedly delay repolarization. The reduction in potassium current found in heart failure could therefore explain the prolongation of the APD and the increase in surface ECG repolarization time. The similarities between heart failure and the long QT syndrome suggest that alterations in the potassium current underlie the arrhythmic risk, and therefore that correcting the potassium channel deficit may reduce the incidence of sudden death. An extensive discussion of the ion channel alterations in heart failure can be found in a review by Tomaselli and Marban [7].

Atrial Fibrillation

Unlike VT associated with heart failure, where the rhythm quickly causes hemodynamic instability and death, atrial fibrillation (AF) can be sustained for long periods of time. As such, the cellular adaptive processes that occur with AF are completely different than those seen with heart failure. During sustained AF, there is a shortening of the APD and refractory period (figure 1C). Clinical and experimental studies have shown a 70% downregulation of the Ca^{2+} current, $I_{Ca.L}$, and the transient outward current, I_{to}, to account for the observed changes in the AP morphology. The inward rectifier and adenosine/acetylcholine activated potassium currents (I_{K1} and $I_{K,Ach}$) are upregulated. The end result of these changes is an improved ability of the atrial myocytes to sustain the rapid and chaotic impulses characteristic of atrial fibrillation. This situation creates a cycle where the rapid rate causes a shortened refractory period which allows the continuation of the rapid rate, an idea that has been termed “AF begets AF”. The maladaptive nature of the ion channel alterations suggests that interrupting these changes on a molecular level is a potential treatment for AF. A comprehensive discussion of the molecular changes in atrial fibrillation can be found in an extensive review by Bosch et al. [8].

Gene Therapy For The Treatment Of Cardiac Arrhythmias

The investigation of gene therapeutic strategies to treat cardiac arrhythmias is in its infancy. There is only a single report in the literature of *in vivo* utilization of this concept [9], and there are a limited number of reports using *in vitro* gene transfer to investigate arrhythmia mechanisms on a cellular level. The general hypothesis guiding investigation of arrhythmia gene therapy is that an understanding of the basic principles of cellular electrophysiology can be used to target specific genes. Initial investigations in isolated cardiac ventricular myocytes demonstrated the ability to affect the AP morphology [10]. These early efforts were directed toward the I_{to} current, shown to be reduced in heart failure, but overexpression of this current led to a bizarre and overly shortened AP. Refining this method *in vitro* by overexpressing the I_{Kr} current led to a more normal appearing AP [11], because the kinetics of I_{Kr} include a later onset of action, giving more time for normal phases 1-3 of the AP than is possible with the early acting I_{to} current. A conclusion from this series of experiments is that the kinetics of the ion channels may be more important than the replacement of specifically downregulated currents. If our ability to control gene expression reaches the point where expression levels can be fine tuned to exactly replicate the normal level of current, then the strategy of replacing downregulated currents may become viable.

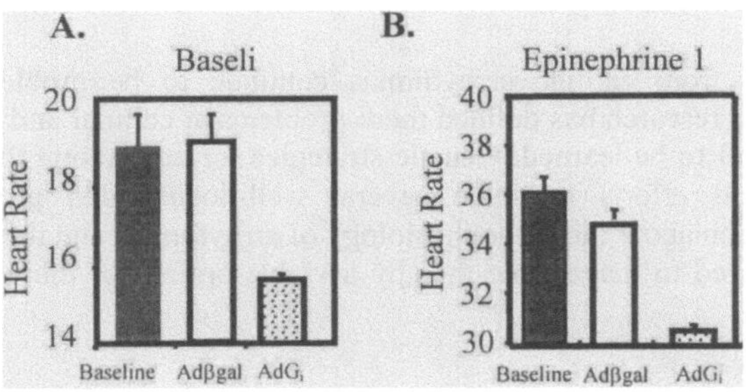

Figure 2. *Ventricular rate after acute induction of atrial fibrillation in a porcine model of gene therapy. Adenovirus was perfused into the AV nodal artery at baseline. Heart rate was measured after burst-pacing induction of atrial fibrillation at baseline and 7 days later.* ***A.*** *drug-free state.* ***B.*** *after 1 mg epinephrine IV.*

In the only *in vivo* report of gene transfer-mediated treatment of an arrhythmia, we targeted overexpression in the AV node of $G\alpha_{i2}$, the α-subunit of an inhibitory G protein, and showed a 15-20% reduction in the ventricular rate during AF, both in the drug-free state and after administration of epinephrine 1 mg IV (figure 2). In a porcine model of acute AF, we infused adenoviruses into the AV nodal artery. Adβgal, an adenovirus encoding E. Coli β-galactosidase, was used as a reporter to document the extent of gene transfer, and it was used as a control to identify non-specific effects of adenovirus infection and gene transfer in the AV node. AdGi was the therapeutic virus, encoding $G\alpha_{i2}$. Prior to virus perfusion, vascular endothelial growth factor (VEGF) and nitroglycerin (TNG) were infused into the AV nodal artery to increase the efficiency of gene transfer [12,13]. Adβgal-infected hearts showed gene transfer to 45% of AV nodal myocytes and no effect on AV nodal function. There was a limited mononuclear

inflammatory infiltrate noted. AdGi-infection slowed conduction through the AV node, ultimately leading to a reduction in the heart rate during acutely-induced AF.
Future efforts at gene therapy for arrhythmias will require solutions to a number of problems that face the field, including those related to homogeneous delivery of the vector to the target tissue, control of gene expression, evaluation of potentially toxic effects of the vector or the transgene, and elimination of non-target organ gene transfer and of host immune responses. Efforts to solve these problems are taking place on multiple fronts. Recent advances in vector design have documented the ability of adeno-associated virus and helper-dependent adenovirus vectors to sustain long-term gene expression and to reduce host immune responses [14,15]. Discovery of methods to increase microvascular permeability to vectors have improved efficacy and homogeneity of gene delivery [13]. New situation-specific promoters or response elements that activate in response to hypoxia, temperature, and steroid or drug exposure have been identified [16], lending hope to the possibility that the timing and amount of gene expression can be controlled.

Conclusions

Morbidity and mortality from cardiac arrhythmias continue to be problems in the developed world. Ongoing research has defined these problems at cellular and subcellular levels, althoug much is left to be learned. Genetic strategies for arrhythmia therapy hold promise, but concentrated efforts to solve several well-documented problems are necessary. More information about the pathophysiology of arrhythmias and the mechanics of gene therapy are required to make gene therapy a viable option for the treatment of arrhythmias.

Acknowledgements
This work was supported by the Richard S. Ross Clinician-Scientist Award, Johns Hopkins University and by the NIH (P50 HL52307).

References

1. Domanski MJ, Zipes DP, Schron E. Treatment of sudden cardiac death: current understandings from randomized trials and future research directions. Circulation 1997;95:2694-9.
2. Beyer E, Veenstra R, Kanter H, Saffitz J. Molecular structure and patterns of expression of cardiac gap junction proteins, in Zipes D, Jalife J (eds): Cardiac Electrophysiology: from cell to bedside. ed2. Philadelphia, PA, W.B. Saunders Co., 1995, pp 31-7.
3. Priori SG, Bloise R, Crotti L. The long QT syndrome. Europace 2001;3:16-27.
4. Brugada P, Brugada J. Right bundle branch block, persistent ST segment elevation and sudden cardiac death: a distinct clinical and electrocardiographic syndrome. A multicenter report. J Am Coll Cardiol 1992;20:1391-6.
5. Alings M, Wilde A. "Brugada" Syndrome: clinical data and suggested pathophysiological mechanism. Circulation 1999;99:666-73.
6. Yan GX, Antzelevitch C. Cellular basis for the Brugada syndrome and other mechanisms of arrhythmogenesis associated with ST-segment elevation. *Circulation* 1999;100:1660-6.
7. Tomaselli G,Marban E. Electrophysiological remodeling in hypertrophy and heart failure. Cardiovasc Res 1999;42:270-83.
8. Bosch R, Zeng X, Grammer J, et al. Ionic mechanisms of electrical remodeling in human atrial fibrillation. Cardiovasc Res 1999;44:121-31.
9. Donahue JK, Heldman A, Fraser H, et al. Focal modification of electrical conduction in the heart by viral gene transfer. Nature Med 2000;6:1395-8.
10. Nuss HB, Johns D, Kaab S, et al. Reversal of potassium channel deficiency in cells from failing hearts by adenoviral gene transfer: a prototype for gene therapy for disorders of cardiac excitability and contractility. Gene Therapy 1996;3:900-12.
11. Nuss B, Marban E, Johns D. Overexpression of a human potassium channel suppresses cardiac hyperexcitability in rabbit ventricular myocytes. J Clin Invest 1999;103:889-96.
12. Nagata K, Marban E, Lawrence J, Donahue JK. Phosphodiesterase inhibitor-mediated potentiation of adenovirus delivery to myocardium. J Mol Cell Cardiol 2001;33:575-80.
13. Donahue JK, Kikkawa K, Thomas AD, Marban E, Lawrence J. Acceleration of widespread adenoviral gene transfer to intact rabbit hearts by coronary perfusion with low calcium and serotonin.Gene Therapy 1998;5:630-4.
14. Monahan P, Samulski R. AAV vectors: is clinical success on the horizon? Gene Therapy 2000;7:24-30.
15. Chen HH, Mack LM, Kelly R, et al. Persistence in muscle of an adenoviral vector that lacks all viral genes. Proc Natl Acad Sci USA 1997;94:1645-50.
16. Fussenegger M. The impact of mammalian gene regulation concepts on functional genomic research, metabolic engineering, and advanced gene therapies. Biotechnol Prog 2001;17:1-51.
17. Wang Q, Curran ME, Splawski I, et al. Positional cloning of a novel potassium channel gene: KVLQT1 mutations cause cardiac arrhythmias. Nat Genet 1996;12:17-23.
18. Curran ME, Splawski I, Timothy KW, et al. A molecular basis for cardiac arrhythmia: HERG mutations cause long QT syndrome. Cell 1995;80:795-803.
19. Wang Q, Shen J, Splawski I, et al. SCN5A mutations associated with an inherited cardiac arrhythmia, long QT syndrome. Cell 1995;80:805-11.
20. Schott JJ, Charpentier F, Peltier S, et al. Mapping of a gene for long QT syndrome to chromosome 4q25-27. Am J Hum Genet 1995;57:1114-22.
21. Schulze-Bahr E, Wang Q, Wedekind H, et al. KCNE1 mutations cause Jervell and Lange-Nielsen syndrome. Nat Genet 1997;17:267-8.
22. Abbott GW, Sesti F, Splawski I, et al. MiRP1 forms IKr potassium channels with HERG and is associated with cardiac arrhythmia. Cell 1999;97:175-

15. DIFFERENTIAL EXPRESSION AND FUNCTIONAL REGULATION OF DELAYED RECTIFIER CHANNELS

M. Stengl, P.G.A. Volders, M.A. Vos

Introduction

The terms differential expression and functional regulation have broad meanings, which can overlap and be explained in several ways. In this chapter, functional regulation is considered an acute modulation of the system (delayed rectifier channels in this case) and as such it will be discussed. Differential expression, on the one hand, reflects heterogeneity of delayed rectifiers expression in myocytes from different regions of the healthy heart; on the other hand it can also mean an altered (usually decreased) expression in pathological conditions.

The delayed rectifier potassium current was first described by Noble and Tsien [1] in sheep cardiac Purkinje fibers. Since then it has been demonstrated in a variety of species and cell types [2-7]. This chapter will focus only on ventricular myocytes. Originally, it was thought that it is a single current (referred as I_K). Gradually, however, it became obvious that two different currents, I_{Kr} with relatively rapid activation kinetics and I_{Ks} with slow activation, can be distinguished on basis of kinetics, sensitivity to drugs, and modulation by cellular signaling messengers. In ventricular myocytes, both currents can be found in most species so far studied, with the exception of cat where only I_{Kr} has been demonstrated [8].

Characteristics of delayed rectifiers

I_{Kr}

Molecular structure

The channel consists of four main, pore-forming, homologous α-subunits. Each α-subunit is made of six transmembrane segments. Segment 4 is highly positively charged and acts as a voltage sensor (for review see [9]). The α-subunit is encoded by ERG, the *ether-a-go-go*-related gene, mutations of which lead to an inherited cardiac arrhythmia, long QT syndrome, type 2 (LQT2 [10]). Although the channel formed by α-subunits is functional and shows properties consistent with those of the native cardiac I_{Kr}, there is one major difference. The ERG current is not blocked by methanesulfonanilide drugs (E-4031 and MK-499), potent and specific blockers of native, cardiac I_{Kr} [2,10], suggesting that an additional subunit responsible for drug sensitivity is present in the native I_{Kr} channel. Recently, a novel potassium channel gene was cloned and characterized [11]. The gene (KCNE2) encodes MinK-related peptide 1 (MiRP1), a subunit that coassembles with ERG protein and these complexes resemble the native cardiac I_{Kr} channels in all aspects including pharmacological sensitivity. Furthermore, mutations in KCNE2 were shown to predispose to arrhythmia (LQT6 [11]).

Table 1 Activation and deactivation time constants in adult ventricular myocytes

	I_{Kr}		I_{Ks}	
Species	Activation	Deactivation	activation	deactivation
Guinea pig	20[1a]	160[1a]	λ_f 420; λ_s 1650[2b]	800[1c]
Rabbit[1d]	40	λ_f 640; λ_s 6500	900	160
Dog[1e]	50	λ_f 360; λ_s 3300	1000	90
Human[1]	30[f]	λ_f 600; λ_s 6800[f]	1000[g]	120[g]

[1]activation at +30 mV, deactivation at –40 mV; [2]activation at +50 mV
[a][2]; [b][38]; [c][30]; [d][12]; [e][14]; [f][13]; [g][7]

Kinetics

Activation of I_{Kr} is relatively rapid (see table 1). Time constants of activation are voltage dependent and vary between species. In guinea pig, the time constant is the longest at –30 mV (~175 ms) and shortens at more positive potentials [2]. In rabbit the time constant also shortens at more positive voltages [12]. In all species studied, activation of I_{Kr} is faster than of I_{Ks}. This is not true for deactivation. Deactivation of I_{Kr} is fast in guinea pig, but much slower in human [13], rabbit [12] or dog [14].

I_{Kr} shows inward rectification, i.e. at positive potentials the amplitude of the current is rather small [15]. On hyperpolarization, however, the tail current transiently increases before the exponential decline starts (so called "hook"). This issue was investigated in Xenopus oocytes or mammalian cells with expression of ERG [10,15] and in the mouse atrial myocyte cell line AT-1 [16], demonstrating that the process responsible for inward rectification is the inactivation of I_{Kr}. At positive potentials, development of the inactivation is faster than activation, therefore only a small current appears. Upon hyperpolarization, recovery from inactivation leads to an increase of current before the deactivation starts. In addition, decreased external K^+ was found to speed up the inactivation thus explaining decreased I_{Kr} at low external K^+. Furthermore, block of the channel by external Na^+ also contributes to the fall of I_{Kr} conductance at low external K^+ [17].

Blockers

I_{Kr} is blocked by external bivalent and trivalent cations, for instance Co^{2+} and La^{3+} [18]. Selective organic blockers of I_{Kr} are the methanesulfonanilide derivatives dofetilide, MK-499, E-4031, and other compounds like almokalant (for review, see [19]). Many other drugs intended to block other channels or receptors, both cardiovascular and non-cardiovascular were also shown to block I_{Kr} (for complete overview see Dr. Woosley's web site www.torsades.org).

The role of I_{Kr} in the action potential

Specific block of I_{Kr} leads to a significant prolongation of APD. I_{Kr} was therefore postulated to play a prominent role in the repolarization of ventricular action potential in most mammals including human [5,20,21]. The prolongation is rate-dependent, being more pronounced at low than high rates of stimulation [20]. A possible mechanism of this rate-dependence could be use-dependent block of I_{Kr} channel during diastole and unblock during depolarization, perhaps by preferential block of the channel in the closed state. Methanesulfonanilides, however, block the channel in the activated, open state (for review, see [19]) and they still prolong APD with this "reverse" use dependence. Jurkiewicz and Sanguinetti [20] addressed this question with dofetilide in guinea pig myocytes. They showed that the magnitude of I_{Kr} as well as the sensitivity to block of I_{Kr} by dofetilide is rate independent. However, rapid pacing increases the magnitude of I_{Ks} as a result of incomplete deactivation and subsequent accumulation. Therefore, in guinea pig the rate dependence of APD prolongation by dofetilide reflects the relatively minor contribution of I_{Kr} to the repolarization at fast rates when I_{Ks} is increased.

Despite the relatively rapid activation kinetics of I_{Kr}, its contribution during the action potential plateau is rather limited by inward rectification, which has been attributed to fast inactivation [10,15,16]. Upon repolarization, a fast recovery from inactivation occurs leading to a transient increase of the current before the deactivation prevails. This phenomenon was, however, observed with square voltage-clamp pulses where the repolarization (as well as depolarization) is abrupt. During the ventricular action potential the repolarization is much slower. Therefore an important question arises, whether under conditions of slow repolarization the recovery from inactivation with subsequent increase in I_{Kr} can still occur. To answer this question and further characterize the involvement of native I_{Kr} in ventricular repolarization, Gintant [22] applied voltage-clamp protocols simulating the plateau and repolarization phases of the action potentials to canine ventricular myocytes and Purkinje fibers. The results demonstrate that I_{Kr} is minimal at plateau potentials but it transiently increases during repolarization. I_{kr} thus provides a transient pulse of outward current later in the action potential plateau consistent with its rapid recovery from inactivation along with slower deactivation kinetics. Action-potential-clamp (the recorded action potential serves as the command waveform) showed that I_{Kr} provides a transient late outward current during the terminal part of the action potential plateau, that promotes the initiation of terminal (phase 3) repolarization and activation of inward rectifier K^+ current (I_{K1}).

I_{Ks}

Molecular structure

The structure of I_{Ks}, in general, is similar to I_{Kr}: four pore-forming α-subunits, each of them consisting of six transmembrane segments. The α-subunit is encoded by KCNQ1 (KvLQT1) gene. KvLQT1 expression induces a current that does not resemble the native, cardiac I_{ks} in many aspects, the most important being fast activation and existence of inactivation (native I_{Ks} shows very slow activation and no inactivation). This suggests that KvLQT1 coassembles with an auxiliary subunit to form a channel with biophysical properties similar to I_{Ks}. MinK encoded by KCNE1 gene was identified to be this subunit [23-25]. Mutations of KvLQT1 or MinK cause an inherited cardiac arrhythmia, LQT1 [26] or LQT5 [27], respectively. Wang et al. [28] investigated the stoichiometry of these two subunits using several fusion channel proteins: proteins consisting of a) 1 KvLQT1 and 1 MinK, b) of 2 KvLQT1, c) of 2 KvLQT1 and 1 MinK. The data revealed that addition of MinK slows significantly the activation, thus making the current more similar to native I_{Ks}. The current characteristics were similar to native I_{Ks}, when the stoichiometry was 1 KvLQT1:1 MinK, 2 KvLQT1:1 MinK, or when KvLQT1 and MinK were coexpressed independently. This observation suggests that the stoichiometry of the two proteins is not fixed, however there is a minimal stoichiometry requirement for incorporation of MinK that modifies KvLQT1 gating in such a way that the resulting current resembles native I_{Ks}.

Kinetics

The hallmark of I_{Ks} is a very slow activation (see table 1). I_{Ks} activates at depolarized potentials (≥-10 mV) and it does not show saturation with increasing depolarization

[4,29]. I_{Ks} only shows activation and deactivation, but no inactivation. Deactivation is slow in guinea pig but relatively fast in dog, rabbit and human [4,7,12,30]. Deactivation shows a voltage dependence being faster at more negative potentials. In contrast to I_{Kr}, extracellular potassium has no direct effect on the gating of the channel, however it will influence the current amplitude through changes in the (intra- versus extracellular) concentration gradient. In zero external potassium I_{Ks} is thus increased (whereas I_{Kr} decreased, see above).

Blockers

Only recently selective blockers of I_{Ks} appeared: chromanol 293B [14,31], L-735,821 [32,33], L-768,673 [34], and HMR 1556 [35]. According to Sun et al. [36] chromanol 293B blocks I_{Ks} with IC_{50} of 2 μM in canine ventricular myocytes. Higher concentrations of chromanol 293B also inhibit transient outward current (I_{to}; IC_{50} 40 μM) but exert little or no effect on I_{Kr}, I_{K1}, or L-type Ca^{2+} current (I_{CaL}; [36]). Even more potent are compounds L-735,821, L-768,673, and HMR 1556 that suppress I_{Ks} with IC_{50} of 6 nM (both L compounds) and 34 nM (HMR1556). Other drugs such as quinidine, azimilide, ambasilide or amiodarone block I_{Ks} but also other K^+ currents (for review see [19]). I_{Ks} can also be blocked by Co^{2+} (half maximal inhibition at 1 mM [37]).

The role of I_{Ks} in the action potential

The role of I_{Ks} in the action potential remains still a controversial issue. Although the absence of specific blockers of I_{Ks} until recently disallowed a direct evaluation, I_{Ks} was generally assumed to be a significant contributor to the repolarization. This view was supported by a number of studies where crippled or reduced I_{Ks} in various pathological states, both congenital and acquired, was associated with a significant prolongation of APD (see below). The magnitude of I_{Ks} in guinea pig myocytes was shown to be increased by rapid pacing, as a result of incomplete deactivation of I_{Ks} during short interpulse intervals and subsequent accumulation [20]. This would suggest that the block of I_{Ks} would prolong APD especially at high rates when I_{Ks} is increased; an ideal effect against reentry-based tachyarrhythmia. In large mammals including human, however, the kinetics of deactivation are much faster than in guinea pig, thus making the accumulation of I_{Ks} at physiologically relevant high rates unlikely [21]. The results with specific blockers of I_{Ks}, chromanol 293B and L-735,821 [31,33] are somewhat contradictory. Bosch et al. [38] reported increased APD in guinea pig and human ventricular myocytes after treatment with chromanol 293B (1 μM), while using different groups of cells (treated with chromanol 293B versus non-treated). Chromanol 293B also induced a prolongation of APD in epi-, mid-, and epicardial tissues [39] as well as in an arterially perfused wedge of canine left ventricle [40]. The concentration necessary to produce an APD prolongation was 30 μM. At such a concentration, however, chromanol 293B blocks I_{to} as well, at least in isolated myocytes [36,38]. Application of L-768,673 in guinea pig ventricular myocytes also elicited an APD prolongation, which was self-limiting, maximal 30% [34]. The same inhibitor, when applied in vivo in 2 canine models of ischemically induced malignant ventricular arrhythmias, showed a significant antiarrhythmic efficacy associated with modest increases in ventricular refractory periods (3% to 10%) and QTc interval (4% to 6% [41]).

On the other hand, there are reports showing that under basal conditions I_{Ks} blockers fail to prolong APD. This was the case in rabbit right ventricle papillary muscles [12, 42] as well as in dog right ventricular papillary muscle [14]. The concentration of chromanol 293B in these studies was 10 μM, which is less than in the studies mentioned above, however, it should be enough to block a substantial portion of I_{Ks} without influencing other currents (e.g. I_{to}). Furthermore, similar results were obtained with L-735,821 [12,14]. Varro et al. [14] noted that although chromanol 293B had no effect under basal conditions, it lengthened the repolarization significantly when APD was artificially pre-prolonged (by E-4031 and veratridine), thus allowing I_{Ks} to activate for a sufficient time. Schreieck et al. [43] in guinea pig papillary muscles and Han et al. [44] in canine cardiac Purkinje cells reported no effect of chromanol 293B on APD under basal conditions. However, in both studies, a significant prolongation of APD by chromanol 293B was observed in the presence of the β-adrenergic agonist, isoproterenol.

The data indicate that I_{Kr} is a primary repolarizing current in ventricular tissue, I_{Ks} (especially in large mammals) plays only a minor role under basal conditions, however in conditions of adrenergic stimulation or an abnormal APD prolongation I_{Ks} increases and becomes an important safety factor that reduces arrhythmia risk.

Functional regulation of delayed rectifiers

I_{Kr}

β-Adrenergic modulation of I_{Kr} is still a matter of discussion. On the one hand, β-adrenergic stimulation was shown to have no effect on I_{Kr} [45] in guinea pig ventricular cells. I_{Kr} was also shown to be independent on intracellular Ca^{2+} [46]. On the other hand, different results were obtained when this issue was assessed using the perforated patch-clamp technique [47], which has some advantages over conventional ruptured patch clamping used in previous work. The most important one is that complete dialysis of the cell and subsequent washout of important (signaling) molecules such as cAMP or protein kinases, is prevented. Under these conditions, the authors could demonstrate an increase in I_{Kr} following stimulation of protein kinase A (PKA) by isoproterenol (β-adrenergic agonist) or forskolin (activator of adenylyl cyclase). The increase was inhibited when Ca^{2+} entry was reduced or intracellular Ca^{2+} strongly buffered. The isoproterenol- and forskolin-induced increases were also inhibited by inhibitors of protein kinase C (PKC), whereas direct activation of PKC lead to an increase in I_{Kr}. The data were interpreted that activation of PKA leads through phosphorylation of L-type

Ca^{2+} channels to an increase in intracellular Ca^{2+} concentration which in turn activates PKC, that stimulates I_{Kr}, possibly through a reduction in the C-type inactivation.
External K^+ is another important regulator of I_{Kr}. At low external K^+ the conductance falls, a phenomenon explained by more pronounced inactivation and/or block by external Na^+ [16,17]. Increased external K^+, on the other hand, enhances I_{Kr} and it was shown that intravenous infusion of K^+ corrects repolarization abnormalities in LQT2 [48]. A long-lasting rise of serum K^+, however, is not achievable, since in the presence of normal renal function the increase of serum K^+ is limited by renal homeostatic mechanisms [49].
Angiotensin II increases I_{Kr} in guinea pig ventricular myocytes [50]. Cell swelling decreases amplitude of I_{Kr} [51], while, in contrast, it increases I_{Ks}.

I_{Ks}

β-adrenergic stimulation leads through activation of PKA to a significant increase of I_{Ks} [45, 52]. Activation of PKC, either directly or through stimulation of α1-adrenoreceptors increases I_{Ks} in guinea-pig myocytes [53], however, it decreases the current obtained when mouse KvLQT1 and MinK are coexpressed in COS cells [23]. Varnum et al. [54] showed that species-specific differences in MinK account for the different responses to PKC. The time factor is also of importance. Lo and Numann [55] showed that stimulation of PKC with phorbol ester induced an increase in I_{Ks}, that peaked after 20 minutes and then, however, subsequently decreased to approximately 50% of the control level after 1 hour. Interestingly, premodulation by PKC prevented I_{Ks} current modulation by PKA, and PKC had no effect on I_{Ks} current after potentiation by PKA suggesting that the I_{Ks} current is modulated by PKC and PKA in a mutually exclusive manner. An interesting and complex role in the modulation of I_{Ks} is played by intracellular Ca^{2+} ions. Ca^{2+} ions contribute to the stimulation of PKC, thus increasing I_{Ks} indirectly [56]. Increase in intracellular Ca^{2+} concentration, however, enhances I_{Ks} also in the presence of specific inhibitors of PKC, PKA or calmodulin kinase II [57]. This enhancement was abolished by a calmodulin antagonist. Thus, intracellular Ca^{2+} regulates I_{Ks} also via a calmodulin-dependent pathway, which does not involve phosphorylation.
Endothelin suppressed I_{Ks}, especially when stimulated by isoproterenol (β-adrenergic agonist). The effect was mediated via a pathway involving endothelin type A receptors, pertussis toxin-sensitive G-protein and adenylyl cyclase [58]. Angiotensin II decreases amplitude of I_{Ks} in guinea pig myocytes [50]. Hypotonic-induced stretch significantly increases amplitude of I_{Ks} [51,59].

Differential expression of I_{Kr} and I_{Ks}

Expression levels

When studying gene expression, 3 levels can be distinguished: a) mRNA expression, b) protein expression, and c) functional expression, i.e. the currents. Changes in one expression level can be accompanied by corresponding changes in other two levels, e.g. decrease in I_{Ks} in dogs with chronic atrioventricular block [60] was matched by decreased levels of KCNQ1 (KvLQT1) and KCNE1 (MinK) mRNA [61]. On the other hand, such a congruity between functional changes (i.e. decrease of the current) and changes of mRNA or protein expression has not always been recovered. Such a discrepancy was shown by Rose et al. [62], who reported a significant decrease in the density of both I_{Kr} and I_{Ks} in dogs with tachycardia-induced heart failure, which, however, was not matched by a reduced mRNA expression. A large discrepancy between the current and the mRNA expression was shown to occur even in individual normal cell [63]. In rat ventricular myocytes, measurement of tetraethylammonium-sensitive delayed rectifier current was combined with qualitative assessment of the presence or absence of Kv2.1 mRNA transcript (which is thought to underlie the channel protein) by a single cell reverse transcriptase-polymerase chain reaction technique. The results revealed a substantial discrepancy, when there was no difference in mean amplitude or inactivation kinetics of the current between cells positive and negative for Kv2.1 mRNA transcripts.

At this point in time, most information about differential expression, both in normal heart and in pathology, comes from functional studies showing various changes in currents (altered density, kinetics, etc.), there is relatively less information available on other two expression levels (mRNA and protein expression). To fill this gap in our knowledge and further elucidate relationships between expression levels will be a major task for near future.

Differential expression in normal heart

Electrical behavior of cardiac cells is by far not homogeneous. Differential expression of I_{Kr} and/or I_{Ks}, shown in various cell types, is likely to contribute to this electrical heterogeneity. Furukawa et al. [64] studied delayed rectifier K^+ current in endocardial and epicardial feline ventricular cells and distinguished fast and slow component (corresponding to I_{Kr} and I_{Ks}) on basis of different kinetics. Both of them were significantly smaller in the endocardial layer. The transmural heterogeneity in I_{Kr} and/or I_{Ks} was also investigated in the canine heart [4,65]. The currents and action potentials were measured in ventricular myocytes from three regions: endocardial, midmyocardial, and epicardial. I_{Kr} density was found to be similar in all three cell types. I_{Ks} density was similar in endo- and epicardial myocytes, but it was significantly smaller in midmyocardial cells. It was suggested that smaller I_{Ks} density could in part account for longer APD in midmyocardial cells as well as for their high pharmacological responsiveness. Brahmajothi et al. [66] studied the localization of ERG in the ferret heart using fluorescence in situ hybridization and immunofluorescence methods. In the ventricle, the highest ERG transcript- and protein-expression levels were found in epicardial layers. In rabbit left ventricle, I_{Ks} density was larger in epi- than in endocardial cells, whereas I_{Kr} density was similar in both endo- and epicardial myocytes [67].

Next to the transmural heterogeneity, significant differences can also be found when apical and basal regions are compared. In the rabbit left ventricle, Cheng et al. [68] showed that the total I_K (sum of I_{Kr} and I_{Ks}) as well as I_{Ks} tail densities were significantly smaller in apical than in basal myocytes, whereas I_{Kr} tail density was significantly larger in the apex than in the base. The ratio of I_{Ks}/I_{Kr} was, therefore, significantly smaller in the apex than in the base. These differences were accompanied by significantly longer APDs in the apex than in the base, which was further pronounced by application of the I_{Kr} blocker E-4031.

Volders et al. [29] investigated the ionic basis of interventricular dispersion in canine heart. Action potentials and ionic currents were recorded in midmyocardial cells from both ventricles. APD was significantly shorter in right ventricular cells. I_{to} as well as I_{Ks} were found to be significantly larger in right ventricle. I_{Kr}, in contrast, had a similar amplitude in both ventricles.

Differential expression in pathology

Hereditary disorders

Mutations in subunits of I_{Ks} or I_{Kr} (KvLQT1, MinK, ERG, MiRP1; see above) form the molecular basis of long QT syndromes (LQTS). The LQTS are characterized by abnormally prolonged ventricular repolarization, manifested as a prolongation of the QT interval, and by a high susceptibility to malignant ventricular tachyarrhythmias. Till now, six subtypes of LQTS, LQT1-6, have been described and four of them have been linked to defects in I_{Ks} or I_{Kr}: LQT1 – defective KCNQ1 (KvLQT1) gene encoding the α-subunit of I_{Ks} channel [26]; LQT2 – defective ERG gene encoding the α-subunit of I_{Kr} channel [10]; LQT5 – defective KCNE1 gene encoding MinK, a β-subunit of I_{Ks} channel [27]; LQT6 - defective KCNE2 gene encoding MiRP1, a β-subunit of I_{Kr} channel [11]. Mutations in genes for I_{Ks} or I_{Kr} account for approximately 90% of LQTS [69]. In general, defects in K^+ channels related to LQTS lead to a loss of function and to a reduction in the outward current during the repolarization phase of the cardiac action potential, thus prolonging APD and QT interval.

Acquired disorders

Changes in I_{Kr} and/or I_{Ks} have been studied in several pathophysiological models of ventricular hypertrophy or heart failure. Volders et al. [60] studied I_{Kr} and I_{Ks} in dogs with chronic (more than 6 weeks) atrioventricular block (CAVB). Long-term bradycardia and atrioventricular asynchrony lead to the volume overload and to the development of biventricular hypertrophy in this model [70]. The hypertrophy is accompanied by electrical remodeling: QT intervals and ventricular endocardial monophasic action potentials are much longer than expected on the basis of bradycardia alone [70]. Anesthetized CAVB dogs show an increased interventricular dispersion of repolarization, early afterdepolarizations and an enhanced susceptibility to the polymorphic tachyarrhythmia torsades de pointes [70]. Furthermore, CAVB dogs are often prone to sudden cardiac death (~15%; [71]). No significant differences were found, when I_{to} and I_{K1} in midmyocardial myocytes from CAVB and control dogs were compared. I_{Ks} in CAVB dogs was significantly reduced by ~50% in both left and right ventricle. I_{Kr} measured as the almokalant-sensitive current was decreased in right (by 45% [60]), as well as in left ventricle myocytes [72]. In conclusion, a significant downregulation of I_{Ks} and I_{Kr} occurs in the heart of CAVB dogs, which could contribute to torsades de pointes in vivo. However, when making a connection between reduction in I_{Ks} (and I_{Kr}) and arrhythmias, one should realize that other important cellular factors could also contribute to the proarrhythmia. One of them could be an altered Ca^{2+} homeostasis and it was recently reported that this is indeed the case and that in CAVB dogs, the Na^+/Ca^{2+} exchange is upregulated [73]. Enhanced Na^+/Ca^{2+} exchange activity most likely contributes to the increased sarcoplasmic reticulum loading and subsequently to the increased Ca^{2+} release in CAVB at lower heart rates. This adaptive mechanism improves the contractile function, but, on the other hand, it may also contribute to proarrhythmia: 1) propensity to Ca^{2+} overload increases; 2) at the time of Ca^{2+} release the increased Na^+/Ca^{2+} exchange inward current tends to prolong and destabilize repolarization.

Decreases in I_{Ks} (as well as in other repolarizing currents I_{K1} and I_{to}) associated with APD prolongation were described in left ventricular cells from dogs with heart failure induced by rapid ventricular pacing (240/min for 5 weeks; [74]). Similar results (i.e. APD prolongation, reduced I_{Ks}, I_{to}, and I_{K1}) were obtained also in right ventricular epicardial cells isolated from human failing hearts [75]. Also myocardial infarction leads to alterations in I_{Kr} and I_{Ks} [76]. Analysis of the currents recorded in infarct zone myocytes, which were isolated from a thin layer of surviving epicardium overlying the infarct, revealed decreased densities of both I_{Kr} and I_{Ks} as well as changes in kinetics of currents: an acceleration of I_{Kr} activation and I_{Ks} deactivation. These functional changes were accompanied by decreased mRNA levels of I_{Kr} and I_{Ks} channel subunits ERG, MinK and KvLQT1.

In a rabbit model of cardiac hypertrophy induced by aortic banding a significant APD prolongation was found, which was associated with reduction of I_{K1} and I_{to} densities [77]. I_{Kr}, however, was not changed in this model. Similarly, in a rabbit model of left ventricular hypertrophy due to renovascular hypertension I_{Kr} was not influenced in either epi- or endocardial cells whereas I_{Ks} density was reduced in both epi- and endocardial myocytes to a similar extent (~40% [67]). Consequently, I_{Kr} was expected to play relatively more important role in the repolarization of hypertrophied cells. This hypothesis was confirmed by experiments, where specific block of I_{Kr} prolonged APD significantly more in hypertrophied myocytes than in control cells. In tachycardia-pacing-induced chronic heart failure Tsuji et al. [78] found, in addition to reduction of I_{to} and I_{CaL}, also I_{Ks} and I_{Kr} to be significantly decreased in left ventricle myocytes.

Table 2 I_{Kr} and I_{Ks} in pathology

Species	Pathology	I_{Kr}			I_{Ks}			Ref.
		mRNA	prot	curr	mRNA	prot	curr	
Human	CHF						↓	[75]
Dog	BVH^{CAVB}						↓	[60]
					↓	↓		[61]
				↓				[60, 72]
Dog	CHF^{TP}						↓	[74]
Dog	CHF^{TP}	=		↓	=	=	↓	[62]
Dog	MI^{LADO}	↓		↓	↓		↓	[76]
Rabbit	LVH^{AB}			=				[77]
Rabbit	LVH^{RH}			=			↓	[67]
Rabbit	CHF^{TP}			↓			↓	[78]
Rabbit	BVH^{CAVB}			↓				[79]
Cat	RVH^{PAB}			↓*				[80]
Cat	LVH^{AB}			↓*				[81]
Hamster	CHF^{IC}			↓				[82]
Rat	LVH^{AB}			↓*				[83]
Guinea pig	LVH^{AB}			=			=	[84]

BVH, biventricular hypertrophy; LVH, left ventricular hypertrophy; RVH, right ventricular hypertrophy; CHF, congestive heart failure; MI, myocardial infarction. In superscript methods used to induce the pathology: CAVB, chronic atrioventricular block; TP, tachycardia pacing; LADO, left anterior descending coronary artery occlusion; AB, aortic banding; RH, renovascular hypertension; PAB, pulmonary artery banding; IC, inherited cardiomyopathy

*; delayed rectifier studied as single current I_K

↓, decrease; =, no change; empty field – parameter not studied

Consistent with downregulation of repolarizing currents APD was prolonged in cells from failing hearts. A progressive QT-interval prolongation together with a marked reduction of I_{Kr} in left ventricular cells was found in rabbits with compensated complete heart block [79].
Right ventricular hypertrophy in the cat led to a downregulation of I_K in right ventricular myocytes, whereas I_{K1} was increased [80]. In another feline model left ventricular hypertrophy was induced by chronic pressure overload [81]. Endocardial myocytes studies revealed a significantly longer APD and decreased I_K in hypertrophied cells. I_{CaL}, on the other hand, was not different. In addition, the responsiveness to metabolic inhibition (i.e. APD shortening) was significantly more pronounced in hypertrophied cells.
The BIO TO-2 strain of cardiomyopathic hamster provides a model of dilated low output heart failure. Changes in potassium currents were identified in this model [82], but at a rather late stage. The densities of I_{to}, I_{Kr} and I_{K1} in 8-month-old myopathic hamsters are not significantly different from their age-matched controls, they become different at 10 months of age when the densities of all 3 currents become significantly lower in myopathic hamsters. A reduction of delayed rectifier current (I_K) was also reported in a rat model of left ventricular hypertrophy induced by pressure overload following the stenosis of the ascending aorta [83]. In this model, I_K was decreased in left ventricular endocardial and midmyocardial myocytes, but not in epicardial ones. In contrast to the previous findings, Ahmmed et al. [84] reported a guinea pig model with pressure overload-induced cardiac hypertrophy and failure where I_{Kr} and I_{Ks} densities remained unchanged. APD, however, was prolonged and this prolongation was attributed to the changes in Ca^{2+} handling.
In conclusion, most studies show a clear decrease in at least one component of delayed rectifier (see table 2) under conditions of hypertrophy or heart failure.
Whether the decrease in current(s) is or is not accompanied by changed expression of mRNA and/or protein, remains to be determined in most models. Another question is, given the fact that both I_{Kr} and I_{Ks} channels consist of two subunits, whether both subunits downregulate to the same extent (and simultaneously) or whether one of them downregulates preferentially? Completely unclear are the pathways leading to the decreased expression of delayed rectifiers. Are they activated as a secondary response to hypertrophy (or other pathology) or are they independent? In this respect, recently an interesting study [85] addressed potential mechanisms of Kv4.3 gene (major contributor to I_{to}) downregulation in cardiac hypertrophy. In this study, angiotensin II as well as phenylephrine induced a hypetrophic response in neonatal rat cardiac myocytes. Each of them also downregulated Kv4.3 mRNA and protein. The mechanisms, however, were different. Whereas angiotensin II decreased Kv4.3 expression through destabilization of mRNA, phenylephrine inhibited the Kv4.3-promoter activity and thus the gene transcription. Similar studies on downregulation of delayed rectifiers are necessary.

I_{Kr} and I_{Ks} as therapeutic target

The "Sicilian gambit" paper [86] introduced the concept of the vulnerable parameter. The vulnerable parameter is an electrophysiological property, arrhythmogenic in nature, modification of which will terminate or prevent the arrhythmia. Two major groups of arrhythmogenic mechanisms can be distinguished: abnormal impulse initiation and abnormal impulse conduction. Delayed rectifier blockers are routinely used only in arrhythmias based on reentry with a short excitable gap (the group of abnormal conduction). The vulnerable parameter here is the refractory period. Delayed rectifier blockers will delay repolarization, subsequently prolong refractoriness and thus inhibit arrhythmia efficiently. It should be emphasized, however, that the currently available delayed rectifier blockers are far from ideal. Pure I_{Kr} blockers prolong APD with a reverse rate-dependence [87]. At high rates, when the prolongation would be the most beneficial, they are less effective, while at low frequencies they induce an excessive APD prolongation with the danger of early afterdepolarizations and accompanying torsades de pointes arrhythmias. Pure I_{Ks} blockers for clinical use are not currently available. In the light of experimental work showing only a minor role of I_{Ks} in the repolarization, it is not clear, how much beneficial they could be. From these reasons, non-specific drugs that influence a variety of channels and receptors appear to be safer and more efficient than pure I_{Kr} (or I_{Ks}) blockers at this point in time (e.g. amiodarone [88]).

Conclusion

In a variety of pathological states, both congenital and acquired, with reduction of delayed rectifiers, arrhythmias with the mechanism of triggered activity based on early afterdepolarizations (the group of abnormal impulse initiation) are often encountered. The vulnerable parameter in this case is the prolonged APD due to (at least in part) reduction of delayed rectifiers. Therefore an enhancement of delayed rectifiers, which will shorten APD back and eliminate early afterdepolarizations, is desirable. Such enhancement can be induced by β-agonists. β-Agonists, however, influence also many other cellular processes, especially the Ca^{2+} homeostasis, which makes the outcome rather uncertain. An interesting possibility to selectively enhance I_{Kr} or I_{Ks} (or other ionic currents) and correct for the reduction, appeared recently: viral-mediated gene transfer of ion channel and/or molecular subunits into cardiac cells. Successful approaches were reported in vitro [89, 90] and in vivo [91].

References

1. Noble D, Tsien RW. Outward membrane currents activated in the plateau range of potentials in cardiac Purkinje fibres. J Physiol 1969;200:205-31
2. Sanguinetti MC, Jurkiewicz NK. Two components of cardiac delayed rectifier K+ current. Differential sensitivity to block by class III antiarrhythmic agents. J Gen Physiol 1990;96:195-215
3. Veldkamp MW, van Ginneken AC, Opthof T, Bouman LN. Delayed rectifier channels in human ventricular myocytes. Circulation 1995;92:3497-504
4. Liu DW, Antzelevitch C. Characteristics of the delayed rectifier current (IKr and IKs) in canine ventricular epicardial, midmyocardial, and endocardial myocytes. A weaker IKs contributes to the longer action potential of the M cell. Circ Res 1995;76:351-65
5. Li GR, Feng J, Yue L, Carrier M, Nattel S. Evidence for two components of delayed rectifier K+ current in human ventricular myocytes. Circ Res 1996;78:689-96
6. Salata JJ, Jurkiewicz NK, Jow B, Folander K, Guinosso PJ Jr, Raynor B, Swanson R, Fermini B. IK of rabbit ventricle is composed of two currents: evidence for IKs. Am J Physiol 1996;271:H2477-89
7. Virag L, Iost N, Opincariu M, Szolnoky J, Szecsi J, Bogats G, Szenohradszky P, Varro A, Papp JG. The slow component of the delayed rectifier potassium current in undiseased human ventricular myocytes. Cardiovasc Res 2001;49:790-7
8. Follmer CH, Colatsky TJ. Block of delayed rectifier potassium current, IK, by flecainide and E-4031 in cat ventricular myocytes. Circulation 1990;82:289-93
9. Nerbonne JM. Molecular basis of functional voltage-gated K+ channel diversity in the mammalian myocardium. J Physiol 2000;525:285-98
10. Sanguinetti MC, Jiang C, Curran ME, Keating MT. A mechanistic link between an inherited and an acquired cardiac arrhythmia: HERG encodes the IKr potassium channel. Cell 1995;81:299-307
11. Abbott GW, Sesti F, Splawski I, Buck ME, Lehmann MH, Timothy KW, Keating MT, Goldstein SA. MiRP1 forms IKr potassium channels with HERG and is associated with cardiac arrhythmia. Cell 1999;97:175-87
12. Lengyel C, Iost N, Virag L, Varro A, Lathrop DA, Papp JG. Pharmacological block of the slow component of the outward delayed rectifier current (I(Ks)) fails to lengthen rabbit ventricular muscle QT(c) and action potential duration. Br J Pharmacol 2001;132:101-10
13. Iost N, Virag L, Opincariu M, Szecsi J, Varro A, Papp JG. Delayed rectifier potassium current in undiseased human ventricular myocytes. Cardiovasc Res 1998;40:508-15
14. Varro A, Balati B, Iost N, Takacs J, Virag L, Lathrop DA, Csaba L, Talosi L, Papp JG. The role of the delayed rectifier component IKs in dog ventricular muscle and Purkinje fibre repolarization. J Physiol 2000;523:67-81
15. Smith PL, Baukrowitz T, Yellen G. The inward rectification mechanism of the HERG cardiac potassium channel. Nature 1996;379:833-6
16. Yang T, Snyders DJ, Roden DM. Rapid inactivation determines the rectification and [K+]o dependence of the rapid component of the delayed rectifier K+ current in cardiac cells. Circ Res 1997;80:782-9
17. Scamps F, Carmeliet E. Delayed K+ current and external K+ in single cardiac Purkinje cells. Am J Physiol 1989;257:C1086-92
18. Sanguinetti MC, Jurkiewicz NK. Lanthanum blocks a specific component of IK and screens membrane surface change in cardiac cells. Am J Physiol 1990;259:H1881-9
19. Carmeliet E, Mubagwa K. Antiarrhythmic drugs and cardiac ion channels: mechanisms of action. Prog Biophys Mol Biol 1998;70:1-72

20. Jurkiewicz NK, Sanguinetti MC. Rate-dependent prolongation of cardiac action potentials by a methanesulfonanilide class III antiarrhythmic agent. Specific block of rapidly activating delayed rectifier K+ current by dofetilide. Circ Res 1993;72:75-83
21. Gintant GA. Two components of delayed rectifier current in canine atrium and ventricle. Does I_{Ks} play a role in the reverse rate dependence of class III agents? Circ Res 1996;78:26-37
22. Gintant GA. Characterization and functional consequences of delayed rectifier current transient in ventricular repolarization. Am J Physiol Heart Circ Physiol 2000;278:H806-17
23. Barhanin J, Lesage F, Guillemare E, Fink M, Lazdunski M, Romey G. K(V)LQT1 and IsK (minK) proteins associate to form the I_{Ks} cardiac potassium current. Nature 1996;384:78-80
24. Sanguinetti MC, Curran ME, Zou A, Shen J, Spector PS, Atkinson DL, Keating MT. Coassembly of K(V)LQT1 and minK (IsK) proteins to form cardiac I(Ks) potassium channel. Nature 1996;384:80-3
25. Tristani-Firouzi M, Sanguinetti MC. Voltage-dependent inactivation of the human K+ channel KvLQT1 is eliminated by association with minimal K+ channel (minK) subunits. J Physiol 1998;510:37-45
26. Wang Q, Curran ME, Splawski I, Burn TC, Millholland JM, VanRaay TJ, Shen J, Timothy KW, Vincent GM, de Jager T, Schwartz PJ, Toubin JA, Moss AJ, Atkinson DL, Landes GM, Connors TD, Keating MT. Positional cloning of a novel potassium channel gene: KVLQT1 mutations cause cardiac arrhythmias. Nat Genet 1996;12:17-23
27. Splawski I, Tristani-Firouzi M, Lehmann MH, Sanguinetti MC, Keating MT. Mutations in the hminK gene cause long QT syndrome and suppress IKs function. Nat Genet 1997;17:338-40
28. Wang W, Xia J, Kass RS. MinK-KvLQT1 fusion proteins, evidence for multiple stoichiometries of the assembled IsK channel. J Biol Chem 1998;273:34069-74
29. Volders PG, Sipido KR, Carmeliet E, Spatjens RL, Wellens HJ, Vos MA. Repolarizing K+ currents ITO1 and IKs are larger in right than left canine ventricular midmyocardium. Circulation 1999;99:206-10
30. Chinn K. Two delayed rectifiers in guinea pig ventricular myocytes distinguished by tail current kinetics. J Pharmacol Exp Ther 1993;264:553-60
31. Busch AE, Suessbrich H, Waldegger S, Sailer E, Greger R, Lang H, Lang F, Gibson KJ, Maylie JG. Inhibition of IKs in guinea pig cardiac myocytes and guinea pig IsK channels by the chromanol 293B. Pflugers Arch 1996;432:1094-6
32. Salata JJ, Jurkiewicz NK, Sanguinetti MC, Siegl PKS, Claremon DC, Remy DC, Elliott JM, Libby BE. The novel class III antiarrhythmic agent L-735,821 is a potent and selective blocker of I_{Ks} in guinea pig ventricular myocytes. Circulation 1996;94:I-529
33. Cordeiro JM, Spitzer KW, Giles WR. Repolarizing K+ currents in rabbit heart Purkinje cells. J Physiol 1998;508:811-23
34. Selnick HG, Liverton NJ, Baldwin JJ, Butcher JW, Claremon DA, Elliott JM, Freidinger RM, King SA, Libby BE, McIntyre CJ, Pribush DA, Remy DC, Smith GR, Tebben AJ, Jurkiewicz NK, Lynch JJ, Salata JJ, Sanguinetti MC, Siegl PK, Slaughter DE, Vyas K. Class III antiarrhythmic activity in vivo by selective blockade of the slowly activating cardiac delayed rectifier potassium current IKs by (R)-2-(2,4-trifluoromethyl)-N-[2-oxo-5-phenyl-1-(2,2,2-trifluoroethyl)- 2, 3-dihydro-1H-benzo[e][1,4]diazepin-3-yl]acetamide. J Med Chem 1997;40:3865-8
35. Gogelein H, Bruggemann A, Gerlach U, Brendel J, Busch AE. Inhibition of IKs channels by HMR 1556. Naunyn Schmiedebergs Arch Pharmacol 2000;362:480-8
36. Sun ZQ, Thomas GP, Antzelevitch C. Chromanol 293B inhibits slowly activating delayed rectifier and transient outward currents in canine left ventricular myocytes. J Cardiovasc Electrophysiol 2001;12:472-8

37. Fan Z, Hiraoka M. Depression of delayed outward K+ current by Co2+ in guinea pig ventricular myocytes. Am J Physiol 1991;261:C23-31
38. Bosch RF, Gaspo R, Busch AE, Lang HJ, Li GR, Nattel S. Effects of the chromanol 293B, a selective blocker of the slow, component of the delayed rectifier K+ current, on repolarization in human and guinea pig ventricular myocytes. Cardiovasc Res 1998;38:441-50
39. Burashnikov A, Antzelevitch C. Block of I(Ks) does not induce early afterdepolarization activity but promotes beta-adrenergic agonist-induced delayed afterdepolarization activity. J Cardiovasc Electrophysiol 2000;11:458-65
40. Shimizu W, Antzelevitch C. Differential effects of beta-adrenergic agonists and antagonists in LQT1, LQT2 and LQT3 models of the long QT syndrome. J Am Coll Cardiol 2000;35:778-86
41. Lynch JJ Jr, Houle MS, Stump GL, Wallace AA, Gilberto DB, Jahansouz H, Smith GR, Tebben AJ, Liverton NJ, Selnick HG, Claremon DA, Billman GE. Antiarrhythmic efficacy of selective blockade of the cardiac slowly activating delayed rectifier current, I(Ks), in canine models of malignant ischemic ventricular arrhythmia. Circulation 1999;100:1917-22
42. Pham TV, Sosunov EA, Gainullin RZ, Danilo P Jr, Rosen MR. Impact of sex and gonadal steroids on prolongation of ventricular repolarization and arrhythmias induced by I(k)-blocking drugs. Circulation 2001;103:2207-12
43. Schreieck J, Wang Y, Gjini V, Korth M, Zrenner B, Schomig A, Schmitt C. Differential effect of beta-adrenergic stimulation on the frequency-dependent electrophysiologic actions of the new class III antiarrhythmics dofetilide, ambasilide, and chromanol 293B. J Cardiovasc Electrophysiol 1997;8:1420-30
44. Han W, Wang Z, Nattel S. Slow delayed rectifier current and repolarization in canine cardiac Purkinje cells. Am J Physiol Heart Circ Physiol 2001;280:H1075-80
45. Sanguinetti MC, Jurkiewicz NK, Scott A, Siegl PK. Isoproterenol antagonizes prolongation of refractory period by the class III antiarrhythmic agent E-4031 in guinea pig myocytes. Mechanism of action. Circ Res 1991;68:77-84
46. Sanguinetti MC, Jurkiewicz NK. Delayed rectifier outward K+ current is composed of two currents in guinea pig atrial cells. Am J Physiol 1991;260:H393-9
47. Heath BM, Terrar DA. Protein kinase C enhances the rapidly activating delayed rectifier potassium current, IKr, through a reduction in C-type inactivation in guinea-pig ventricular myocytes. J Physiol 2000;522:391-402
48. Compton SJ, Lux RL, Ramsey MR, Strelich KR, Sanguinetti MC, Green LS, Keating MT, Mason JW. Genetically defined therapy of inherited long-QT syndrome. Correction of abnormal repolarization by potassium. Circulation 1996;94:1018-22
49. Tan HL, Alings M, Van Olden RW, Wilde AA. Long-term (subacute) potassium treatment in congenital HERG-related long QT syndrome (LQTS2). J Cardiovasc Electrophysiol 1999;10:229-33
50. Daleau P, Turgeon J. Angiotensin II modulates the delayed rectifier potassium current of guinea pig ventricular myocytes. Pflugers Arch 1994;427:553-5
51. Rees SA, Vandenberg JI, Wright AR, Yoshida A, Powell T. Cell swelling has differential effects on the rapid and slow components of delayed rectifier potassium current in guinea pig cardiac myocytes. J Gen Physiol 1995;106:1151-70
52. Walsh KB, Kass RS. Regulation of a heart potassium channel by protein kinase A and C. Science 1988;242:67-9
53. Tohse N, Nakaya H, Kanno M. Alpha 1-adrenoceptor stimulation enhances the delayed rectifier K+ current of guinea pig ventricular cells through the activation of protein kinase C. Circ Res 1992;71:1441-6
54. Varnum MD, Busch AE, Bond CT, Maylie J, Adelman JP. The min K channel underlies the cardiac potassium current IKs and mediates species-specific responses to protein kinase C. Proc Natl Acad Sci U S A 1993;90:11528-32

55. Lo CF, Numann R. Independent and exclusive modulation of cardiac delayed rectifying K+ current by protein kinase C and protein kinase A. Circ Res 1998;83:995-1002
56. Tohse N, Kameyama M, Irisawa H. Intracellular Ca2+ and protein kinase C modulate K+ current in guinea pig heart cells. Am J Physiol 1987;253:H1321-4
57. Nitta J, Furukawa T, Marumo F, Sawanobori T, Hiraoka M. Subcellular mechanism for Ca(2+)-dependent enhancement of delayed rectifier K+ current in isolated membrane patches of guinea pig ventricular myocytes. Circ Res 1994;74:96-104
58. Washizuka T, Horie M, Watanuki M, Sasayama S. Endothelin-1 inhibits the slow component of cardiac delayed rectifier K+ currents via a pertussis toxin-sensitive mechanism. Circ Res 1997;81:211-8
59. Groh WJ, Gibson KJ, Maylie JG. Hypotonic-induced stretch counteracts the efficacy of the class III antiarrhythmic agent E-4031 in guinea pig myocytes. Cardiovasc Res 1996;31:237-45
60. Volders PG, Sipido KR, Vos MA, Spatjens RL, Leunissen JD, Carmeliet E, Wellens HJ. Downregulation of delayed rectifier K(+) currents in dogs with chronic complete atrioventricular block and acquired torsades de pointes. Circulation 1999;100:2455-61
61. Ramakers C, Doevedans PA, Vos MA, Antzelevitch C, Dumaine R. KCNQ1 and KCNE1 expression is reduced in dogs with chronic AV block. Biophys J 2000;78:220A (abstract)
62. Rose J, Zheng MQ, Juang G, Kong W, O'Rourke B, Tomaselli GF. Delayed rectifier current in heart failure: Depressed function without changes in protein expression or protein level. Circulation 1999;100:I-425 (abstract)
63. Schultz JH, Volk T, Ehmke H. Heterogeneity of Kv2.1 mRNA expression and delayed rectifier current in single isolated myocytes from rat left ventricle. Circ Res 2001;88:483-490
64. Furukawa T, Kimura S, Furukawa N, Bassett AL, Myerburg RJ. Potassium rectifier currents differ in myocytes of endocardial and epicardial origin. Circ Res 1992;70:91-103
65. Sicouri S, Antzelevitch C. A subpopulation of cells with unique electrophysiological properties in the deep subepicardium of the canine ventricle. The M cell. Circ Res 1991;68:1729-41
66. Brahmajothi MV, Morales MJ, Reimer KA, Strauss HC. Regional localization of ERG, the channel protein responsible for the rapid component of the delayed rectifier, K+ current in the ferret heart. Circ Res 1997;81:128-35
67. Xu X, Rials SJ, Wu Y, Salata JJ, Liu T, Bharucha DB, Marinchak RA, Kowey PR. Left ventricular hypertrophy decreases slowly but not rapidly activating delayed rectifier potassium currents of epicardial and endocardial myocytes in rabbits. Circulation 2001;103:1585-90
68. Cheng J, Kamiya K, Liu W, Tsuji Y, Toyama J, Kodama I. Heterogeneous distribution of the two components of delayed rectifier K+ current: a potential mechanism of the proarrhythmic effects of methanesulfonanilideclass III agents. Cardiovasc Res 1999;43:135-47
69. Splawski I, Shen J, Timothy KW, Lehmann MH, Priori S, Robinson JL, Moss AJ, Schwartz PJ, Towbin JA, Vincent GM, Keating MT. Spectrum of mutations in long-QT syndrome genes. KVLQT1, HERG, SCN5A, KCNE1, and KCNE2. Circulation 2000;102:1178-85
70. Vos MA, de Groot SH, Verduyn SC, van der Zande J, Leunissen HD, Cleutjens JP, van Bilsen M, Daemen MJ, Schreuder JJ, Allessie MA, Wellens HJ. Enhanced susceptibility for acquired torsade de pointes arrhythmias in the dog with chronic, complete AV block is related to cardiac hypertrophy and electrical remodeling. Circulation 1998;98:1125-35
71. van Opstal JM, Verduyn SC, Leunissen HD, de Groot SH, Wellens HJ, Vos MA. Electrophysiological parameters indicative of sudden cardiac death in the dog with chronic complete AV-block. Cardiovasc Res 2001;50:354-61
72. Thomas GP, Vos MA, Antzelevitch C. The effect of volume overload hypertrophy on transmural distribution of the delayed rectifier (I_{Kr} and I_{Ks}) and transient outward (I_{to}) currents in the canine heart. Pace 2001;24:597 (abstract)

73. Sipido KR, Volders PG, de Groot SH, Verdonck F, Van de Werf F, Wellens HJ, Vos MA. Enhanced Ca(2+) release and Na/Ca exchange activity in hypertrophied canine ventricular myocytes: potential link between contractile adaptation and arrhythmogenesis. Circulation 2000;102:2137-44
74. Li GR, Sun H, Nattel S. Action potential and ionic remodeling in a dog model of heart failure. Pace 1998a;21:877 (abstract)
75. Li GR, Sun H, Feng J, Nattel S. Ionic mechanisms of the action potential prolongation in failing human ventricular cells. Pace 1998b;21:877 (abstract)
76. Jiang M, Cabo C, Yao J, Boyden PA, Tseng G. Delayed rectifier K currents have reduced amplitudes and altered kinetics in myocytes from infarcted canine ventricle. Cardiovasc Res 2000;48:34-43
77. Gillis AM, Geonzon RA, Mathison HJ, Kulisz E, Lester WM, Duff HJ. The effects of barium, dofetilide and 4-aminopyridine on ventricular repolarization in normal and hypertrophied rabbit heart. J Pharmacol Exp Ther 1998;285: 262-270
78. Tsuji Y, Opthof T, Kamiya K, Yasui K, Liu W, Lu Z, Kodama I I. Pacing-induced heart failure causes a reduction of delayed rectifier potassium currents along with decreases in calcium and transient outward currents in rabbit ventricle. Cardiovasc Res 2000;48:300-9
79. Suto F, Cahill SA, Greenwald I, Gross GJ. Early onset of QT interval prolongation and I_{Kr} downregulation in rabbits with compensated complete heart block. J Am Coll Cardiol 2001;37:91A (abstract)
80. Kleiman RB, Houser SR. Outward currents in normal and hypertrophied feline ventricular myocytes. Am J Physiol 1989;256:H1450-61
81. Furukawa T, Myerburg RJ, Furukawa N, Kimura S, Bassett AL. Metabolic inhibition of ICa,L and IK differs in feline left ventricular hypertrophy. Am J Physiol 1994;266:H1121-31
82. Lodge NJ, Normandin DE. Alterations in Ito1, IKr and Ik1 density in the BIO TO-2 strain of syrian myopathic hamsters. J Mol Cell Cardiol 1997;29:3211-21
83. Volk T, Nguyen TH, Schultz JH, Faulhaber J, HE H. Regional alterations of repolarizing K+ currents among the left ventricular free wall of rats with ascending aortic stenosis. J Physiol 2001;530:443-55
84. Ahmmed GU, Dong PH, Song G, Ball NA, Xu Y, Walsh RA, Chiamvimonvat N. Changes in Ca(2+) cycling proteins underlie cardiac action potential prolongation in a pressure-overloaded guinea pig model with cardiac hypertrophy and failure. Circ Res 2000;86:558-70
85. Zhang TT, Takimoto K, Stewart AF, Zhu C, Levitan ES. Independent regulation of cardiac Kv4.3 potassium channel expression by angiotensin II and phenylephrine. Circ Res 2001;88:476-482
86. The Sicilian gambit. A new approach to the classification of antiarrhythmic drugs based on their actions on arrhythmogenic mechanisms. Task Force of the Working Group on Arrhythmias of the European Society of Cardiology. Circulation 1991;84:1831-51
87. Hondeghem LM, Snyders DJ. Class III antiarrhythmic agents have a lot of potential but a long way to go. Reduced effectiveness and dangers of reverse use dependence. Circulation 1990;81:686-90
88. van Opstal JM, Schoenmakers M, Verduyn SC, de Groot SHM, Leunissen HDM, Van der Hulst FF, Molenschot MMC, Wellens HJJ, Vos MA. Chronic amiodarone evokes no torsade de pointes arrhythmias despite QT lengthening in an animal model of acquired long QT syndrome. Circulation. In the press.
89. Hoppe UC, Marban E, Johns DC. Molecular dissection of cardiac repolarization by in vivo Kv4.3 gene transfer. J Clin Invest 2000;105:1077-84
90. Hoppe UC, Marban E, Johns DC. Distinct gene-specific mechanisms of arrhythmia revealed by cardiac gene transfer of two long QT disease genes, HERG and KCNE1. Proc Natl Acad Sci U S A 2001;98:5335-40

91. Donahue JK, Heldman AW, Fraser H, McDonald AD, Miller JM, Rade JJ, Eschenhagen T, Marban E. Focal modification of electrical conduction in the heart by viral gene transfer. Nat Med 2000;6:1395-8

16. GENETIC POLYMORPHISMS AND THEIR ROLE IN VENTRICULAR ARRHYTHMIAS

S. Kääb, M. Näbauer, A. Pfeufer

Introduction

Potentially life threatening ventricular arrhythmias may occur as a consequence of genetic variants affecting key proteins that directly or indirectly alter myocardial electrical properties. Diseases such as Long-QT-Syndrome, Brugada Syndrome, catecholaminergic polymorphic ventricular tachycardia in structurally normal hearts and hypertrophic, dilated, and right ventricular cardiomyopathy in structurally abnormal hearts are the best known examples of monogenic conditions causing ventricular arrhythmias and sudden cardiac death (SCD).
With the human genome sequence being available the role of minor gene effects such as the contribution of single nucleotide polymorphisms (SNPs) to arrhythmia predisposition is currently under investigation. Whether these common DNA variants can create a significant disadvantageous profile in some patients towards the susceptibility to common symptomatic ventricular arrhythmias or SCD in the sense of a polygenic disease predisposition remains to be investigated.

Monogenic diseases causing ventricular arrhythmias in structurally normal hearts

Long-QT-Syndrome
Long-QT-Syndrome is an inheritable disease characterized by abnormal repolarization typically manifested in a prolonged QT interval and stress mediated life threatening polymorphic ventricular arrhythmias with syncope occurring for the first time during childhood and adolescence. A rare autosomal recessive form with congenital deafness (Jervell and Lange-Nielsen Syndrome) and the more frequent autosomal dominant form (Romano Ward Syndrome) have been described. At present five genes encoding subunits

of cardiac ion channels have been linked to Long-QT-Syndrome [1]. For three common Long-QT-Syndromes genotype-phenotype correlations have been established allowing for gene-specific epidemiology and risk stratification. (table 1) [2,3,4]

Congenital LQT-Syndromes

Syndrome	Chromsomal Localisation	Gene/-Product	(autosomal)	Prevalance
Romano-Ward Syndrome (1963/64)				1 : 10.000
LQT 1	Chr. 11p15.5	KVLQT1 / α-I_{Ks}	dominant	50-60%
LQT 2	Chr. 7q35-36	HERG / α-I_{Kr}	dominant	30%
LQT 3	Chr. 3ß21-24	SCNA5 / I_{Na}	dominant	5-10%
LQT 4	Chr. 4q25-27	?	dominant	
LQT 5	Chr. 21q22.1-22.2	KCNE1 / ß- I_{Ks} (minK)	dominant	
LQT 6	Chr. 21q22.1-22.2	KCNE2 / ß- I_{Kr} (Mirp)	dominant	
LQT X	no LQT1-LQT6	?		
Jervell and Lange-Nielsen Syndrome (1957)				1-6 : 1 Mio.
JLN 1	Chr. 11p15.5	KCNQ1 / ß-I_{Ks}	recessive	
JLN 2	Chr. 21q22.1-22.2	KCNE1 / ß- I_{Kr}	recessive	

Genetic defects in Long-QT-Syndrome cause altered function of myocardial ion channels leading to prolonged repolarization either by increased depolarizing current (I_{Na}) or decreased repolarizing currents (I_{Ks}, HERG). LQT3 appears to be the most malignant variant generally presenting with first symptoms later in adolescence and the one less effectively managed by betablockers. Arrhythmic events in LQT3 are more likely to occur during bradycardia and during sleep and rest. LQT1 and LQT2 seem to have a higher frequency of syncopal events starting in early childhood but their lethality is lower and the protection afforded by betablockers especially in LQT1 is much higher [5]. I_{Ks} the current affected in LQT1 is augmented by adrenergic stimulation causing the physiologic QT shortening during exercise. Failure to shorten QT with exercise may reflect calcium current dominating over defective I_{Ks}. Subjects with HERG mutations (LQT2) display near normal QT shortening during exercise. Diagnosis of Long-QT-Syndrome is established ideally with characteristic clinical manifestation and ECG findings together with a family history of syncope and SCD. Genetic testing should be encouraged in all cases with reasonable clinical evidence to establish a better database for genotype-phenotype correlation with the goal to establish gene-specific therapy.
At present pharmacological therapy relies on the use of betablockers. Additional therapeutic options include left-sided cardiac sympathectomy, cardiac pacing and implantable cardioverter defibrillators in selected high-risk patients. The occurrence of torsades de pointes arrhythmias with potentially lethal outcome has been identified as a major risk of antiarrhythmic as well as of some non-antiarrhythmic drugs. Despite rapidly

expanding knowledge of genetic and molecular aspects of the pathophysiology of congenital Long-QT-Syndrome, predisposing factors and strategies for identifying patients at risk for acquired Long-QT-Syndrome remain difficult to identify.

Genotype - Phenotype - Correlation of Common Long-QT-Syndromes

	LQT 1	LQT 2	LQT3
affected gene	KvLQT1	HERG	SCN5A
chromosome	Chr. 11p 15.5	Chr. 7q 35-36	Chr. 3p21-24
affected current	$\downarrow I_{Ks}$	$\downarrow I_{Ks}$	$\uparrow I_{Na}$
prevalance (%LQTS)	50-60%	20-30%	5-10%
ECG	broad T-wave	biphasic or small T-wave	sharp pointed T-wave long isoelectric ST-segment
	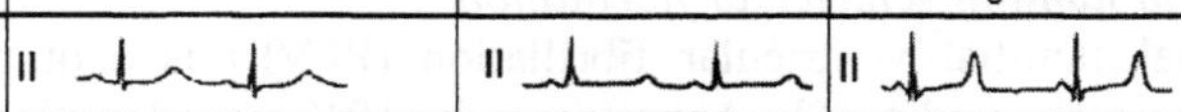		
trigger	physical or emotional stress	stress oder rest acustic signals(?)	rest, sleep
mean age at time of first symptoms	9	12	16
QT-changes with exercise	no change (or increase)	normal shortening	supra - normal shortening
arrhythmic events	++	+++	+
SCD	++	+	+++

\+ rare
++ occasional
+++ common

Current understanding postulates extrinsic factors to destabilize repolarization, while an unknown number of intrinsic myocardial variables may predispose to an increased lability of the repolarization process. Evidence for an increased susceptibility to QT-prolongation upon challenge of repolarization led to the concept of altered repolarization reserve as a proarrhythmic substrate. Disproportionate QT-prolongation during therapy with class I or III antiarrhythmic agents followed by arrhythmias of the torsades de pointes type have long been described as unpredictable for an individual patient. More recently, increasing awareness of drug induced arrhythmias has pointed to the QT-prolonging and arrhythmogenic potential of a wide variety of non-antiarrhythmic drugs, expanding the population at risk and demonstrating the need for understanding factors determining susceptibility for drug induced Long-QT-Syndrome in the individual patient.

Experimental evidence demonstrates that QT-prolongation by class III antiarrhythmic agents as well as non-antiarrhythmic drugs is primarily due to block of the rapidly activating component of the delayed rectifier potassium current, I_{Kr}. Attempts to explain the individual predisposition to acquired Long-QT-Syndrome by mutations in HERG and MIRP, the alpha and beta subunit encoding the human I_{Kr}, or by mutations in other genes known to cause congenital Long-QT-Syndrome, revealed apparent genetic predisposition only in a small fraction of patients. Clinical and experimental evidence led to the hypothesis of a variable impact of a number of "modifier" genes that affect both susceptibility and severity of acquired Long-QT-Syndrome. In addition, altered drug metabolism due to renal or hepatic insufficiency, cytochrome P-450 polymorphism, or

drug-drug interactions with an unpredictable increase of the plasma concentration of QT-prolonging drugs have been implicated to be crucial for the occurrence of acquired Long-QT-Syndrome.
Among the many risk factors for drug induced torsades de pointes, none has been rigorously validated with respect to its actual role for the predisposition to acquired Long-QT-Syndrome. The preponderance of e.g. female gender and heart failure in patients with acquired Long-QT-Syndrome suggests that intrinsic electrophysiologic properties of the myocardium may be of major importance. Additional factors such as hypokalemia and exposure to QT-prolonging drugs may cause disproportionate QT-prolongation and torsades de pointes arrhythmias primarily in patients with intrinsic myocardial predisposition, manifested by reduced repolarization reserve.

Paroxysmal familial ventricular fibrillation
Paroxysmal familial ventricular fibrillation (PFVF) is another primary form of ventricular tachycardia [6]. Approximately 10% of all patients with ventricular fibrillation have no demonstrable cardiac or noncardiac causes for the disease, including no prolongation of the QT interval, and it is therefore by exclusion classified as a primary condition.
Clinical hallmarks of PFVF are syncopal attacks and sudden cardiac death typically triggered by emotional stimuli. Important electrocardiographic changes at rest are a shortened PR interval. Fibrillation can be induced by stressful emotional stimuli. Exercise stress tests were notably often unrewarding in revealing VF as opposed to catecholaminergic polymorphic ventricular tachycardia (CPVT see below).
Available figures suggest an 11% rate of sudden death with 1 year of diagnosis [7]. Class IA antiarrhythmic agents as well as Propranolol were found to be effective for prophylaxis and also appeared to improve prognosis.
In some but not in all of the PFVF patients class IA antiarrhythmic agents also led to a significant change in the electrocardiogram by inducing a pattern of right bundle branch block and elevation of ST segment in leads V1 to V3 as usually seen in the Brugada Syndrome (see below) [8]. This feature led to the notion that both conditions share in part a common pathophysiology.
PFVF has been described to occur as an autosomal dominant trait. In some families it has been shown to be caused by one mutation in the cardiac sodium channel gene SCN5A, namely Ser1710Leu [9]. Therefore this form of PFVF is not only pathophysiologically related to but also allelic to Brugada Syndrome. The alterations of the SCN5A channel characteristics at the molecular level due to this mutation yet have to be determined.

Brugada Syndrome
Brugada Syndrome is closely related to PFVF in molecular, pathophysiological and clinical terms. Patients with this syndrome suffer from PFVF and in addition present at rest with a characteristic transient electrocardiographic pattern consisting of right bundle branch block and elevation of ST segment [10]. This disorder can therefore be viewed as a specific subgroup of PFVF and may account for 40 to 60% of all PFVF cases.

Typical symptoms are syncope and SCD caused by rapid polymorphic ventricular arrhythmias mainly occurring at rest or during sleep with an incidence of 30% over 3 years both in symptomatic and asymptomatic patients. Symptoms occur mainly in males at a mean age of 38 years.
Administration of class I antiarrhythmics unmasks ST segment elevation in concealed forms of the disease, which are in essence variants of PFVF [11] (see above). In contrast to PFVF no drug has proven efficacy in the prevention of SCD in Brugada syndrome and therefore cardiac arrest survivors and patients with a history of syncope or a family history of juvenile SCD are recommended to receive an implantable cardiodefibrillator.
On the molecular side Brugada Syndrome is also caused by specific alterations in the SCN5A Gene, in which other mutations lead to PFVF and to the LQT3 form of inherited long QT syndrome.

Catecholaminergic polymorphic ventricular tachycardia
Catecholaminergic polymorphic VT (CPVT) is a primary form of ventricular tachycardia occurring in the absence of structural heart disease. It has been initially recognized as a distinct disease entity with autosomal dominant inheritance in 1978 [12] and was extensively described in 1985 [13].
Its main features are ventricular bigemini or polymorphic bidirectional VT that can in affected patients be reproducibly triggered by increased sympathotonic activity, e.g. by physical activity or the administration of sympathomimetics. The electrocardiographic pattern of CPVT closely resembles the arrhythmias associated with intracellular calcium overload due to digitalis toxicity or inadequate release of calcium into the sarcoplasm. This feature points to delayed afterdepolarizations and triggered activity as the likely arrhythmogenic mechanism of the condition. Typical symptoms of patients include unexplained syncope, rapid and sustained runs of VT and a positive family history for SCD. The administration of class IC antiarrhythmic drugs does not induce electrocardiographic abnormalities in CPVT as opposed to Brugada syndrome. The only apparently effective therapy is betablocker treatment based on a retrospective analysis which shows SCDs in 4/38 (10,5%) of patients with and 10/21(48%) of patients without this therapy [14]. The relatively high mortality despite pharmacotherapy (10·5%) may indicate ICD implantation at least for secondary prevention in those patients with early onset of symptoms or a positive family history of SCD.
CPVT disease may follow an autosomal dominant inheritance pattern in some families. Mutations of the cardiac ryanodine receptor gene (RYR2) were described in four unrelated families with one or several affected individuals [15].

Monogenic diseases causing ventricular arrhythmias in structurally abnormal hearts

Right ventricular cardiomyopathy
Right Ventricular Cardiomyopathy (RVC, initially termed Arrhythmogenic Right Ventricular Dysplasia or ARVD) is a genetically heterogeneous disease due to mutations in at least 7 genomic loci [16]. The disease is familial in at least 30% of cases

with dominant as well as recessive modes of inheritance.

The distinguishing electrocardiographic (ECG) pattern of RVC is that of inverted T-waves and prolonged QRS complex with epsilon waves in the right precordial leads. Morphologically RVC is characterized by regional or global fibro-fatty replacement of the right ventricular myocardium, with or without left ventricular involvement and with relative sparing of the septum. The disease typically manifests in young adults with syncope or cardiac arrest due to ventricular arrhythmias. Monomorphic ventricular tachycardia with left bundle branch block morphology is most common but any arrhythmia from asymptomatic premature ventricular beats to polymorphic ventricular tachycardia has been documented.

Recent data, suggest that like the other inherited cardiomyopathies, RVC is one of the major causes of SCD in the age group under 35 years accounting for approximately 25% of deaths in young athletes [17]. Only limited information is available on risk assessment of SCD in RVC. Recent data suggests, that if SCD can be prevented, life expectancy may be normal or near normal suggesting a minor role of right ventricular failure for disease severity.

Six genomic loci, namely ARVD1-6, featuring autosomal dominant RVC have been determined. ARVD2 is caused by mutations in the RYR2 gene and is therefore allelic to CPVT. In addition a recessive form of ARVC with coinherited skin and hair abnormalities called Naxos disease has been recognized in which a 2 base pair deletion in the plakoglobin gene is causal [18].

Dilated cardiomyopathy

Dilated cardiomyopathy (DCM) is a monogenic form of congestive heart failure (CHF) caused mainly by mutations in cardiac cytoskeleton protein genes. Inheritance is autosomal dominant in most cases but also recessive forms of the disease have been described [19].

Various arrhythmogenic features such as atrial flutter or fibrillation, AV-conduction disturbances of varying degree and paroxysmal VT or ventricular fibrillation often but not always accompany DCM. The two different entities have been termed arrhythmogenic DCM as opposed by pure DCM. The affected disease gene is regarded as the main determinant of the DCM subform. In second line it is the kind of mutation within that gene. Arrhythmogenic DCM is ascribed to mutations in the lamin A/C gene (CMD1A, 1q21.2 , allelic to Emery Dreifus muscular dystrophy), and the genomic loci CMD1E (3p25-p22, formerly termed CDCD2), CMD1F (6q23) and CMD1H (2q14-q22). Pure forms of DCM are presumably caused by mutations in the cardiac actin gene (15q14), the desmin gene (CMD1I, 2q35), the δ- Sarcoglycan gene (5q33) as well as the genomic loci CMD1B (9q13), CMD1C (10q21-q23), CMD1D (1q32), CMD1G (2q31), CMD1J (6q23-q24, where the disease cosegregates with deafness) and CMD1G (6q12-q16). Recently DCM causing mutations have also been identified in sarcomere protein genes having previously thought to cause only HCM. Some β-myosin heavy chain (MYH7) and cardiac troponin T (TNNT2) mutation which caused early-onset ventricular dilatation and frequently resulted in heart failure without antecedent hypertrophy [20]. Autosomal dominant familial atrial fibrillation (FAF) without CHF has also been linked to the genomic locus CMD1C known to cause a pure form of

DCM. Mutations leading to FAF therefore might be allelic variants in a yet to be identified DCM gene. In this case, the cause of the disease might prove to be rather dependent on the specific mutation.

Nevertheless in DCM the influence of the specific mutation and of potential co-acting modifier genes on the subform is especially difficult to quantify or to rule out in the case of DCM at the current stage of investigation as the number and sizes of the investigated pedigrees are often limited, the affected families share common genetic background and because most sporadic cases have not received a molecular diagnosis yet. Investigation of genotype phenotype correlations including predisposition to arrhythmia is in its infancy in DCM research whereas it is already somewhat advanced in the monogenic diseases LQT and HCM.

Hypertrophic Cardiomyopathy

Hypertrophic cardiomyopathy (HCM) is a monogenic form of cardiac hypertrophy. It has first been recognized as a novel disease entity with autosomal dominant inheritance in 1958 [21] but also sporadic forms of the disease are often observed suggesting de novo mutations or frequent cases of incomplete penetrance. Its population prevalence has been estimated between 1:500 [22] and 1:5000.

HCM has a highly characteristic pathology (myocardial hypertrophy, myocyte disarray and fibrosis probably due to replacement scarring) which contributes to a broad spectrum of functional abnormalities including myocardial ischemia, diastolic dysfunction, left ventricular outflow obstruction supraventricular and ventricular tachyarrhythmia. The pathophysiological process may result in congestive heart failure, clinically important arrhythmias and SCD in some patients.

Since the initial elucidation of its molecular cause HCM has been viewed as a disease of the sarcomere where mutations in either one of several sarcomeric protein genes trigger the disease process [23]. Among the known disease genes are the β-myosin heavy chain gene (CMH1, 14q11.2), the cardiac troponin T2 gene (CMH2, 1q32) [24], the α-tropomyosin gene (CMH3, 15q2.4), the cardiac myosin binding protein C gene (CMH4, 11p11.2) [25,26], the cardiac troponin I gene (CMH7, 19q13.4) [27], the cardiac troponin C gene (3p21.3-p14.3) [28], the essential myosin light chain gene MYL3 (CMH8, 3p) and the regulatory myosin light chain gene MYL2 (12q23-q24). Mutations in the cardiac actin gene (15q14) have also been described to cause HCM [29]

Recent investigation has demonstrated, that also mutations in the gene for the AMP dependent protein kinase γ2 subunit can lead to a form of HCM that cosegregates with WPW syndrome and has previously been linked to the chromosomal locus CMH6 (7q36) [30]. The identification of a regulatory enzyme component of central role in responding to intracellular changes in ATP levels and energy load as a cause of HCM has together with the observation that mutations in sarcomere proteins lead to ATP wasting rather than impaired contractility led to a change in the understanding of HCM. Intracellular ATP depletion under certain circumstances, especially those of increased cardiac workload, appear to be the hallmark change of the condition and sarcomere mutations are only one of several possible ways towards it.

As opposed to the previous one this new concept of viewing HCM as an ATP wasting disease does easily incorporate some other diseases, previously regarded as independent from HCM. These conditions also share cardiac hypertrophy as a significant feature and their common molecular cause is impaired mitochondrial energy metabolism. These diseases include Friedreich Ataxia caused by Frataxin mutation, deficiency of very long chain fatty acid CoA-dehydrogenase (VLCAD deficiency) and mitochondrial myopathy and cardiomyopathy caused by mutations in the Chondriome or genes of nuclear encoded mitochondrial proteins [31].

At present it is not clear how this new concept will affect our understanding of arrhythmogenesis in HCM. It is currently also unknown how the development of two conditions previously thought to be unrelated such as HCM and WPW, which also is a primarily arrhythmogenic disease, is orchestrated at the molecular level presumably by a single genetic variant. A more detailed understanding of this condition may also shed light on arrhythmogenic mechanisms in other forms of HCM.

Polygenic diseases causing ventricular arrhythmias

Research into arrhythmias caused by single genetic variants has seen a tremendous accumulation of knowledge and insight into disease within the last decade. Yet little is known about potential genetic variants with higher population frequencies (commonly termed "polymorphisms"), which might confer an increased susceptibility to or risk of arrhythmias, although some lines of evidence suggest that such variants unrelated to a monogenic disease may exist.

This evidence has in part emerged from large-scale epidemiological studies that have demonstrated familial associations of SCD. A recent case control study of more than 500 SCD survivors demonstrated that family history was a relevant risk factor and independent predictor of SCD with an odds ratio 1.6 [32]. A second study followed a cohort of 7000 subjects for 23 years including 118 SCD patients [33] and also confirmed family history was a strong independent predictor of SCD susceptibility with an odds ratio between 1·8 when only one other family member was affected and a remarkable odds ratio of 9.4 when a positive history for SCD was present in both parental lines. These two studies provide evidence for the existence of factors in addition to the established risk predictors making common genetic variants a likely candidate.

One emerging polygenic risk factor for SCD in the special case of postinfarction is the well-studied ACE I/D polymorphism. In a population-based study of myocardial infarction survivors the ACE DD-genotype was associated with longer QT dispersion ($p<0.001$) but was not so in healthy subjects without infarction [34]. The ACE DD-genotype therefore appears to confer a recessive genetic predisposition towards increased myocardial damage or unfavorable postinfarction remodeling resulting in a more than average decrease in repolarization homogeneity of the postinfarct heart.

A polygenic recessive predisposition for congestive heart failure has also been described. In the Japanese mainland population the homozygous state of the relatively frequent variant Asn654Lys of the cardiac Nebulette gene (10p12) appears to increase risk of DCM (35). The authors found an odds ratio of 6,25 of the Lys/Lys homozygous genotype for the development of DCM but did not find this genotype associated with any of the specific clinical subphenotypes of DCM or CHF including different forms of arrhythmia presence or absence. The finding of Nebulette involvement in polygenic DCM awaits independent confirmation.

Conclusion

At present it seems appropriate to encourage the assessment of family history of patients with symptomatic ventricular arrhythmias and especially in all survivors of SCD. In the presence of familial clustering of symptomatic ventricular arrhythmias and SCD the possibility of monogenic arrhythmic disorders such as Long-QT-Syndrome, Brugada Syndrome and others should be carefully evaluated.

Functional expression studies and investigations probing genotype-phenotype correlations in the rare monogenic arrhythmic diseases will continue to help understanding the pathophysiology of ventricular arrhythmogenesis and may guide a genome wide search for single nucleotide polymorphisms that may ultimately be useful for risk stratification in common polygenic ventricular arrhythmias.

List of disease genes

Locus	Name	Disease	OMIM	Gene	Symbol	OMIM
Long-QT-Syndrome						
11p15.5	LQT1	LQT		IK(s)-a,KCNQ1	KCNQ1	
7q35-q36	LQT2	LQT		IK(r)-a,HERG	KCNH2	
3p21	LQT3	LQT		Sodium Channel V-a	SCN5A	
4q25-q27	LQT4	LQT+SSS	*600919	?	?	
21q22.12	LQT5	LQT		IK(s)-b,MinK	KCNE1	
21q22.12	LQT6	LQT		IK(r)-b,MiRP1	KCNE2	
Paroxysmal familial ventricular fibrillation						
3p21	LQT3	PFVF		Sodium Channel V-a	SCN5A	
Brugada Syndrome						
3p21	LQT3	Brugada Sy		Sodium Channel V-a	SCN5A	
Catecholaminergic polymorphic ventricular tachycardia						
1q42-q43	CPVT	CPVT	#604772	Ryanodinreceptor2	RYR2	*180902

Locus	Name	Disease	OMIM	Gene	Symbol	OMIM
Right ventricular Cardiomyopathy						
17q21	ARVD	ARVD+PPK	#601214	Plakoglobin	JUP	173325
14q23-q24	ARVD1	ARVD	*107970	?	?	
1q42-q43	ARVD2	ARVD	#600996	Ryanodinreceptor2	RYR2	*180902
14q12-q22	ARVD3	ARVD	*602086	?	?	
2q32-q32	ARVD4	ARVD+LSB	*602087	?	?	
3p23	ARVD5	ARVD	*604400	?	?	
10p14-p12	ARVD6	ARVD	*604401	?	?	
Dilated Cardiomyopathy						
1q21.2	CMD1A	DCM+Arrh	#115200	Lamin A/C	LMNA	*150330
9q13	CMD1B	DCM	*600884	?	?	
10q21-q23	CMD1C	DCM	*601493	?	?	
1q32	CMD1D	DCM	*601494	?	?	
3p25-p22	CMD1E	DCM+Arrh	*601154	?	?	
6q23	CMD1F	DCM+Arrh	*602067	?	?	
2q31	CMD1G	DCM	*604145	?	?	
2q14-q22	CMD1H	DCM+Arrh	*604288	?	?	
2q35	CMD1I	DCM	#604765	Desmin	DES	*125660
6q23-q24	CMD1J	DCM+Deaf	*605362	?	?	
6q12-q16	CMD1K	DCM	*605582	?	?	
5q33	CMD	DCM		Sarcoglycan-d	SGCD	601411
15q14	CMD	DCM	#115200	Actin-a,cardiac	ACTC	*102540
10p12	CMD	DCM		Nebulette	NEBL	
14q11.2	CMD	DCM		Myosin-b HC	MYH7	160760
1q32	CMD	DCM		Troponin T2,cardial	TNNT2	191045
6p24	CMD	DCM+PPK	#605676	Desmoplakin	DSP	*125647
10q22-q24	Afib	Afib+Brady	*163800	?	?	
Hypertrophic Cardiomyopathy						
15q14	CMH1	HCM	#192600	Actin-a,cardiac	ACTC	*102540
14q11.2	CMH1	HCM		Myosin-b HC	MYH7	160760
1q32	CMH2	HCM		Troponin T2,cardial	TNNT2	191045
15q22.1	CMH3	HCM		Tropomyosin-a	TPM1	191010
11p11.2	CMH4	HCM		Myosinb.Protein C	MYBPC3	600958
7q36.1	CMH6	HCM+WPW	600858	ProtKinase-AMP-g2	PRKAG2	*602743
19q13.4	CMH7	HCM		Troponin I,cardial	TNNI3	191044
19pq	CMH	HCM		Myosin bind prot C	MYBPC2	
12q23-q24.3	CMH	HCM		Myosin LC2,reg	MYL2	160781
3p21.3-p21.2	CMH	HCM		Myosin LC3,v,ess	MYL3	160790
9q13	FRDA	FRDA	*229300	Frataxin	FRDA	*229300

References

1. Schwartz PJ, Priori SG, Napolitano C. The Long QT Syndrome. In: Zipes DP, Jalife J, eds. Cardiac Electro-physiology. From Cell to Bedside. Philadelphia: WB Saunders Co, 2000:597–615.
2. Priori SG, Napolitano C, Schwartz PJ. Low penetrance in the long-QT syndrome: clinical impact. Circulation 1999;99:529–33.
3. Zareba W, Moss AJ, Schwartz PJ et al.Influence of genotype on the clinical course of the long-QT syndrome. International Long-QT Syndrome Registry Research Group. N Engl J Med 1998;339:960–5.
4. Schwartz PJ, Priori SG, Spazzolini C.et al. Genotype-phenotype correlation in the long-QT syndrome : gene-specific triggers for life-threatening arrhythmias. Circulation 2001;103: 89–95.
5. Moss AJ, Zareba W, Hall WJ et al.Effectiveness and limitations of beta-blocker therapy in congenital long-QT syndrome. Circulation 2000;101:616–23.
6. McRae J.R,Wagner G.S,Rogers M.C, Canent R.V. Paroxysmal familial ventricular fibrillation. J. Pediat.1974;84:515-8.
7. Viskin S, Lesh MD, Eldar M, Fish R, Setbon I, Laniado S, Belhassen B. Mode of onset of malignant ventricular arrhythmias in idiopathic ventricular fibrillation. J.Cardiovasc. Electrophysiol. 1997;8:1115-20.
8. Akai J, Makita N, Sakurada H, Shirai N, Ueda K, Kitabatake A, Nakazawa K, Kimura A, Hiraoka M. A novel SCN5A mutation associated with idiopathic ventricular fibrillation without typical ECG findings of Brugada syndrome. FEBS Lett. 2000;479: 29-34.
9. Chen Q, Kirsch G.E, Zhang D. et al. Genetic basis and molecular mechanism for idiopathic ventricular fibrillation. Nature 1998;392:293-5.
10. Brugada P, Brugada J. Right bundle branch block, persistent ST segment elevation and sudden cardiac death: a distinct clinical and electrocardiographic syndrome. A multicenter report. J Am Coll Cardiol 1992; 20:1391–6.
11. Brugada R, Brugada J, Antzelevitch C et al. Sodium channel blockers identify risk for sudden death in patients with ST-segment elevation and right bundle branch block but structurally normal hearts. Circulation 2000;101:510–15.
12. Coumel P, Fidelle J, Lucet V, Attuel P, Bouvrain Y. Catecholaminergic-induced severe ventricular arrhythmias with Adams-Stokes syndrome in children: report of four cases. Br Heart J 1978; 40: 28–37
13. Leenhardt A, Lucet V, Denjoy I, Grau F, Ngoc DD, Coumel, P. Catecholaminergic polymorphic ventricular tachycardia in children. A 7-year follow-up of 21 patients. Circulation 1995;91: 1512–9.
14. de Paola AA, Horowitz LN, Marques FB et al. Control of multiform ventricular tachycardia by propranolol in a child with no identifiable cardiac disease and sudden death. Am Heart J 1990; 119:1429–32.
15. Priori SG, Napolitano C, Tiso N et al. Mutations in the cardiac ryanodine receptor gene (hRyR2) underlie cate-cholaminergic polymorphic ventricular tachycardia. Circulation 2001; 103: 196–200.
16. Fontaine G, Fontaliran F, Hebert JL, Chemla D, Zenati O, Lecarpentier Y, Frank R. Arrhythmogenic right ventricular dysplasia. Annu. Rev. Med. 1999;50:17-35.
17. Tabib A, Miras A, Taniere P, Loire R. Undetected cardiac lesions cause unexpected sudden cardiac death during occa-sional sport activity. A report of 80 cases. Eur Heart J 1999;20: 900–3.
18. McKoy G, Protonotarios N, Crosby A et al. Identification of a deletion in plakoglobin in arrhythmogenic right ventricular cardiomyopathy with palmoplantar keratoderma and woolly hair (Naxos disease). Lancet. 2000;355 : 2119-24.

19. Keeling PJ, Gang Y, Smith G et al. Familial dilated cardiomyopathyin the United Kingdom. Br Heart J 1995; 73: 417–21
20. Kamisago M, Sharma SD, DePalma SR, et al. Mutations in sarcomere protein genes as a cause of dilated cardiomyopathy. New Eng. J. Med. 2000; 343: 1688-1696.
21. Teare D. Asymmetrical hypertrophy of the heart in young adults. Brit heart J 1958; 20: 1-8
22. Maron BJ, Mathenge R, Casey SA, Poliac LC, Longe TF. Clinical profile of hypertrophic cardiomyopathy identified de novo in rural communities. J Am Coll Cardiol 1999;33:1590-5
23. Geisterfer-Lowrance AAT, Kass S, Tanigawa G, Vosberg HP, McKenna W, Seidman CE, Seidman JG. A molecular basis for familial hypertrophic cardiomyopathy: a beta cardiac myosin heavy chain gene missense mutation. Cell 1990;62:999-1006
24. Watkins H, MacRae C, Thierfelder L, Chou YH, Frenneaux M, McKenna W, Seidman JG, Seidman CE. A disease locus for familial hypertrophic cardiomyopathy maps to chromosome 1q3. Nature Genet 1993;3:333-7
25. Watkins H, Conner D, Thierfelder L, Jarcho JA, MacRae C, McKenna WJ, Maron BJ, Seidman JG, Seidman CE. Mutations in the cardiac myosin binding protein-C gene on chromosome 11 cause familial hypertrophic cardiomyopathy. Nature Genet 1995;11:434-7
26. Bonne G, Carrier L, Bercovici J, Cruaud C, Richard P, Hainque B, Gautel M, Labeit S, James M, Beckmann J, Weissenbach J,Vosberg HP, Fiszman M, Komajda M, Schwartz K. Cardiac myosin binding protein C gene splice acceptor site mutation is associated with familial hypertrophic cardiomyopathy. Nature Genet 1995;11:438-40
27. Kimura A, Harada H, Park JE et al. Mutations in the cardiac troponin I gene associated with hypertrophic cardiomyopathy. Nature Genet 1997;16:379-82
28. Hoffmann B, Schmidt-Traub H, Perrot A, Osterziel KJ, Gessner R. First mutation in cardiac troponin C, L29Q, in a patient with hypertrophic cardiomyopathy. Hum Mutat. 2001; 17: 524.
29. Mogensen J, Klausen IC, Pedersen AK et al. Alpha-cardiac actin is a novel disease gene in familial hypertrophic cardiomyopathy. J. Clin. Invest 1999;103:R39-R43
30. Gollob MH, Green MS, Tang AS et al. Identification of a gene responsible for familial Wolff-Parkinson-White syndrome. N Engl J Med. 2001; 344 :1823-31.
31. Blair E, Redwood C, Ashrafian H, Oliveira M, Broxholme J, Kerr B, Salmon A, Ostman-Smith I, Watkins H. Mutations in the gamma(2) subunit of AMP-activated protein kinase cause familial hypertrophic cardiomyopathy: evidence for the central role of energy compromise in disease pathogenesis. Hum Mol Genet. 2001; 10: 1215-20.
32. Friedlander Y, Siscovick DS, Weinmann S et al. Family history as a risk factor for primary cardiac arrest. Circulation1998; 97: 155–60.
33. Jouven X, Desnos M, Guerot C, Ducimetiere P. Predicting sudden death in the population: the Paris Prospective Study I. Circulation 1999; 99: 1978–83.
34. Jeron A, Hengstenberg C, Engel S, Lowel H, Riegger GA, Schunkert H, Holmer S. The D-allele of the ACE polymorphism is related to increased QT dispersion in 609 patients after myocardial infarction. Eur Heart J. 2001; 22: 663-8.
35. Arimura T, Nakamura T, Hiroi S et al. Characterization of the human nebulette gene: a polymorphism in an actin-binding motif is associated with nonfamilial idiopathic dilated cardiomyopathy. Hum Genet. 2000; 107: 440-51.

17. MOLECULAR MECHANISMS OF REMODELING IN HUMAN ATRIAL FIBRILLATION

B.J.J.M. Brundel, R.H. Henning, H.H. Kampinga, I.C. van Gelder, H.J.G.M. Crijns

Introduction

Atrial fibrillation (AF) is the most common sustained cardiac arrhythmia in humans. Most frequently, AF occurs in conjunction with other cardiovascular disease, such as hypertension, ischemic heart disease, valve disease or cardiac failure. However, in 20-50% of the patients AF is not associated with any underlying disease [1]. One of the most intriguing properties of AF is its tendency to become more persistent over time [2]. Consequently, a large percentage of patients with paroxysmal AF will develop persistent AF [2]. Also, conversion to and maintenance of sinus rhythm by pharmacological or electrical methods becomes increasingly difficult the longer the arrhythmia exists [3].

Experimental and clinical studies point at two major mechanisms involved in the intrinsic progressive nature of AF. The first consists of a change in the electrical properties of the atrium, notably a shortening of the atrial effective refractory period (AERP) and a loss of rate adaptation [4,5], and hence was named *electrical remodeling*. Furthermore, it has been considered that AF is also associated with elaborate adaptive and maladaptive changes in tissue and cellular architecture [6,7]. By parallel, this type of changes was denominated *structural remodeling*. Together, these mechanisms will increase the probability of generating multiple atrial wavelets by enabling rapid atrial activation and dispersion of refractoriness [8].

Having identified electrical and structural remodeling as general pathophysiological mechanisms in the progressive nature of AF, recent studies have attempted to detect the underlying molecular mechanisms. So far, the molecular research has been focussed mainly at various ion-channels and at proteins involved in calcium homeostasis. The purpose of this review is to discuss the observed molecular changes in relation with the pathophysiological adaptations in human AF and point out future directions of research.

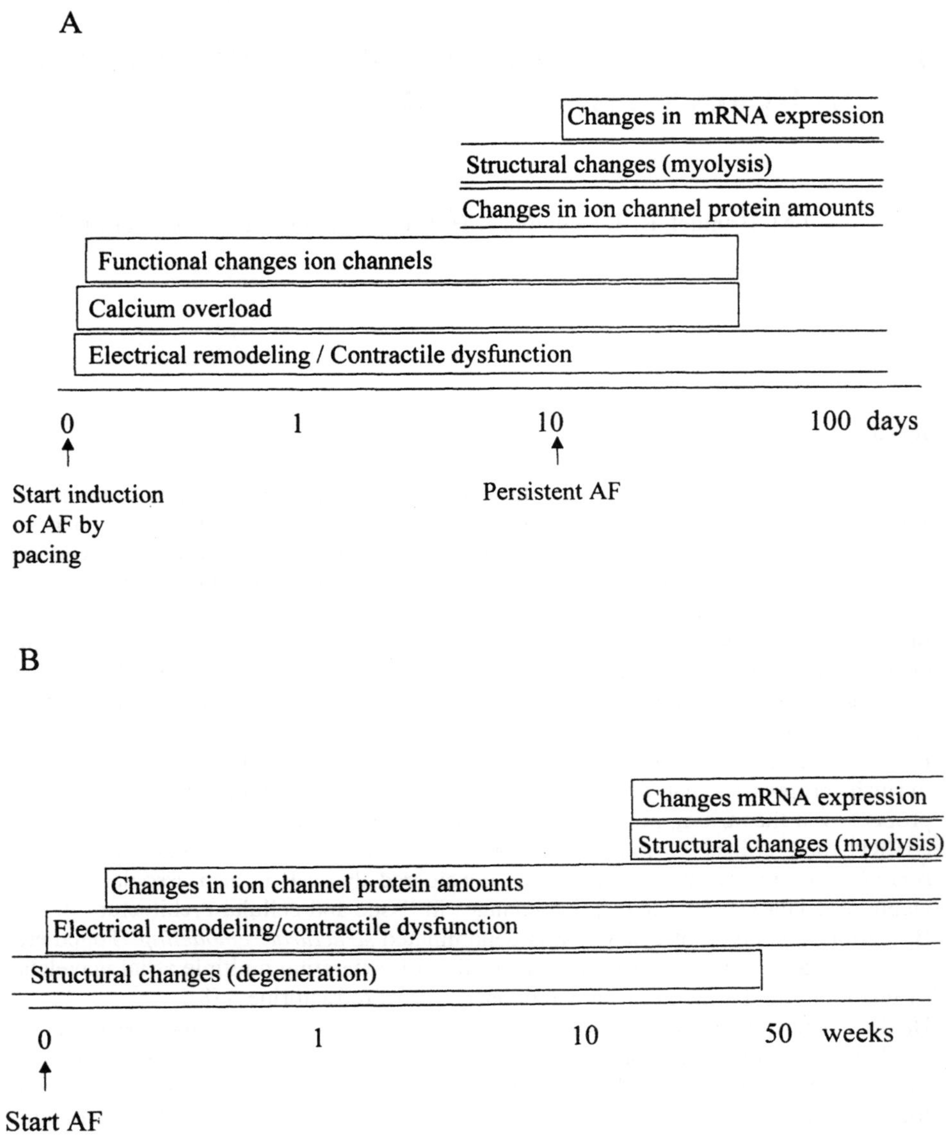

Figure 1 *A. Overview AF induced remodeling processes in experimental AF.*
B. Overview AF induced remodeling processes in human AF. Figures reproduced from thesis Brundel [36].

Protein remodeling

Ion-channels

It is well known that an abrupt increase in heart frequency, like in AF, causes an immediate (within one action potential) and then a gradual decrease (reaching steady state over several minutes) in action potential duration (APD) [9]. The decrease in APD reduces the AERP and shortens the wavelength for reentry, thus facilitating the occurrence and maintenance of reentrant arrhythmias like AF. This short-term adaptation mainly involves functional changes in the L-type Ca^{2+} channel following calcium overload [10-12]. In addition, longer periods of sustained atrial tachycardia induce further changes in electrical properties over a course of hours to days [4,13-15]. As opposed to short-term adaptations, long-term changes appear to be caused by regulation of ion channel density, which are related to modified protein expression
(figure 1A,B) [10,16].

In human AF, the relationship between changes in AERP and ion channel protein expression was investigated by studying the regulation of L-type Ca^{2+} channel and several K^+ channels. A study in patients with persistent and paroxysmal AF demonstrated a positive correlation between the ion-channel protein expression of L-type Ca^{2+} channel and the AERP, but also with the rate adaptation to AERP (figure 2A,B) [5]. Patients with reduced L-type Ca^{2+} channel protein levels were associated with short AERP and poor rate adaptation. L-type Ca^{2+} channel amounts and activity in persistent AF have also been investigated with experimental binding and electrophysiological studies. In figure 3 it is reflected that the L-type Ca^{2+} channel gene expression and function is reduced by 70% after six weeks of experimental AF. By comparison with human AF, these changes are postponed. A 70% reduction in current density was found after at least 18 months of AF. This result indicates that other (protective) adaptation mechanisms seem to play a role in human AF.

Since shortening of AERP can also be explained by increased K^+ conductance, a number of studies evaluated the expression levels of K^+ channels, notably Kv4.3, Kv1.5, HERG, minK/KvLQT1, Kir3.1/Kir3.4 and Kir6.2, in paroxysmal and persistent AF [5,17-19]. Remarkably, a reduction in mRNA and protein levels were found for several K^+ channels in patients with persistent AF, which is in plain contradiction with the shortening of AERP. The general interpretation of these results is that the electrophysiological changes in AF are primarily caused by the reduction in L-type Ca^{2+} channel [5,16]. Secondary to this process, the reduced expression of K^+ channels may serve to adapt the myocardial cell to the high rate and counteract the shortening of AERP [5].

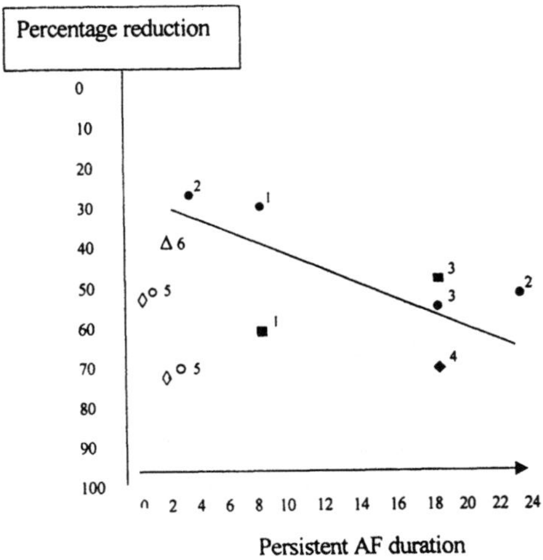

Figure 2.
Summary changes in L-type Ca² channel gene expression and function. (•), mRNA amount human study (1[5], 2[24], 3[25]) (o), mRNA amount experimental study (5[10,16]; (■), protein amount human study (1[5], 3[24]); (♦), amount current human study (4[75]); (□), amount current experimental study (5[10]); (Δ), experimental binding study (6[20]) Figure reproduced from [36].

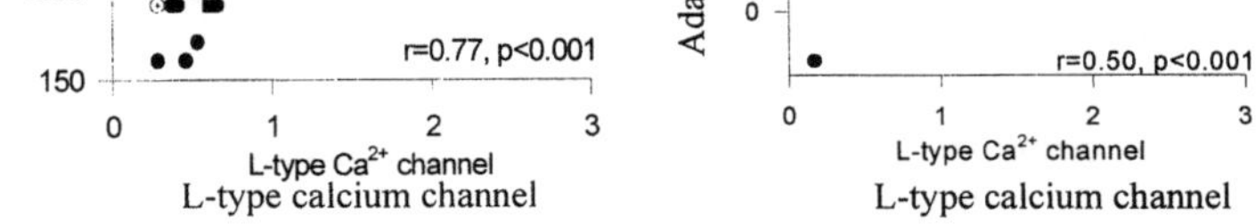

Figure 3 *A Correlation between the L-type Ca² channel protein expression and the AERP measured at basic cycle length of 500 ms in RAA and LAA. B. Correlation between the L-type Ca² channel protein expression and the rate adaptation coefficient. (o) represents control patients in sinus rhythm undergoing CABG, (⊙) patients with lone paroxysmal AF, (•) patients with lone persistent AF, (□) patients in sinus rhythm with underlying mitral valve disease, (□) patients with paroxysmal AF and mitral valve disease, (■) patients with persistent AF and mitral valve disease (reproduced from Ref. [5]).*

Of note is the observation from one electrophysiological study, in which increased I_{KACh} and I_{K1} currents were found in isolated human atrial cells of patients with persistent AF due to different underlying heart diseases [19]. The apparent inconsistency between the reduction in K^+ channel protein levels and increased current densities can only be explained by assuming changed single channel properties, such as an increase of mean open-time or channel conductance, or a change in voltage dependency. Such changes in single-channel properties of K^+ channels await experimental conformation.

While it is clear that protein expression of various ion-channels is modified during human AF, estimation of its time course depends largely on data from experimental studies. A study in dogs subjected to rapid atrial pacing for 7 and 42 days [10,16] demonstrated that high rate atrial stimulation does not change a variety of currents, including inward and delayed rectifier K^+ currents, T-type Ca^{2+} current, and Ca^{2+} dependent Cl^- current. Yet the approximately 70% reduction in L-type Ca^{2+} current (I_{CaL}) and transient outward K^+ current (I_{To}) observed after 6 weeks of rapid atrial pacing can be totally explained by reductions in channel protein levels [16,20]. In accordance, channel properties of I_{CaL} and I_{To}, like voltage, time and frequency dependence are unchanged. Despite the quantitatively similar reduction in I_{CaL} and I_{To}, it has become clear that reduction in I_{CaL} plays the central role in the changes in APD caused by the atrial tachycardia [10].

Finally, studies on I_{Na} in atrial myocytes of dogs subjected to rapid atrial pacing indicate a gradual reduction in current density over time [21]. This reduction parallels those in conduction velocity, suggesting that alterations in I_{Na} contribute to the conduction changes associated with sustained atrial tachycardia. Changes in the expression of the Na^+-channel in human AF have only been documented on the mRNA level so far. The mRNA level was found unaffected in persistent AF and 35% upregulated in paroxysmal lone AF [5]. Clearly, the determination of the role of the Na^+-channel in human AF awaits determination of changes on the protein level.

In recent years it has become clear that electrical remodeling is paralleled by a more or less general reduction in ion-channel protein levels as part of the adaptation mechanisms during AF. Since the reduction in ion-channel protein expression was supposed to occur due to the AF, this phenomenon was called ion-channel remodeling. It seems likely that ion-channel remodeling plays an important role in the susceptibility to AF after restoration of sinus rhythm, especially the occurrence of IRAF (immediate reinitiation of AF) and subacute relapses. It may be hypothesized that reduced ion-channel protein expression persist for some time after restoration of sinus rhythm, thereby creating a highly arrhythmogenic substrate due to changes in cellular electrophysiological and possibly also mechanical function. In the presence of such a substrate only a single atrial premature beat may induce a relapse of AF.

Proteins influencing calcium homeostasis

Animal experimental studies show that electrical remodeling and the accompanying cessation of contractile properties of the atrium occur with several days after the onset of AF (figure 1A) [4,22,23]. Both are largely attenuated by blocking the

L-type Ca^{2+} channel, indicating that changes in the calcium homeostasis play a pivotal role in the induction of atrial electrical remodeling and contractile dysfunction. This observation has prompted research on the molecular remodeling of proteins which influence calcium homeostasis [24-27]. One of the main findings of these human studies was the reduction in mRNA and protein expression of the L-type calcium channel (see above). Examination of additional gene products only revealed a decrease in sarcoplasmatic reticulum Ca^{2+} ATPase (SR Ca^{2+} ATPase) mRNA and protein levels, predominantly in patients with persistent AF. However, no changes in mRNA expression were found for phospholamban, sodium-calcium exchanger, ryanodine receptor 2, and calsequestrin [24-27]. Taken together these studies show that changes occur in the protein expression of proteins influencing the calcium homeostasis in persistent human AF, although they are so far limited to the L-type Ca^{2+} channel and SR Ca^{2+} ATPase. However, given their importance in contractile function, such changes probably represent a contributory factor for the atrial contractile dysfunction in AF.

Connexins

Since AF is promoted by slow conduction [13,14,28,29] a number of studies have focussed on gap-junction proteins, which play an important role in rapid and homogenous propagation of the wavefront in the heart [13,30-32]. Gap-junctions are clusters of connexin-channels, which span the closely apposed plasma membranes forming cell-to-cell pathways. Connexins are permeable to ions and small molecules up to 1 kDa in molecular mass, including second messengers such as inositol triphosphate, cyclic AMP and calcium. The initial data presented on changes in intercellular connexins seemed contradictory. One study in the dog showed that AF increases connexin43 expression, the most abundant connexin [13], while another study in goat suggested that connexin43 is unaltered, but that the distribution of its atrial isoform, connexin40, was altered [31]. In a recent goat study, the gap junctional changes in relation to stabilization of AF were studied [30]. While in sinus rhythm a homogeneous distribution of connexin40 was found, a marked heterogeneity was observed after 2 weeks of AF, by the time intracellular Ca^{2+} is deposited [33] and just before AF became sustained. Moreover, the connexin40 distribution pattern correlated with the occurrence of structural changes (myolysis) in atrial myocytes and might be involved in the pathogenesis of sustained AF [30]. Thus, these animal experimental results suggest that heterogeneity of connexin distribution, rather than simple up- or down-regulation, may play an important role in the susceptibility to AF.

Underlying molecular mechanisms

As described in the previous sections, AF induced changes in the expression of multiple proteins are the likely mechanisms which promote the susceptibility of the human atrium to the arrhythmia. Often, the level of protein expression is controlled by the rate of transcription of its mRNA from the genetic material. Consequently, changes in protein levels will be accompanied by similar changes in mRNA.

However, this appears not to be consistent in AF. A marked discrepancy between changes in mRNA and protein levels were found in various studies in experimental and human AF [5,17,34,35]. Whereas ion-channel protein levels were substantially decreased, the mRNA levels were essentially unaffected. A number of mechanisms may be implicated, such as activation of mRNA silencers, a general reduction in translation efficiency or activation of protein degradation (proteolysis). The latter seems the most likely mechanism, as very recently increased proteolytic activity due to activation of calpains was found in atrial tissue of patients with paroxysmal and persistent lone AF [36]. In human AF, the increased proteolytic activity was predominantly due to increased calpain I activity and expression. Furthermore, the increased calpain I protein was localized at the nucleus and intercalated discs of atrial myocytes. Finally, a good correlation between calpain activity and protein levels of L-type Ca^{2+} channel was found. In addition, calpains have been reported to mediate cell death in metabolically inhibited cultured rat cardiomyocytes and are involved in troponin proteolysis and cross-linking following cardiac stunning and calcium overload [37-39].

Although a causal relationship between calpain activation and regulation of protein expression controlling the remodeling in AF has not been proven yet, it seems an attractive hypothesis. Since calpains are activated by an increase in cytosolic calcium, this hypothesis would link calcium overload, which is perceived as one of the most important features of AF [33,40], to the molecular changes observed in AF. Calpains are calcium-activated neutral proteases [41], which cleave mainly cytoskeletal and membrane-associated proteins into 'limited fragments' without further degradation [42]. Calpains have been implicated in the degradation of the L-type Ca^{2+} channel [43-45], cytoskeletal proteins [46] but also proteins directly involved in excitation-contraction coupling [47]. At the nucleus [48] calpain can induce degenerative features leading to cell death, which is also observed in human AF [7,49]. Since extreme cellular stress, in combination with sustained elevated cytosolic calcium levels, as in experimental AF [33] often results in necrosis, calpain may play an important role in this condition [50]. Whether gap junctional remodeling may be caused by calpain induction is unknown, but it is known that at least proteasome activity underlies a connexin43 degradation [51].

On the other hand, the crucial role of cytosolic calcium increase, in particular calcium overload, in the changes induced by AF has been recognized widely. Atrial contractile dysfunction occurs both after short-term and after chronic AF [52-54] (Figure 1A,B). The explanation for the atrial dysfunction after short-term AF might be an increase in cytosolic Ca^{2+} due to the high rate of atrial activation. Fast successive action potentials inhibit a proper sarcoplasmic reticulum Ca^{2+} re-uptake, resulting in elevated cytosolic Ca^{2+}, possibly impairing the excitation-contraction coupling and contractile function [11,15,22,23,55-57]. Contractile dysfunction after chronic AF is most likely related to the cellular alterations in atrial myocytes [24,25] reflected by structural alterations, probably induced by proteolysis. In experimental AF, Ausma and coworkers showed sarcolemma-bound Ca^{2+} and Ca^{2+} deposits in mitochondria to increase markedly up to 2 weeks and tends to regress towards normal levels at 4 and 8 weeks of AF (figure 1A) [33]. Unfortunately, the methods used limit the visualization of Ca^{2+} load at subcellular sites like the sarcoplasmic reticulum. Additional data, however, show that atrial tachycardia causes an immediate increase in cytoplasmic Ca^{2+} concentration, which results in impaired Ca^{2+} release and cellular contractile dysfunction after the cessation of tachycardia [40]. In addition to the L-type Ca^{2+} channel, T-type Ca^{2+} channels may be implicated in atrial tachycardia-induced electrical remodeling, because the T-type Ca^{2+} channel blocker mibefradil limits both the ERP changes and AF promotion caused by one week of rapid atrial pacing. Also in this case calcium overload would be prevented by blocking a calcium channel [58].

Apart from calcium overload, atrial ischemia has been put forward as an underlying mechanism explaining the cellular ultrastructural changes caused by sustained human AF because of its resemblance with the hibernating myocardium [6]. Whether ischemia occurs in AF is still debatable, but a reduced atrial blood flow in dogs with rapid pacing induced AF was found and may result in atrial ischemia [59]. A potential role for atrial ischemia is consistent with the protective effect by blockade of the Na^+/H^+ exchanger in short term tachycardia-induced atrial electrophysiological remodeling [60] and contractile dysfunction [61]. In this model, ischemia would give rise to a decrease in intracellular pH, which leads to an exchange of intracellular hydrogen ions for extracellular Na^+ ions. Such an increase in intracellular Na^+ results in 'reverse-mode' functioning of the Na^+/Ca^{2+} exchanger and therefore an influx of Ca^{2+} ions [62]. Alternatively, inhibition of the Na^+/H^+ exchanger may alter cellular ionic homeostasis and combat calcium overload induced by ischemia.

Thus, several lines of evidence point to a central role for intracellular calcium overload in AF induced remodeling. Recent results reveal the potential role of calcium sensitive processes that lead to changes in protein expression and structural changes [36]. Because elevated levels of intracellular Ca^{2+} are known to activate proteolysis, this could result in increased breakdown of myofilaments [39,63] and ion-channel proteins [5]. In turn this could be responsible for decreased contractility as well as for the remodeling processes that convey the vulnerability to AF.

Possible clinical relevance

The probability of successful chemical or electrical cardioversion is dependent on the duration of AF. The clinically observed diminished efficacy of cardioversion after long term AF cannot only by explained by the occurrence of electrical remodeling. The ion-channel protein remodeling and structural remodeling probably also affect the electrophysiological function of the atrial myocardium. In patients with persistent AF, there is a correlation between the duration of AF and the time needed to recover atrial contractile function after cardioversion [53,64]. The increase in calpain activity which could lead to structural remodeling of the atrial myocytes might offer an explanation for the delay in recovery of contractile function in the atria after conversion to sinus rhythm as seen in patients with persistent AF.

Interference with the calpain pathway by pharmacological means may represent a novel therapeutic strategy to decrease protein degradation and thereby reduce the vulnerability to AF. Calpain inhibitors are as therapeutic agents already used in nerve and muscle degeneration [65], but their potential benefit in heart diseases is not studied yet. After restoration of normal sinus rhythm it may require the cardiomyocytes a certain period to rebuild a normal amount of sarcomeres, if that is still possible [66]. Unfortunately, data on the recovery of ion-channel protein expression after cessation of AF are still lacking. However, a few reports describe reversal of electrical remodeling in human AF after cardioversion, in contrast to the absence of reversal of structural changes [67,68]. Since AF induces remodeling in the atria it is essential to restore sinus rhythm as soon as possible, thereby preventing the continuation of the atrial structural, ion-channel protein and electrical remodeling. On the other hand, identification of the mechanisms underlying protein remodeling in AF may enable successful "priming" of a patient prior to cardioversion.

Future research

Research on AF focussed mainly on describing electrical remodeling, contractile dysfunction and structural changes. Functional experiments were designed to block electrical remodeling and contractile dysfunction. Possibly studying the intracellular pathways which are activated by electrical stress lead to the adaptive and maladaptive, degenerative structural changes [69,69,70]. One of the possible important roles might be for the calcium overload induced proteolytic activity. This postulation is supported by recent views on the dominant role of calpain in forms of apoptosis and necrosis through the degradation of cytoskeletal and contractile proteins, and activation of caspases [71-74]. These pathways may open new avenues for understanding the pathophysiology of AF and perhaps additional pharmacological interventions.

References

1. Murgatroyd FD and Camm AJ. Atrial arrhythmias. The Lancet 1993;341:1371-22.
2. Godtfredsen J. Etiology, course and prognosis. A follow-up study of 1212 cases. Copenhagen: University of Copenhagen. Thesis 1975.
3. Van Gelder IC, Crijns HJGM, Tieleman RG, De Kam PJ, Gosseling ATM, Verheugt FWA, and Lie KI. Value and limitation of electrical cardioversion in patients with chronic atrial fibrillation - importance of arrhythmia risk factors and oral anticoagulation. Arch Intern Med 1996;156:2585-92.
4. Wijffels MC, Kirchhof CJ, Dorland R, and Allessie MA (1995) Atrial fibrillation begets atrial fibrillation. A study in awake chronically instrumented goats. Circulation 1995;92:1954-68.
5. Brundel BJJM, Van Gelder IC, Henning RH, Tuinenburg AE, Tieleman RG, Wietses M, Grandjean JG, Van Gilst WH, and Crijns HJGM. Ion channel remodeling is related to intra-operative atrial refractory periods in patients with paroxysmal and persistent atrial fibrillation. Circulation 2001;103:684-90.
6. Ausma J, Wijffels M, Thone F, Wouters L, Allessie M, and Borgers M. Structural changes of atrial myocardium due to sustained atrial fibrillation in the goat. Circulation 1997;96:3157-63.
7. Thijssen VLJL, Ausma J, Liu GS, Allessie M, Eys GJJM, and Borgers M. Structural changes of atrial myocardium during chronic atrial fibrillation. Cardiovasc Pathol 2000;9:17-28.
8. Moe GK and Abildskov JA. Experimental and laboratory reports. Atrial fibrillation as a self-sustained arrhythmia independent of focal discharge. Am Heart J 1959;58:59-70.
9. Boyett MR and Jewell BR. Analysis of the effects of changes in rate and rhythm upon electrical activity in the heart. Prog Biophys Molec Biol 1980;36:903-23.
10. Yue L, Feng J, Gaspo R, Li GR, Wang Z, and Nattel S. Ionic remodeling underlying action potential changes in a canine model of atrial fibrillation. Circ Res 1997;81:512-25.
11. Daoud EG, Knight BP, Weiss R, Bahu M, Paladino W, Goyal R, Man KC, Strickberger SA, and Morady F. Effect of verapamil and procainamide on atrial fibrillation-induced electrical remodeling in humans. Circulation 1997;96:1542-50.
12. Yu WC, Chen SA, Lee SH, Tai CT, Feng AN, Kuo BIT, Ding YA, and Chang MS. Tachycardia-induced change of atrial refractory period in humans. Rate dependency and effects of antiarrhythmic drugs. Circulation 1998;97:2331-7.
13. Elvan A, Wylie K, and Zipes DP. Pacing-induced chronic atrial fibrillation impaires sinus node function in dogs: electrophysiological remodeling. Circulation 1996;94:2953-60.
14. Gaspo R, Bosch RF, Talajic M, and Nattel S. Functional mechanisms underlying tachycardia-induced sustained atrial fibrillation in a chronic dog model. Circulation 1997;96:4027-35.
15. Nattel S. Atrial electrophysiological remodeling caused by rapid atrial activation: underlying mechanisms and clinical relevance to atrial fibrillation. Cardiovasc Res 1999;42:298-308.
16. Yue L, Melnyk P, Gaspo, Wang Z, and Nattel S. Molecular mehanisms underlying ionic remodeling in a dog model of atrial fibrillation. Circ Res 1999;84:776-84.
17. Brundel BJJM, Van Gelder IC, Henning RH, Tuinenburg AE, Tieleman RG, Wietses M, Grandjean JG, Wilde AAM, Van Gilst WH, and Crijns HJGM. Alterations in potassium channel gene expression in atria of patients with persistent and paroxysmal atrial fibrillation. J Am Coll Cardiol 2001;37:926-32.
18. Van Wagoner DR, Pond AL, McCarthy PM, Trimmer JS, and Nerbonne JM. Outward K^+ current densities and Kv1.5 expression are reduced in chronic human atrial fibrillation. Circ Res 1997;80:1-10.

19. Bosch RF, Zeng X., Grammer JB, Popovic K, Lewis C, and Kühlkamp V. Ionic mechanisms of electrical remodeling in human atrial fibrillation. Cardiovasc Res 1999;44:121-31.
20. Gaspo R, Sun H, Fareh S, Levi M, Yue L, Allen BG, Hebert TE, and Nattel S. Dihydropyridine and beta adrenergic receptor binding in dogs with tachycardia-induced atrial fibrillation. Cardiovasc Res 1999;42:434-42.
21. Gaspo R, Bosch RF, Bou-Abboud E, and Nattel S. Tachycardia-induced changes in Na+ current in a chronic dog model of atrial fibrillation. Circ Res 1997;81:1045-52.
22. Tieleman RG, De Langen CDJ, Van Gelder IC, De Kam PJ, Grandjean JG, Bel KJ, Wijffels MC, Allessie MA, and Crijns HJGM. Verapamil reduces tachycardia-induced electrical remodeling of the atria. Circulation 1997;95:1945-53.
23. Leistad E, Aksnes G, Verburg E, and Christensen G. Atrial contractile dysfunction after short-term atrial fibrillation is reduced by verapamil but increased by BAY K8644. Circulation 1996;93:1747-54.
24. Brundel BJJM, Van Gelder IC, Henning RH, Tuinenburg AE, Deelman LE, Tieleman RG, Grandjean JG, Van Gilst WH, and Crijns HJGM. Gene expression of proteins influencing the calcium homeostasis in patients with persistent and paroxysmal atrial fibrillation. Cardiovasc Res 1999;42:443-54.
25. Van Gelder IC, Brundel BJJM, Henning RH, Tuinenburg AE, Tieleman RG, Deelman LE, Grandjean JG, De Kam PJ, Van Gilst WH, and Crijns HJGM. Alterations in gene expression of proteins involved in the calcium handling in patients with atrial fibrillation. J Cardiovasc Electrophysiol 1999;10:552-60.
26. Lai LP, Su MJ, Lin J, Tsai CH, Chen YS, Huang SK, Tseng YZ, and Lien WP. Down-regulation of L-type calcium channel and sarcoplasmic reticular Ca^{2+} -ATPase mRNA in human atrial fibrillation without significant change in the mRNA of ryanodine receptor, calsequestrin and phospholamban: an insight into the mechanism of atrial electrical remodeling. J Am Coll Cardiol 1999;33:1231-7.
27. Ohkusa T, Ueyama T, Yamada J, Yano, Fujumura Y, Esato K, and Matsuzaki M. Alterations in cardiac sarcoplasmic reticulum Ca^{2+} regulatory proteins in the atrial tissue of patients with chronic atrial fibrillation. J Am Coll Cardiol 1999;34:255-63.
28. Morillo CA, Klein GJ, Jones D, and Guiraudom CM. Chronic rapid atrial pacing. Structural, functional, and electrophysiological characteristics of a new model of sustained atrial fibrillation. Circulation 1995;91:1588-95.
29. Franz MR, Karasik PL, Li C, Moubarak J, and Chavez M. Electrical remodeling of the human atrium: similar effects in patients with chronic atrial fibrillation and atrial flutter. J Am Coll Cardiol 1997;30:1785-92.
30. Van der Velden HMW, Ausma J, Rook MB, Hellemons AJ, Van Veen TA, Allessie M, and Jongsma HJ. Gap junctional remodeling in relation to stabilization of atrial fibrillation in the goat. Cardiovasc Res 2000;46:476-86.
31. Van der Velden HMW, Van Kempen MJA, Wijffels MCEF, Van Zijverden M, Groenewegen AW, Allessie MA, and Jongsma HJ. Altered pattern of connexin40 distribution in persistent atrial fibrillation in the goat. J Cardiovasc Electrophysiol 1998;9:596-607.
32. Britz-Cunningham SH, Shah MM, Zuppan CW, and Fletcher W. Mutations of the connexin43 gap-junction gne in patients with heart malformations and defects of laterality. N Engl J Med 1995;332:1323-9.
33. Ausma J, Dispersyn GD, Duimel H, Thone F, Ver Donck L, Allessie M, and Borgers M. Changes in ultrastructural calcium distribution in goat atria during atrial fibrillation. J Mol Cell Cardiology 2000;32:355-64.
34. Goette A., Arndt M, Röcken C, Spiess A, Staack T, Geller C, Huth C, Ansorge S, Klein H, and Lendeckel U. Regulation of angiotensin II receptor subtypes during atrial fibrillation in humans. Circulation 2000;101:2678-781.

35. Brundel BJJM, Van Gelder IC, Tuinenburg AE, Wietses M, Van Veldhuisen DJ, Van Gilst WH, Crijns HJGM, and Henning RH. Endothelin system in human persistent and paroxysmal atrial fibrillation. J Cardiovasc Electrophysiol 2000;12:737-42.
36. Brundel BJJM. Molecular adaptations in human atrial fibrillation: mechanisms of protein remodeling. Thesis University Groningen, The Netherlands 2000.
37. Atsma DE, Bastiaanse EM, Jerzewski A, Van Der Valk LJ, and Van Der Laarse A. Role of calcium-activated neutral protease (calpain) in cell death in cultured neonatal rat cardiomyocytes during metabolic inhibition. Circ Res 1995;76:1071-8.
38. Gorza L, Menabo R, Vitadello M, Bergamini CM, and Di Lisa F. Cardiomyocyte troponin T immunoreactivity is modified by cross-linking resulting from intracellular calcium overload. Circulation 1996;93:1896-904.
39. Gao WD, Atar D, Liu Y, Perez NG, Murphy AM, and Marban E. Role of troponin I proteolysis in the pathogenesis of stunned myocardium. Circ Res 1997;80:393-9.
40. Sun H, Chartier D, Leblanc N, and Nattel S. Intracellular calcium changes and tachycardia-induced contractile dysfunction in canine atrial myocytes. Cardiovasc Res 2001;49:751-61.
41. Matsumura Y, Saeki E, Otsu K, Morita T, Takeda H, Kuzuya T, Hori M, and Kusuoka H. Intracellular calcium level required for calpain activation in a single myocardial cell. J Mol Cell Cardiol 2001;33:1133-42.
42. Suzuki K, Imajoh S, Emori Y, Kawasaki, Minami Y, and Ohno S. Calcium-activated neutral protease and its endogenous inhibitor. Activation at the cell membrane and biological function. FEBS Letters 1987;220:271-7.
43. Hao LY, Kameyama A, Kuroki S, Takano J, Takano E, Maki M, and Kameyama M. Calpastatin domain L is involved in the regulation of L-type Ca^{2+} channels in guinea pig cardiac myocytes. Biochem Biophys Res Commun 2000;279:756-61.
44. De Jongh KS, Colvin AA, Wang KK, and Catterall WA. Differential proteolysis of the full-length form of the L-type calcium channel alpha 1 subunit by calpain. J Neurochem 1994;63:1558-64.
45. Belles B, Hescheler J, Trautwein W, Blomgren K, and Karlsson JO. A possible physiological role of the Ca-dependent protease calpain and its inhibitor calpastatin on the Ca current in guinea pig myocytes. Pflugers Arch 1998;412:554-6.
46. Papp Z, Van Der Velden J, and Stienen G. Calpain-I induced alterations in the cytoskeletal structure and impaired mechanical properties of single myocytes of rat heart. Cardiovasc Res 2000;45:981-93.
47. Laflamme MA and Becker PL. G(s) and adenylylcyclase in transverse tubules of heart: implications for cAMP-dependent signaling. Am.J.Physiol.1999;277:H1841-8.
48. Lane RD, Allan DM, and Mellgren R. A comparison of the intracellular distribution of mu-calpain, m-calpein and calpastatin in proliferating human A431 cells. Exp Cell Res 1992;203:5-16.
49. Aime-Sempe C, Folliguet T, Rucker-Martin, Krajewska M, Krajewska S, Heimburger M, Aubier M, Mercardier JJ, Reed JC, and Hatem SN. Myocardial cell death in fibrillating and dilated human right atria. J Am Coll Cardiol 1999;34:1577-86.
50. Majno G and Joris I. Apoptosis, oncosis and necrosis. An overview of cell death. Am J Pathol 1995;146:3-15.
51. Laing JG, Tadros PN, Saffitz J, and Beyer EC. Proteolysis of connexin43-containing gap junctions in normal and heat-stressed cardiac myocytes. Cardiovasc Res 1998;38:711-8.
52. Manning WJ, Leeman DE, Gotch P, and Come PC. Pulsed evaluation of atrial mechanical function after electrical cardioversion of atrial fibrillation. J Am Coll Cardiol 1989;13:617-23.
53. Manning WJ, Silverman DI, Katz SE, Riley MF, Come PC, Doherty RM, Munson JT, and Douglas PS. Impaired left atrial mechanical function after cardioversion: relation to the duration of atrial fibrillation. J Am Coll Cardiol 1994;23:1535-40.

54. Daoud EG, Marcovitz P, Knight B, Goyal R, Ching Man K, Strickberger A, Armstrong WF, and Morady F. Short-term effect of atrial fibrillation on atrial contractile function in humans. Circulation 1999;99:3024-7.
55. Tieleman RG, Van Gelder IC, Crijns HJ, De Kam PJ, Van Den Berg MP, Haaksma J, Van Der Woude HJ, and Allessie MA. Early recurrences of atrial fibrillation after electrical cardioversion: A result of fibrillation induced electrical remodeling of the atria? J Am Coll Cardiol 1998;31:167-73.
56. Goette A., Honeycutt C, and Langberg JJ. Electrical remodeling in atrial fibrillation. Time course and mechanisms. Circulation 1996;94:2968-74.
57. Nattel S, Li D, and Yue L. Basic mechanisms of atrial fibrillation, very new insights into very old ideas. Annu rev Physiol 2000;62:51-77.
58. Fareh S, Bénardeau A, Thibault, and Nattel S. The T-type Ca^{2+} channel blocker mibefradil prevents the development of a substrate for atrial fibrillation by tachycardia-induced atrial remodeling in dogs. Circulation 1999;100:2191-7.
59. Jayachandran JV, Winkle W, Sih HJ, Zipes DP, and Olgin JE. Chronic atrial fibrillation from rapid atrial pacing is associated with reduced atrial blood flow: a positron emission tomography study. Circulation 1998;98: I-209.
60. Jayachandran JV, Zipes DP, Weksler J, and Olgin JE. Role of the Na^{+}/H^{+} exchanger in short-term atrial electrophysiological remodeling. Circulation 2000;101:1861-6.
61. Altemose GT, Zipes DP, Weksler J, Miller JM, and Olgin JE. Inhibition of the Na(+)/H(+) exchanger delays the development of rapid pacing-induced atrial contractile dysfunction. Circulation 2001;103:762-8.
62. Levi AJ, Dalton GR, Hancox JC, Mitcheson JS, Issberner J, Bates JA, Evans SJ, Howarth FC, Hobai IA, and Jones JV. Role of intracellular sodium overload in the genesis of cardiac arrhythmias. J Cardiovasc Electrophysiol 1997;8:700-21.
63. Gorza L, Menabo R, Di Lisa F, and Vitadello M. Troponin T cross-linking in human apoptotic cardiomyocytes. Am J Pathol 1997;150:2087-97.
64. Van Gelder IC, Crijns HJGM, Van Gilst WH, Verwer R, and Lie KI. Prediction of uneventful cardioversion and maintenance of sinus rhythm from direct-current electrical cardioversion of chronic atrial fibrillation and flutter. Am.J.Cardiol. 1991;86:41-6.
65. Stracher A. Calpain inhibitors as therapeutic agents in nerve and muscle degeneration. Ann N Y Acad Sci 1999;28:52-9.
66. Ausma J, Duimel H, Borgers M, and Allessie M. Recovery of structural remodeling after cardioversion of chronic atrial fibrillation. Circulation 2000;102:II-153.
67. Hobbs WJC, Fynn S, Todd D, Wolfson P, Galloway M, and Garrett CJ. Reversal of atrial electrical remodeling after cardioversion of persistent AF in humans. Circulation 2000;101:1145-51.
68. Pandozi C, Bianconi L, Villani M, Gentilucci G, Castro A, Altamura G, Jesi AP, Lamberti F, Ammirati F, and Santini M. Electrophysiological characteristics of the human atria after cardioversion of persistent atrial fibrillation. Circulation 1998;98:2860-5.
69. Lendeckel U, Arndt M, Wrenger S, Neppie K, Huth C, Ansorge S, Klein HU, Goette A. Expression and activity of ectopeptidases in fibrillating human atria. J Mol Cell Cardiol. 2001;33:1273-81.
70. Goette A., Staack T, Röcken C, Arndt M, Geller JC, Huth C, Ansorge S, Klein HU, and Lendeckel U. Increased expression of extracellular signal-regulated kinase and angiotensin-converting enzyme in human atria during atrial fibrillation. J Am Coll Cardiol 2000;35:1669-77.
71. Beere HM and Green DR. Stress management - heat shock protein-70 and the regulation of apoptosis. Trends Cell Biol 2001;11:6-10.
72. Wolf BB, Goldstein JC, Stennicke HR, Beere HM, Amarante-Mendes GP, Salvesen GS, and Green DR. Calpain functions in a caspase independent manner to promote apoptosis-like events during platelet activation. Blood 1999;94:1683-92.

73. Leist M and Jäättelä M. Four deaths and a funeral: from caspases to alternative mechanisms. Nat Rev Mol Cell Biol 2001;2:589-98.
74. Lee MS, Kwon YT, Li M, Peng J, Friedlander RM, and Tsai LH. Neurotoxicity induces cleavage of p35 to p25 by calpain. Nature 2000;405:360-4.
75. Van Wagoner DR, Pond AL, Lamorgese M, Rossie SS, McCarthy PM, and Nerbonne JM Atrial L-Type Ca^{2+} Currents and Human Atrial Fibrillation. Circ Res 1999;85:428

18. G-PROTEIN β3-SUBUNIT POLYMORPHISM AND ATRIAL FIBRILLATION

U. Ravens, E. Wettwer, T. Christ and D. Dobrev

Introduction

In search for genetic causes of hypertension, Siffert and coworkers [1] have recently identified a single nucleotide polymorphism in a gene encoding for pertussis-toxin (PTX) sensitive G_i-proteins. The authors had noted that a subgroup of patients with "essential hypertension" displayed in their blood platelets high activity of the Na^+/H^+ exchanger that could be traced back to enhanced signal transduction in this ubiquitously expressed ion transport system. Signal transduction between receptors and effectors was also enhanced in immortalized cell lines from these individuals and PTX-sensitive G_i-proteins were likely to be involved [2,3]. The altered activity of G_i-proteins was related to a thymine-for-cytosine polymorphism (C825T-polymorphism) in exon 10 of the gene encoding for the β3-subunit of the heterotrimeric complexes (*GNB3*) [3]. The 825T-allele is associated with alternative splicing of exon 9 resulting in an additional Gβ3-s-subunit which is shorter by 41 amino acids but more active than the full-size Gβ3-subunit [3]. Clinically, the 825T-allele was first associated with essential hypertension [3,4,5,6,7,8] though recently several other clinical conditions have also been associated (see table 1). This is not surprising if one considers the central role of heterotrimeric G-proteins in signal transduction. So far, the enhanced activity of the splice variant Gβ3-s has been demonstrated mostly in blood cells, i.e. in immortalized lymphoblasts or freshly isolated neutrophils [26,27]. Since the 825T-allele is present in all cells, we have studied the putative association between 825T-allele status and the activity of signal transduction in human atrial myocytes.

Signal transduction between cardiac M_2-receptors and acetylcholine-activated K^+ channels.

In human atrial myocytes, stimulation of muscarinic M_2-receptors opens acetylcholine-activated K^+ channels ($I_{K,ACh}$). Drug binding to the receptor will activate PTX-sensitive G_i-proteins that release the βγ-dimers to activate $I_{K,ACh}$ by direct binding to this channel [28,29]. There are numerous isoforms of the α-, β- and γ-subunits. At least 6 isoforms for β- and 12 for γ-subunits have been described so far [30], but the isoform composition of the βγ-subunit involved in activation of

Table 1 Pathophysiological parameters and pharmacodynamic responses associated with the 825T-allele of the *GNB3* gene

Pathophysiological condition	**Reference**
Essential hypertension	Siffert et al., 1998 [3] Benjafield et al., 1998 [4] Hegele et al., 1998 [5] Schunkert et al., 1998 [6] Beige et al., 1999 [7] Dong et al., 1999 [8] Hengstenberg et al., 2001 [9]
Diseases associated with hypertension	
Left ventricular filling (↓)	Jacobi et al., 1999 [10]
Left ventricular hypertrophy	Poch et al., 2000 [11]
Ischemic stroke	Morrison et al., 2001 [12]
Myocardial infarction	Naber et al., 2000 [13]
Hypertension associated with GNB3 825T-allele status links to	
Obesity	Hegele et al., 1999 [14] Siffert et al, 1999a, b [15,16]
Low plasma renin level	Schunkert et al., 1998 [6]
Enhanced renal perfusion	Zeltner et al., 2001 [17]
Low birth weight	Hocher et al., 2000 [18]
Weight retention post partum	Gutersohn et al., 2000 [19]
Pharmacodynamic responses	
Positive association	
Thiazide diuretics (+)	Turner et al., 2001 [20]
Antidepressants (+)	Zill et al., 2000 [21]
α_2-AR mediated coronary constriction (+)	Baumgart et al., 1999 [22]
Adrenalin-induced thrombocyte aggregation (+)	Naber et al., 2000 [23]
Lack of association	
Hand vein response to M-receptor stimulation	Grossmann et al., 2001 [24]
Hand vein response to α_2-AR stimulation	Schäfers et al, 2001 [25]

human atrial $I_{K,ACh}$ is not known. $I_{K,ACh}$ can be measured as an inwardly rectifying K^+ current in myocytes isolated from a small biopsy of atrial tissue obtained from patients undergoing open heart surgery. Its voltage dependence is very similar to the inward rectifier current I_{K1}. Both channels are of clinical significance since they substantially contribute to the atrial resting potential and determine the shape of cardiac action potentials during the final phase of repolarization. In addition, $I_{K,ACh}$ is the major effector of vagal stimulation in atrial myocytes [29].
Since βγ-dimers are involved in activation of atrial $I_{K,ACh}$ channels when muscarinic receptors are stimulated with appropriate agonists, we hypothesized that this signal transduction cascade could serve as a sensitive parameter to monitor G_i-protein activity as the phenotype associated with the presence of the 825T-allele leading to expression of the alternatively spliced, more active Gβ3-subunit.

Association between atrial K^+ current activity and GNB3 gene status

In our recently published study, right atrial myocytes were obtained from 70 patients undergoing cardiac surgery with cardio-pulmonary bypass and were investigated with conventional single electrode voltage clamp techniques [31]. Patients with chronic atrial fibrillation (AF) were excluded from this study and are investigated separately (see below). The GNB3 gene status of the patients was assessed by PCR of the relevant region of the genomic DNA, followed by restriction enzyme digestion, gel electrophoresis, and ethidium bromide staining. Thirty patients (43%) had CC-, 31 (44%) had CT-, and 9 (13%) had TT-genotype. Of the various patient-related parameters recorded there were no significant differences with exception of a history of myocardial infarction (20 out of 30 homozygous C825-allele carriers, 14 out of 31 heterozygous and 2 out of 9 homozygous 825T-allele carriers, [31]). Cell size was estimated from membrane capacitance (C_M) and did not differ in myocytes from the the three genotype groups. Currents were measured in high extracellular K^+ solution (20 mM) during a depolarizing ramp pulse from –100 to –10 mV and verified as inward rectifier at the end of each experiment by complete block with Ba^{2+} (1 mM). Typical current tracings are depicted in figure 1a (see legend for pulse protocol). When corrected for cell size, I_{K1} amplitude analysed at –100 mV was significantly larger in myocytes from patients with TT-genotype than in the other two groups. $I_{K,ACh}$ was activated by 2 μM carbachol and was measured as the difference between total current in the presence of carbachol and I_{K1}. Mean density of $I_{K,ACh}$ was significantly smaller in myocytes from patients with TT-genotype, whereas the total current amplitudes (I_{K1} + $I_{K,ACh}$) were not significantly different in the three genotype groups (figure 1b). These results suggest a significant association between *GNB3* gene status and amplitude of I_{K1} and $I_{K,ACh}$. The individual values for current amplitudes, however, varied enormously both between myocytes derived from the same biopsy as well as between patients within one genotype group. This is illustrated in figure 2, where we have plotted density of

$I_{K,ACh}$ against density of I_{K1} for a total of 80, 117, and 27 myocytes from patients with CC-, CT-, and TT-genotype, respectively, and indicated the median of all current densities for $I_{K,ACh}$ as horizontal and for I_{K1} as vertical lines, respectively (figure 2). This graphic lay-out allows to allocate any individual myocytes within four fields to attribute "small" current, i.e. below the median value, or as "large" current , i.e. above the median value. The typical current phenotype for myocytes from C825-allele carriers, i.e. "small" I_{K1} and "large" $I_{K,ACh}$, was found in 31.3% cells out of 80 from CC-genotype and in 26.5% out of 117 from CT-genotype, but only in 11.1% out of 27 myocytes from TT-genotype. Conversely, the typical phenotype of "large" I_{K1} and "small" $I_{K,ACh}$ in TT-genotype was observed in 51.9 % of myocytes from TT-genotype but only in 23.8% and 22.2% in cells from CC- and CT-genotype, respectively. These differences were highly significant and confirm that despite the large variability between current amplitudes of myocytes from the same biopsy and from different patients, there is a significant association between *GNB3* gene status and activity of inward rectifier K^+ currents in human atrial myocytes. We considered measurement of $I_{K,ACh}$ as a suitable system for studying activity of G_i-protein-mediated signal transduction events and expected an increased $I_{K,ACh}$ as indicator of enhanced signaling. To our surprise, this current was smaller in TT-genotype, whereas that of background I_{K1} was larger than in C825-allele carriers, though total current (I_{K1} and $I_{K,ACh}$) was similar in all groups. This unexpected finding could be explained, if we suppose that the seemingly large I_{K1} in TT-genotype actually consists of unchanged I_{K1} and an additional fraction of spontaneously active $I_{K,ACh}$, since based on whole cell steady-state current-voltage relations only, the two currents are impossible to distinguish. This would also explain the smaller increase of $I_{K,ACh}$ in response to carbachol because only channels not yet activated spontaneously can still respond to receptor stimulation. The additional Gβ3-s in 825T-allele carriers will form Gβ3-sγ dimers that are not as tightly associated with Gα as other βγ-dimers [32] creating a larger cellular pool of free βγ-dimers which could spontaneously activate $I_{K,ACh}$ in the absence of an agonist and thus explain the increased I_{K1} we observed. Receptor stimulation could still increase this pool of βγ-dimers in all genotypes, since G proteins typically are in excess of the receptor molecules to which they couple [33,34] and in human myocardium $G\alpha_i$ is by a factor 1000-10,000 more abundant than M_2-receptors [35]. The maximum response, however, may be limited by channel density number which could explain the similar levels of maximum currents in cells of patients with different genotypes.

Influence of chronic AF on K^+ current activity in myocytes from different Gβ3-genotypes

Current density measured in atrial tissue from patients is also affected by the underlying heart disease. Thus, we have to consider not only genetic factors but also the disease state of the patients. Right atrial myocytes of patients with chronic AF exhibit also enhanced I_{K1} density [36], although a recent study did not confirm these results [37]. Discrepant findings were published also for $I_{K,ACh}$. While stimulation of

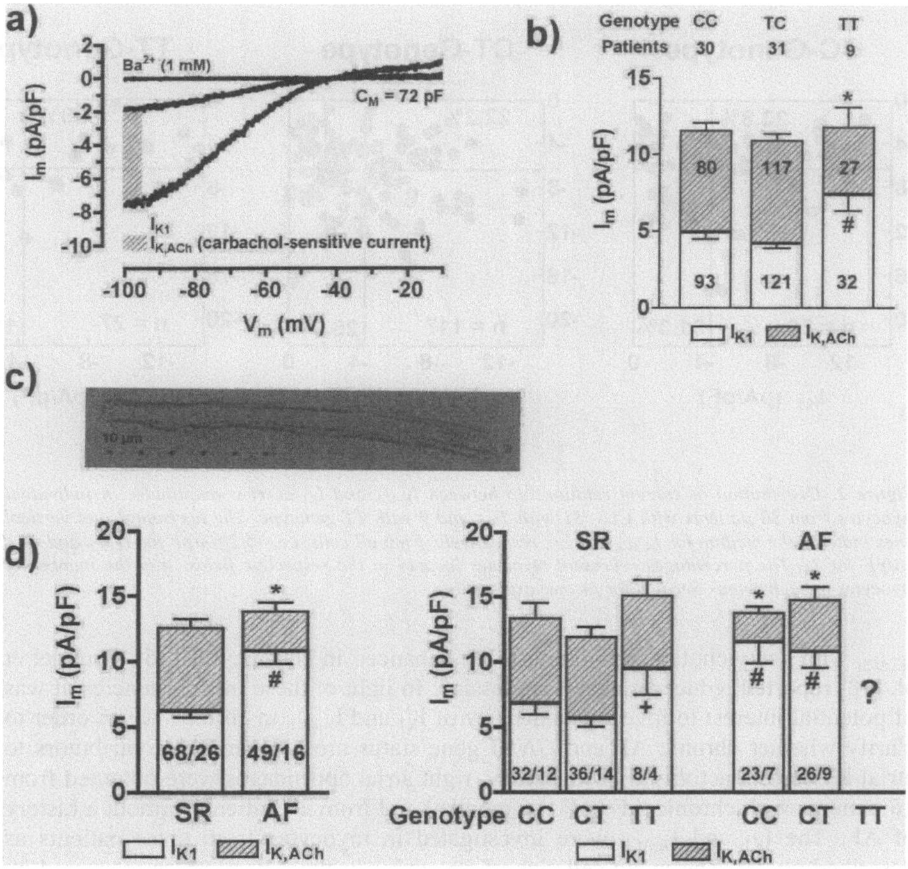

***Figure 1.** GNB3 gene status and inward rectifier K^+ currents in human atrial myocytes from patients in sinus rhythm (SR) and atrial fibrillation (AF). **a.** Top, representative atrial myocyte, calibration as indicated; bottom, representative current tracings of I_{K1} (middle tracing, open column), of current in the presence of 2 µM carbachol (lower tracing), and of leak current in the presence of 1 mM Ba^{2+} (upper tracing). $I_{K,ACh}$ (hatched column) is defined as the difference between total current and I_{K1}. Clamp protocol: holding potential –80 mV, 50-ms pulse to –100 mV, depolarizing ramp of 800 ms duration from –100 mV to –10 mV, 50-ms pulse to –50 mV, return to holding potential; 1 clamp cycle every 2 s; room temperature (22 – 24°C). C_M, membrane capacitance in pF of this myocyte. **b.** Inward rectifier currents I_{K1} (open columns) and $I_{K,ACh}$ (activated by 2 µM carbachol; closed columns) in patients with different Gβ3-genotypes. Numbers in the columns indicate number of cells investigated. **c.** I_{K1} and $I_{K,ACh}$ in atrial myocytes from patients in sinus rhythm (SR) and chronic atrial fibrillation (AF). Numbers in columns indicate number of myocytes / number of patients. Please note, that all patients with TT-genotype were excluded in the SR group. **d.** Current densities of I_{K1} and $I_{K,ACh}$ in relation to genotype. All current amplitudes are expressed in pA/pF. *P < 0.05 for $I_{K,ACh}$, #,+P < 0.05 for I_{K1} when compared to SR or to the other genotypes (one-way ANOVA with post hoc multiple comparison test or Student's t test), respectively.*

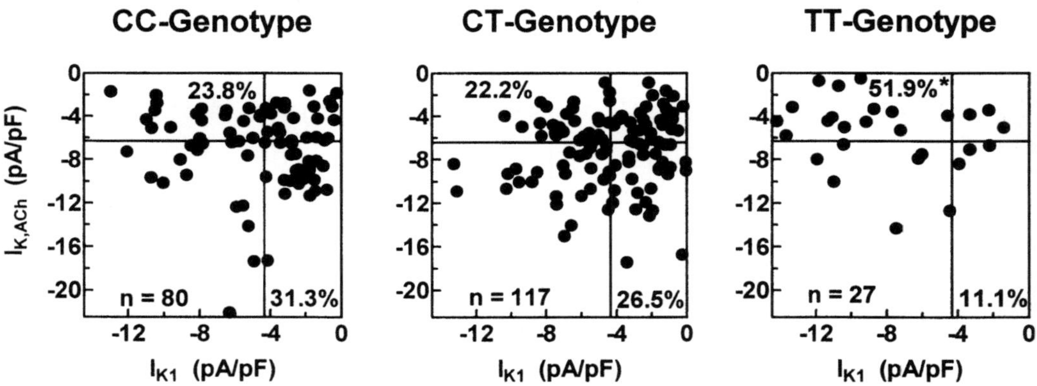

Figure 2. *Distribution of current relationship between $I_{K,ACh}$ and I_{K1} current amplitudes in individual myocytes from 30 patients with CC-, 31 with TC- and 9 with TT-genotype. The horizontal and vertical lines indicate the median for $I_{K,ACh}$ and I_{K1}, respectively, from all cells, i.e. –6.2 pA/pF for $I_{K,ACh}$ and –4.4 pA/pF for I_{K1}. The percentages represent myocytes located in the respective fields. n is the number of myocytes in each group. See text for further explanation.*

$I_{K,ACh}$ with acetylcholine was found to be enhanced in chronic AF [36], Brundel et al. [38] reported reduced channel expression. In light of these inconsistencies, it was of potential interest to investigate activity of I_{K1} and $I_{K,ACh}$ in chronic AF in order to clarify whether chronic AF and *GNB3* gene status are independent contributors to atrial K^+ current activity. For this study, right atrial appendages were obtained from 16 patients with chronic AF (AF > 6 months) and from 26 patients without a history of AF. The I_{K1} and $I_{K,ACh}$ were investigated in myocytes from these patients as described above. The mean cell size of myocytes in the AF group as measured by C_M was larger than that in sinus rhythm ($C_M = 139 \pm 11$ pF, n = 49, for AF *vs.* 84 ± 4 pF, n = 68, for SR; $P < 0.001$). At -100 mV, I_{K1} density was significantly larger in AF than in sinus rhythm (SR, figure 1c). These differences were observed also at other potentials, confirming previous results [36]. In contrast, $I_{K,ACh}$ was found to be reduced by about 50% in AF patients, which corresponded well with the recent report on $I_{K,ACh}$ reduction by 34% of mRNA and by 62% of protein [38]. Interestingly, the AF-related decrease in $I_{K,ACh}$ was observed only in large-sized myocytes (> 100 pF), whereas the increase of I_{K1} was independent of cell size (data not shown). The discrepancy between low (present results) and increased $I_{K,ACh}$ density [36] can be explained by differences between the myocytes investigated: we detected smaller $I_{K,ACh}$ density only in myocytes with membrane capacitance > 100 pF, while Bosch et al. [36] chose to investigate cells of < 80 pF in which he found no change. Finally, these authors defined $I_{K,ACh}$ as *total* current in the presence of acetylcholine, i.e. I_{K1} plus $I_{K,ACh}$. If $I_{K,ACh}$ is calculated as the difference between total current and I_{K1} – as in our study – there was also no change in amplitude of $I_{K,ACh}$ in these small cells from patients with chronic AF [36].

To investigate the impact of *GNB3* gene status on density of I_{K1} and $I_{K,ACh}$ during chronic AF, all patients were genotyped. In the SR group, the distribution was CC-genotype in 12 patients, CT-genotype in 14 patients, and TT-genotype in 4 patients (figure 1d). Like in our first study (see above), I_{K1} was significantly larger in TT-allele carriers than in CC- and CT-allele carriers, although current amplitude of $I_{K,ACh}$ was not different, probably due to the small number of cells. In patients with chronic AF, the distribution was 7 CC- and 9 CT-genotypes, but not a single AF patient was found to be of TT-genotype. Therefore, all data from homozygous 825T-allele carriers in the SR group had to be excluded from analysis for comparison with the AF group (figure 1c). Although myocytes from patients in chronic AF exhibited similar absolute and relative amplitudes of I_{K1} and $I_{K,ACh}$ as myocytes from of homozygous 825T-allele carriers in SR, our results suggest that the AF-related changes are independent of *GNB3* gene status, because the AF-related differences are still present in the respective genotypes. However, possible interaction effects between the 825T-allele and chronic AF on activity of atrial K^+ currents cannot be excluded with certainty due to lack of homozygous 825T-allele carriers in the AF group.

Conclusion

We have demonstrated a significant association between 825T-allele and activity of inward rectifier K^+ currents in human atrial myocytes. Amplitude of I_{K1} was enhanced, whereas activation of $I_{K,ACh}$ produced by muscarinic receptor stimulation was smaller in homozygous 825T-allele carriers than in the other two groups. Similar current phenotypes were observed in cells from patients with chronic AF. Our results indicate that the AF-related changes in K^+ current density are independent of G-protein β3-subunit C825T polymorphism and may be a consequence of or a contributory factor to chronic AF.

Acknowlegements

This work has been supported by grants from DFG Ra 222/9-1 (U.R.) and MeDDrive 2000 (D.D.).

References

1. Siffert W, Rosskopf D, Siffert G, Busch S, Moritz A, Erberl R, Shamrma AM, Ritz E, Wichmann H-E, Jakobs K-H, Horsthemke B. Association of a human G-protein β3 subunit variant with hypertension. Nature Genetics 1998;18:45-8.
2. Rosskopf D, Schröder K-J, Siffert W. Role of sodium-hydrogen exchange in the proliferation of immortalized lymphoblasts from patients with essential hypertension and normotensive subjects. Cardiovasc Res 1995;29:254-59.
3. Siffert W, Düsing R. Sodium-proton exchange and primary hypertension. Hypertension 1995;26:649-55.
4. Benjafield AV, Jeyasingam CL, Nyholt DR, Griffiths LR, Morris BJ. G-Protein β3 subunit (*GNB3*) variant in causation of essential hypertension. Hypertension 1998;32:1094-7.
5. Hegele RA, Harris SB, Hanley AJG, Cao H, Zinman B. G protein β3 subunit gene variant and blood pressure variation in Canadian Oji-Cree. Hypertension 1998;32:688-92.
6. Schunkert H, Hense HW, Döring A, Riegger GAJ, Siffert W. Association between a polymorphism in the G protein β3 subunit gene and lower renin and elevated diastolic blood pressure levels. Hypertension 1998;32:510-3.
7. Beige J, Hohenbleicher H, Distler A, Sharma AM. G-Protein β3 subunit C825T variant and ambulatory blood pressure in essential hypertension. Hypertension 1999;33:1049-51.
8. Dong Y, Zhu H, Sagnella GA, Carter ND, Cook DG, Cappuccio FP. Association between the C825T polymorphism of the G protein β3-subunit gene and hypertension in blacks. Hypertension 1999;34:1193-6.
9. Hengstenberg Ch, Schunkert H, Mayer B, Döring A., Loewel H, Hense HW, Fischer M, Riegger GAJ, Holmer SR. Association between a polymorphism in the G protein β3 subunit gene (*GNB3*) with arterial hypertension but not myocardial infarction. Cardiovasc Res 2001;49:820-7.
10. Jacobi J, Hilgers KF, Schlaich MP, Siffert W, Schmieder RE. 825T allele of the G-protein β3 subunit gene (*GNB3*) is associated with impaired left ventricular diastolic filling in essential hypertension. J Hypertens 1999;17:1457-62.
11. Poch E, Gonzalez D, Gomez-Angelats E, Enjuto M, Pare JC, Rivera F, de La Sierra A. G-Protein β3 subunit gene variant and left ventricular hypertrophy in essentual hypertension. Hypertension 2000;35:214-8.
12. Morrison AC, Doris PA, Folsom AR, Nieto FJ, Boerwinkle E. G-protein β3 subunit and α-adducin polymorphisms and risk of subclinical and clinical stroke. Stroke 2001;32:822-9.
13. Naber CK, Husing J, Wolfhard U, Erbel R, Siffert W. Interaction of the ACE D allele and the GNB3 825T allele in myocardial infarction. Hypertension 2000;36:986-9.
14. Hegele RA, Anderson C, Young TK, Connelly PW. G-protein β3 subunit gene splice variant and body fat distribution in Nunavut Inuit. Genome Res 1999;9:972-7.
15. Siffert W, Forster P, Jockel KH, Mvere DA, Brinkmann B, Naber C, Crookes R, Du P Heyns A, Epplen JT, Fridey J, Freedman BI, Muller N, Stolke D, Sharma AM, Al Moutaery K, Grosse-Wilde H, Buerbaum B, Ehrlich T, Ahmad HR, Horsthemke B, Du Toit ED, Tiilikainen A, Ge J, Wang Y, Rosskopf D, et al. Worldwide ethnic distribution of the G protein beta3 subunit 825T allele and its association with obesity in Caucasian, Chinese, and Black African individuals. J Am Soc Nephrol 1999a;10:1921-30.
16. Siffert W, Naber C, Walla M, Ritz E. G protein beta3 subunit 825T allele and its potential association with obesity in hypertensive individuals. J Hypertens 1999b;17:1095-8.
17. Zeltner R, Delles C, Schneider M, Siffert W, Schmieder RE. G-Protein β_3 subunit gene (*GNB3*) 825T allele is associated with enhanced renal perfusion in early hypertension. Hypertension 2001;37:882-6.
18. Hocher B, Slowinski T, Stolze T, Pleschka A, Neumayer HH, Halle H. Association of maternal G protein β3 subunit 825T allele with low birth weight. Lancet 2000;355;1241-2.

19. Gutersohn A, Naber C, Müller N, Erbel R, Siffert W. G Protein β3 subunit 825 TT genotype and post-pregnancy weight retention. Lancet 2000;355:1240-1.
20. Turner ST, Schwartz GL, Chapman AB, Boerwinkle E. C825T Polymorphism of the G protein β3-subunit and antihypertensive response to a thiazide diuretic. Hypertension 2001;37:739-43.
21. Zill P, Baghai TC, Zwanzger P, Schule C, Minov C, Riedel M, Neumeier K, Rupprecht R, Bondy B. Evidence for an association between a G-protein β3-gene variant with depression and response to antidepressant treatment. Neuroreport 2000;11:1893-7.
22. Baumgart D, Naber C, Haude M, Oldenburg O, Erbel R, Heusch G, Siffert W. G Protein β3 825T allele and enhanced coronary vasoconstriction on α_2-adrenoceptor activation. Circ Res 1999;85: 965-9.
23. Naber C, Hermann BL, Vietzke D, Altmann C, Haude M, Mann K, Rosskopf D, Siffert W. Enhanced epinephrine-induced platelet aggregation in individuals carrying the G protein β3 subunit 825T allele. FEBS Lett 2000;484:199-201.
24. Grossmann M, Dobrev D, Siffert W, Kirch W. Heterogeneity of hand vein responses to acetylcholine is not associated with polymorphisms in the G-protein β_3-subunit (C825T) and endothelial nitric oxide synthase (G894T) genes but with serum LDL cholesterol. Pharmacogenetics 2001;11:307-16.
25. Schafers RF, Nurnberger J, Rutz A, Siffert W, Wenzel RR, Mitchell A, Philipp T, Michel MC. Haemodynamic characterization of young normotensive men carrying the 825T-allele of the G-protein β3 subunit. Pharmacogenetics 2001;11:461-70.
26. Siffert W, Rosskopf D, Moritz A, Wieland T, Kaldenberg-Stasch S, Ketler N, Hartung K, Beckmann S, Jakobs K-H. Enhanced G protein activation in immortalized lymphoblasts from patients with essential hypertension. J Clin Invest 1995;96:759-66.
27. Virchow S, Ansorge N, Rubben H, Siffert G, Siffert W. Enhanced fMLP-stimulated chemotaxis in human neutrophils from individuals carrying the G protein β3 subunit 825 T-allele. FEBS Lett 1998;436:155-8.
28. Krapivinsky G, Krapivinsky L, Wickman K, Clapham DE. Gβγ binds directly to the G protein-gated K^+ channel, $I_{K,ACh}$. J Biol Chem 1995;270:29059-62.
29. Yamada M, Inanobe A, Kurachi Y. G protein regulation of potassium ion channels. Pharmacol Rev 1998;50:723-57.
30. Farfel Z, Bourne HR, Iiri T. The expanding spectrum of G protein diseases. N Engl J Med 1999;340:1012-20.
31. Dobrev D, Wettwer E, Himmel HM, Kortner A, Kuhlisch E, Schüler S, Siffert W, Ravens U. G-protein β_3-subunit 825T-allele is associated with enhanced human atrial inward rectifier potassium currents. Circulation 2000;102: 692-7.
32. Iiri T, Farfel Z, Bourne HR. G-protein diseases furnish a model for the turn-on switch. Nature 1998;394:35-8.
33. Neubig RR, Gantzos RD, Brasier RS. Agonist and antagonist binding to α_2-adrenergic receptors in purified membranes from human platelets. Implications of receptor-inhibitory nucleotide-binding protein stoichiometry. Mol Pharmacol 1985;28:475-86.
34. Milligan G. The stoichiometry of expression of protein components of the stimulatory adenylyl cyclase cascade and the regulation of information transfer. Cell Signal 1996;8:87-96.
35. Böhm M, Ungerer M, Erdmann E. Beta adrenoceptors and m-cholinoceptors in myocardium of hearts with coronary artery disease or idiopathic dilated cardiomyopathy removed at cardiac transplantation. Am J Cardiol 1990;66:880-2.
36. Bosch RF, Zeng XR, Grammer JB, Popovic K., Mewis C, Kühlkamp V.. Ionic mechanisms of electrical remodeling in human atrial fibrillation. Cardiovasc Res 1999;44:121-31.

37. Van Wagoner DR, Pond AL, McCarthy PM, Trimmer JS, Nerbonne JM. Outward K^+ current densities and $K_V1.5$ expression are reduced in chronic human atrial fibrillation. Circ Res 1997;80:772-81.
38. Brundel BJJM, Van Gelder IC, Henning RH, Tieleman RG, Tuinenburg AE, Wietses M, Grandjean JG, Van Gilst WH, Crijns JGM. Ion channel remodeling is related to intraoperative atrial effective refractory periods in patients with paroxysmal and persistent atrial fibrillation. Circulation 2001;103:684-90.

ESF workshop Maastricht 2001: Session 6

Towards cellular transplantion

19. HUMAN STEM CELL GENE THERAPY

A.A.F. de Vries

Introduction

Human embryonic and adult stem cells hold great promise for the treatment of a wide variety of acquired and inherited human diseases including congestive heart failure, neurodegenerative diseases (e.g. Parkinson's disease), liver cirrhosis, cerebral infarction, muscular dystrophy and autoimmune diseases (e.g. type I diabetes) [1-4]. Apart from their use to restore or replace tissues that have been damaged, worn out or lost as a result of disease, senescence or trauma, human stem cells (hSCs) may be excellent tools to study the genes and proteins controlling embryonic development, cell differentiation and oncogenic transformation. Other possible applications of hSCs and their derivatives include the testing of novel drugs, the screening of potential toxins and the production of therapeutic proteins. To exploit the full potential of hSCs it will be important to develop safe and efficient methods for the stable transduction of these cells.

Genetic modification of stem cells

In vitro propagation of human embryonic stem cells (hESCs) typically yields a heterogeneous pool of cells consisting of various cell types even in the presence of specific growth factors [5,6]. The heterogeneous character of hESC derivatives grown in culture has restricted their use in transplantation studies. To solve this problem, hESCs could be transduced with genes encoding tissue-specific transcription factors that direct the development of a single cell type. Furthermore, marker genes under the control of tissue-specific promoters could be introduced in hESCs to allow the selection of lineage-restricted cells from hESC cultures [7]. Another potential hurdle to the clinical application of hESC-derived cells is the possible destruction of the transplanted material by the recipient's immune system. To avoid immune-mediated graft rejection, the hESCs could be provided with a proper set of major histocompatibility complex genes or with genes encoding immunosuppressive proteins. Alternatively, human adult stem cells (hASCs) derived from the patient itself could be used to generate cell or tissue transplants. However, hASCs are rare, may have a smaller developmental repertoire than hESCs and seem to be much more limited in the number of times they can divide compared with hESCs [8]. Moreover, for some types of hASCs (including human hematopoietic stem cells [hHSCs]) researchers are having difficulties to establish culture conditions that allow proliferation without inducing cell death or differentiation. Accordingly, the number of specialized cells than can be produced from hASCs *in vitro* is in many cases insufficient to effectively treat a patient. To expand the proliferation capacity of hASCs they could be transduced with the telomerase gene under the control of a heterogeneous promoter and/or with genes encoding mitogens. In addition, hASCs could be transduced with regulable genes encoding proteins that block apoptosis or suppress differentiation. Although the use of autologous cell transplants greatly reduces the risk of immune-mediated graft rejection or the introduction into the patient of harmful infectious agents, this approach is unsuitable for the treatment of inherited diseases since the donor cells will have a mutant genotype. Successful application of autologous hASCs to cure hereditary diseases will hence depend on the ability to correct the genetic defect in these cells by either gene complementation or gene correction. It may also be important to provide hESCs and hASCs with (regulable) suicide genes as a fail-safe system in case the cell transplants give rise to tumors, cause inflammations or have any other adverse effects in the patient. Other genetic modifications to expand the therapeutic possibilities of hSCs include the transduction of h(H)SCs with genes coding for chimeric T cell receptors to generate T lymphocytes that specifically target tumor cells or infected cells. Alternatively, genes encoding anti-inflammatory proteins to suppress autoimmune reactions could be used. In addition, hSCs could be stably transduced with resistance genes to prevent the infection of their progeny with intracellular pathogens. Finally, the function of specialized cell types could be enhanced by the genetic manipulation of metabolic activities or signal transduction pathways. An example of the latter approach is the introduction in hSCs of genes affecting Ca^{2+}-homeostasis to increase the contractile parameters of cardiomyocytes derived from these cells.

Transduction of stem cells

Since the uptake of unconjugated nucleic acid molecules by cells is a very inefficient process, special gene transfer systems have been developed to facilitate transduction of target cells *in vitro* and *in vivo*. The requirements to which these so-called gene therapy vectors have to conform depend on their specific application. Of special importance is the question whether genetic intervention should occur on a temporary or permanent basis. For applications that only require transient expression of the therapeutic gene, vector integration in the host cell genome is unnecessary or even undesirable. In this case, episomal vectors (i.e. vectors that are present as extrachromosomal genetic elements) are usually preferred since they do not carry the risk of insertional mutagenesis of the target cell genome. However, when the transgene needs to be continuously expressed, nonintegrative and replication-defective vectors are of limited use since in dividing tissues they are diluted out upon cell division. This is particularly relevant for stem cell gene therapy in which the proliferation capacity of stem cells is exploited for the transfer of a novel genetic trait to a large number of descendants. Although different applications call for different gene delivery vehicles, in general terms they must be safe (i.e. nonimmunogenic, noninflammatory, nonmutagenic, nontoxic and replication-deficient) and should be producible in high concentrations, at low costs, and in a stable form. Moreover, gene transfer vectors should efficiently and selectively enter the target cells and should express the transgene in the right amounts, at the proper time and in the right place. An additional requirement of gene transfer systems for *in vivo* applications is that they should not be inactivated by host factors including pre-existing antibodies. This is less important for stem cell gene therapy which usually takes place *ex vivo*.

Viral vectors

Recombinant adenoviruses (Ads) are attractive gene transfer vehicles because they can be produced in large quantities and at high titers. Furthermore, Ad capsid-mediated DNA import into the nucleus is extremely efficient and does not require cell division which may be particularly relevant for the transduction of hHSCs as these cells are mostly quiescent [9,10]. In addition, the life cycle of Ads is very well characterized and they can be relatively easily targeted to specific cell types by mutagenesis of their capsid genes [11]. In an attempt to increase the utility of Ad vectors for therapeutic gene transfer, we decided to exploit the difference in cell tropism between human Ad serotypes which is largely determined by the amino acid sequence of their fiber proteins. We thus constructed a series of first-generation E1-deleted Ad serotype 5 (Ad5) vectors carrying the fiber shafts and knobs of other human Ad subtypes and expressing the *gfp* gene from the cytomegalovirus (CMV) immediate-early promotor. These chimeric Ad5 vectors were subsequently used for the transduction of hHSCs and *gfp* expression was monitored using multi-parameter flow cytometric analysis. In this study, the HSCs were phenotypically characterized as expressing high levels of the CD34 antigen and lacking the early lineage markers (ELM) CD33, CD38 and CD71. This population composes less than 1% of the $CD34^+$ fraction, harbors cells engrafting in NOD/SCID mice, and can be maintained *ex vivo* for several days when cultured on a layer of allogenic human BMSCs (Knaän-Shanzer *et al*., in preparation). Analysis of transduced $CD34^+$ cells after a 5-day culture period showed that vectors carrying Ad subgroup B fiber sequences transduced 70-100% of the most primitive $CD34^{++}ELM^-$ cells at a multiplicity of infection of 1000 virus particles per cell. Importantly, these cells retained the capacity to repopulate the hematopoietic system of sublethally irradiated NOD/SCID mice indicating that neither the transduction procedure nor the *gfp* expression were toxic for these cells. With the nonchimeric Ad5 vector, the maximum transduction level of $CD34^{++}ELM^-$ cells did not exceed 40% whereas Ad5 vectors with shafts and knobs of Ad subgroup D transduced hardly any of these cells [12].

Adenoassociated adenovirus-adenovirus hybrid

Unfortunately, the earlier generations of Ad vectors have a limited cloning capacity, may be toxic for certain target cell types and are highly inflammatory *in vivo* due to the expression of viral genes from the vector genome. To prevent these problems, new Ad vectors have recently been developed that completely lack viral genes and hence can accommodate up to 37 kb of foreign DNA. The production of these so-called high-capacity Ad vectors depends on recombinant helper Ads to provide *in trans* all necessary functions involved in genome replication and packaging. Animal experiments have shown that the *in vivo* performance of helper-dependent Ad vectors is superior to that of previous generations of Ad vectors. However, whereas vectors derived from adeno-associated virus (AAV) and retroviruses can integrate into the host cell genome, Ad vectors normally remain episomal which makes them unsuitable for long-term (trans)gene expression in dividing cell populations. We therefore developed an entirely new vector system in which the highly efficient gene transfer and large cloning capacity of helper-dependent Ad vectors are combined with the integrating potential of AAV. This was achieved by coupling the replication mechanism of AAV to the packaging process of Ad [13]. Our system is based on enlarged rAAV genomes provided with Ad packaging elements. These AAV/Ad chimeric genomes were rescued, replicated and packaged into Ad capsids in the presence of AAV *rep* gene products and an E1-deleted Ad vector. The latter vector provided *in trans* both AAV helper activities and Ad structural proteins. This new AAV/Ad hybrid gene transfer system overcomes the limited packaging capacity of AAV particles and exploits the efficient Ad capsid-mediated nuclear gene delivery into both dividing and resting cells. The large cloning capacity of our AAV/Ad hybrid vectors will be particularly useful for stem cell gene therapy as it allows transduction of genomic DNA fragments for the proper temporal and spatial regulation of gene expression and to counteract transgene inactivation, a phenomenon commonly observed in stem cells [14]. In stringent *in vitro* models based on transduction of proliferating cells we have shown that AAV/Ad hybrid vectors are superior to Ad vectors in establishing prolonged transgene expression and can be used to stably transduce DNA fragments of at least 27 kb. Furthermore, since AAV/Ad hybrid vectors consist of recombinant AAV replicative intermediates packaged into Ad capsids, their tropism is determined by the specific helper Ad vector used for their production. Accordingly, by selecting one of the previously described E1-deleted Ad vectors carrying Ad subgroup B specific fibers as a helper, AAV/Ad hybrid vector particles can be produced that efficiently enter human HSCs. The combined features of the latter vector will hopefully allow us to treat human patients with genetically corrected autologous hHSCs in the near future. Moreover, our vector technology can be easily adjusted to the genetic modification of other hSCs including hESCs by selection of a suitable helper virus from the library of *gfp*-positive fiber-modified E1-deleted Ad5 vectors.

The bottleneck of the AAV/Ad hybrid vector system has been the production of high-titered vector stocks. Until recently, production of AAV/Ad hybrid vectors relied on transfection of Ad E1-complementing producer cells (PER.C6) with (i) an AAV/Ad hybrid shuttle construct, (ii) an AAV *rep* expression construct and, since the Ad E2A protein was found to be limiting for efficient replication of AAV/Ad hybrid vector genomes, (iii) a construct that overexpresses the *E2A* gene. After transfection of these three plasmids, the producer cells were infected with an E1-deleted Ad helper vector. AAV/Ad particles were harvested when the producer cells displayed a complete cytopathic effect. This production system appeared to be very difficult to scale up due to (i) the presence of large amounts of helper Ad vector in the AAV/Ad hybrid vector preparations and (ii) the dependence on plasmid transfections.

To tackle the helper virus problem, we recently developed an improved production system for the AAV/Ad hybrid vectors based on the Cre/loxP recombinase system of bacteriophage P1 (Gonçalves *et al.*, in preparation). In the new production system, the packaging signal of the helper Ad vectors is flanked by loxP sites and the production of the AAV/Ad hybrid vectors is carried out in producer cells (PER.tTA.Cre) that constitutively express the gene for the Cre recombinase. This enzyme catalyzes site-specific recombination between loxP sites which results in the excision of the packaging signal from the helper virus genome. As a consequence, the helper virus DNA can no longer be incorporated in virus particles but retains its ability to replicate and express *in trans* high levels of the Ad functions necessary for the production of AAV/Ad hybrid vectors. In preliminary experiments, the new production system has proven to cause a 5000-fold reduction of the helper virus content of AAV/Ad hybrid vector preparations.

At this moment, we are addressing the transfection-associated limitations by incorporation of regulable expression cassettes of the AAV *rep* and Ad *E2a* gene into the Ad helper vector genome. This will allow us to propagate the AAV/Ad hybrid vectors by simple and scalable infection procedures and to use them to efficiently and stably transduce various stem cell populations for scientific or clinical purposes.

References

1. Asahara, T., C. Kalka, and J. M. Isner. Stem cell therapy and gene transfer for regeneration. Gene Ther. 2000;7:451-7.
2. Fuchs, E., and J. A. Segre. Stem cells: a new lease on life. Cell 2000;100:143-55.
3. McKay, R. Stem cells – hype and hope. Nature (London) 2000;406:361-4.
4. Smith, A. G. Embryo-derived stem cells: of mice and men. Annu. Rev. Cell Dev. Biol. 2001;17:435-62.
5. Schamblott, M. J., J. Axelman, J. W. Littlefield, P. D. Blumenthal, G. R. Huggins, Y. Cui, L. Cheng, and J. D. Gearhart.. Human embryonic germ cell derivatives express a broad range of developmentally distinct markers and proliferate extensively *in vitro*. Proc. Natl. Acad. Sci. U.S.A. 2001;98:113-8.
6. Schuldiner, M., O. Yanuka, J. Itskovitz-Eldor, D. A. Melton, and N. Benvenisty. Effects of eight growth factors on the differentiation of cells derived from human embryonic stem cells. Proc. Natl. Acad. Sci. U.S.A. 2000;97:11307-12.
7. Odorico, J. S., D. S. Kaufman, and J. A. Thomson. Multilineage differentiation from human embryonic stem cell lines. Stem Cells 2001;19:193-204.
8. Weissman, I. L., D. J. Anderson, and F. Gage. Stem and progenitor cells: origins, phenotypes, lineage commitments, and transdifferentiations. Annu. Rev. Cell Dev. Biol. 2001:17:387-403.
9. Knaän-Shanzer, S., D. Valerio, and V. W. van Beusechem. Cell cycle state, response to hemopoietic growth factors and retroviral vector-mediated transduction of human hemopoietic stem cells. Gene Ther. 1996;3:323-333.
10. Knaän-Shanzer, S., S. F. F. Verlinden, V. W. van Beusechem, D. W. van Bekkum, and D. Valerio. Intrinsic potential of phenotypically defined human hemopoietic stem cells to self-renew in short-term in vitro cultures. Exp. Hematol. 1999;27:1440-1450.
11. Krasnykh, V. N., J. T. Douglas, and V. W. van Beusechem. 2000. Genetic targeting of adenoviral vectors. Mol. Ther. 2000;1:391-405.
12. Knaän-Shanzer, S., I. van der Velde, M. J. E. Havenga, A. A. C. Lemckert, A. A. F. de Vries, and D. Valerio. Highly efficient targeted transduction of undifferentiated human hematopoietic cells by adenoviral vectors displaying fiber knobs of subgroup B. Hum. Gene Ther. 2001:12:1989-2005.
13. Gonçalves, M. A .F. V., M. G. Pau, A. A. F. de Vries, and D. Valerio. Generation of a high-capacity hybrid vector: packaging of recombinant adenoassociated virus replicative intermediates in adenovirus capsids overcomes the limited cloning capacity of adenoassociated virus vectors. Virology 2001;288:236-6.
14. Cherry, S. R., D. Biniszkiewicz, L. van Parijs, D. Baltimore, and R. Jaenisch. Retroviral expression in embryonic stem cells and hematopoietic stem cells. Mol. Cell. Biol. 2000;20:7419-26.

20. TOWARDS HUMAN EMBRYONIC STEM CELL DERIVED CARDIOMYOCYTES

C.Mummery, D.Ward, C.E.van den Brink, S.D.Bird, P.A.Doevendans, D.J. Lips, T.Opthof, A.Brutel de la Riviere, L.Tertoolen, M.van der Heyden, M.Pera.

Introduction

Stem cells are the primitive cells present in all organisms that can divide and give rise to more stem cells or switch to become more specialized cells, such as those of the blood, brain, heart or liver. Embryonic stem (ES) cells are present very early in development and can give rise to all tissues of the adult individual. There are therefore termed "pluripotent". Stem cells are also present in some adult tissues capable of regeneration and repair after injury although in very small numbers. Probably exceptionally, the adult heart contains no stem cells and is incapable of repair.

Stem cells from various sources could be used to replace specialized cells lost or malfunctioning as a result of disease. Transplantation of cardiomyocytes could thus be a treatment for cardiac failure. However, much basic research will be required before cell transplantation therapy is transformed from a media item into serious clinical practice. For example, for most potential clinical applications, it is entirely unclear what the most suitable source of stem cells for the derivation of transplantable cells might be. Human embryonic stem cells are one option but it will be necessary to generate large numbers of functionally mature cells either *in vitro* or after transplantation *in vivo,* before, for example, cardiac function can be restored. The first step is to control differentiation. Studies in animals and cell culture have suggested that mechanisms that control stem cell differentiation to particular cell types may be the same whether the stem cell is of adult or embryonic origin.

Molecular pathways that lead to specification and terminal differentiation of cardiomyocytes from embryonic mesoderm during development are not entirely clear but data largely derived from amphibian and chick, and more recently from mouse, have suggested that signals emanating from endoderm in the early embryo may be involved in both processes. Tissue recombination experiments have shown that, for example, in chick, primitive hypoblast (endoderm) induces cardiogenesis in posterior epiblast (ectoderm), while in *Xenopus,* endoderm and Spemann organizer

synergistically induce cardiogenesis in embryonic mesoderm undergoing erythropoiesis [1]. Induction of cardiac myogenesis in avian pregastrula epiblast: the role of the hypoblast and activin [2]. In addition, the zebrafish mutant casanova, which lacks endoderm, also exhibits severe heart anomalies. We have previously isolated various cell lines by cloning from a culture of P19 EC cells treated as aggregates in suspension (embryoid bodies) with retinoic acid before replating [3]. One of these cell lines, END-2, has characteristics of visceral endoderm (VE), expressing alpha fetoprotein and the cytoskeletal protein ENDO-A. When undifferentiated P19 EC cells are plated on to a confluent monolayer of these END-2 cells, they aggregate spontaneously and within a week differentiate to cultures containing areas of beating muscle at high frequency [4]. This was not observed in co-cultures with other differentiated clonal cell lines from P19 EC that did not express characteristics of VE. Differentiation to beating muscle however, was observed when aggregates of P19 EC cells were grown in conditioned medium from END-2 cells, although not in the absence of conditioned medium. This effect was inhibited by activin [5]. More recently we have observed similar effects on mouse embryonic stem (mES) cells, as reported previously by Rohwedel et al [6] using the same END-2 cells. In addition, Dyer et al. [7] have shown that END-2 cells also induce the differentiation of epiblast cells from the mouse embryo to undergo hematopoiesis and vasculogenesis and respecify prospective neurectodermal cell fate, an effect they show to be largely attributable to Indian hedgehog (Ihh), a factor secreted by END-2 cells and VE from the mouse embryo.
Here, we present preliminary results showing similar effects in hES cells [8] co-cultured with END-2 cells, including the appearance of beating muscle. By contrast, in a pluripotent hEC cell line, GCT27X [9], aggregation takes place in the co-culture but we found no evidence of beating muscle. Characterization of the hES-derived cardiomyocytes has been initiated. The myocardial infarction (MI) model in mice, in which we will test cardiomyocyte function after transplantation is described.

Materials and Methods

Co-culture

END-2 cells, P19 EC , hEC and hES cells were cultured as described previously [3-5,8, 10]. To initiate co-cultures, mitogenically inactive END-2 cell cultures replaced mouse embryonic fibroblasts (MEFs) as feeders for hEC, mES and hES. Co-cultures with P19EC, which are feeder-independent, were initiated and maintained as described previously [4]. Cultures were then grown for 2-3 weeks and scored for the presence of areas of beating muscle from 5 days onwards.

Isolation of primary human adult cardiac cells

Human atrial cells from surgical biopsies served as a control for antibody staining, electrophysiology and characterization of ion channels by RT-PCR. Cardiac tissue was obtained with consent from patients undergoing cardiac surgery. Atrial appendages routinely removed during surgery were immediately transferred to ice cold Krebs-Ringer (KR) saline solution. Tissues were trimmed of excess connective and adipose tissue and washed twice with sterile KR solution. Myocardial tissue was minced with sterile scissors, then dissociated to release individual cells by a three step enzymatic isolation procedure using published methods [11]. The first step involved a fifteen minute incubation with 4.0 U/ml protease type XXIV (Sigma, St Louis, MO, USA) at 37° C. Tissues were then transferred to a solution consisting of collagenase 1.0 mg/ml and hyaluronidase 0.5 mg/ml, followed by three further incubations with collagenase (1.0 mg/ml) for twenty minutes each at 37° C. Tissue extracts were combined and the calcium concentration restored to 1.79 mmol/L. Cardiomyocytes were transferred to tissue culture medium M199 enriched with 10% FBS, penicillin (100U/ml) / streptomyocin (100μg/ml), 2.0 mmol/L L-carnitine, 5.0 mmol/L creatine, 5.0 mmol/L taurine and seeded directly onto glass cover-slips coated with 50 μg/ml poly-L-lysine and cultured overnight.

Immunocytochemistry
Attached primary cardiomyocytes, mES (E14 and R1) and hES-derived cardiomyocytes were fixed with 3.0% paraformaldehyde in PBS with Ca^{2+} and Mg^{2+} for 30 minutes at room temperature, then permeablized with 0.1% triton X 100 in PBS for four minutes. Immunocytochemistry was performed by standard methods using monoclonal antibodies directed against sarcomeric proteins including α-actinin and tropomyosin (Sigma). Antibodies specific for isoforms of myosin light chain (MLC2a/2v) were used to distinguish between atrial and ventricular cells (gift of Dr Ken Chien), Table 1. Secondary antibodies were from Jackson Immunoresearch Labs. Cultured cardiac fibroblasts served as a negative control for sarcomeric proteins and cells were visualized using a Zeiss Axiovert 135M epifluorescence microscope (Carl Zeiss, Jena GmbH, Germany). Images were pseudocoloured using image processing software.

Table 1 Antibodies used to stain atrial cardiomyocytes

Primary antibody	**Dilution**	**Secondary antibody**	**Dilution**
Mouse anti α-actinin IgG	1:800	Goat anti mouse IgG-cy3/FITC conjugated	1:250
Mouse anti tropomyosin IgG	1:50	Goat anti mouse IgG-cy3 conjugated	1:250
Polyclonal rabbit anti mouse mlc-2a (atrial)	1:500	Goat anti rabbit IgG-cy3 conjugated	1:250
Hoechst (nucleic acid)	1:500		

Semi-quantitative RT-PCR for ion channel expression
P19EC cells were differentiated into beating muscle by the aggregation protocol in the presence of 1% dimethyl sulphoxide [12]. After 16 days in these culture conditions, beating areas were excised and RNA was isolated using Trizol (Gibco) and reversed transcribed using M-MLV-RT (Gibco). Primers for cardiac actin [13], MLC2v [14], ERG [15], Kir2.1 [16] were used as described previously. Primers for

mouse L-type calcium channel subunit α1c (sense 5'-CCAGATGAG ACCCGCAGCGTAA; antisense 5'- TGTCTGCGGCGTT CTCCATCTC; GenBank accession no. L01776; product size 745 bp), Scn5a (sense 5'-CTTGGCCA AGATCAACCTGCTCT; antisense 5'-CGGACAGGGCCAAA TACTCAATG; AJ271477; 770 bp) and β-tubulin (sense 5'- TCACTGTG CCTGAACTTACC; antisense 5'-GGAACATAGCCGTAAACTGC; X04663; 319 bp) were designed using VectorNTI software (InforMax, North Bethesda, USA).

Patchclamp electrophysiology.

Experiments were performed at 33° C, using the whole cell voltage clamp configuration of the patchclamp technique. After establishment of the gigaseal the action potentials were measured in the current clamp mode. The data were recorded from cells in spontaneously beating areas using an Axopatch 200B amplifier (Axon Instruments Inc., Foster City, CA, U.S.A.). Output signals were digitized at 2 kHz using a Pentium III equipped with an AD/DAC LAB PC+ acquisition board (National Instruments, Austin, TX, U.S.A.). Patch pipettes with a resistance between 2 and 4 MΩ were used. Composition of the bathing medium was 140 mM NaCl, 5mM KCL, 2mM $CaCL_2$, 10 mM HEPES, adjusted to pH 7.45 with NaOH. Pipette composition: 145 mM KCL, 5 mM NaCL, 2 mM $CaCL_2$, 10 mM EGTA, 2 mM $MgCL_2$, 10 mM HEPES, adjusted to pH 7.30 with KOH.

Myocardial infarction model in mice

In order to test the ability of stem cell derived cardiomyocytes to restore cardiac function, a MI model was developed in mice. In pentobarbital anesthetized adult mice, the chest is opened through a midsternal approach. The anterior descending branch is identified and ligated. Successful procedures induce a discoloration of the distal myocardium. The chest is closed with three sutures and the animal is allowed to recover. In total 17 animals have been operated on to date. Seven received a sham procedure including positioning of the suture and 10 were ligated. Four weeks after MI the mice were anaesthetized again using the same medication by intraperitoneal injection. For the hemodynamic study the animals were intubated, and connected to a rodent respirator (Hugo Sachs Electronics, March-Hugstetten Germany). Instrumentation was performed with the chest closed by introducing a catheter into the jugular vein. A 1.4 French conductance-micromanometer (Millar Instruments, Houston USA) was delivered to the left ventricle through the carotid artery. Pressure and conductance measurements were recorded using Sigma SA electronic equipment (CDLeycom, Zoetermeer, The Netherlands) and stored for offline analysis. Typical pressure volume (PV) loop recorded in a normal heart is presented in figure 5a. From the PV-loops many hemodynamic parameters can be deduced including the end systolic PV relationship (ESPVR) and preload recruitable stroke work (PRSW).

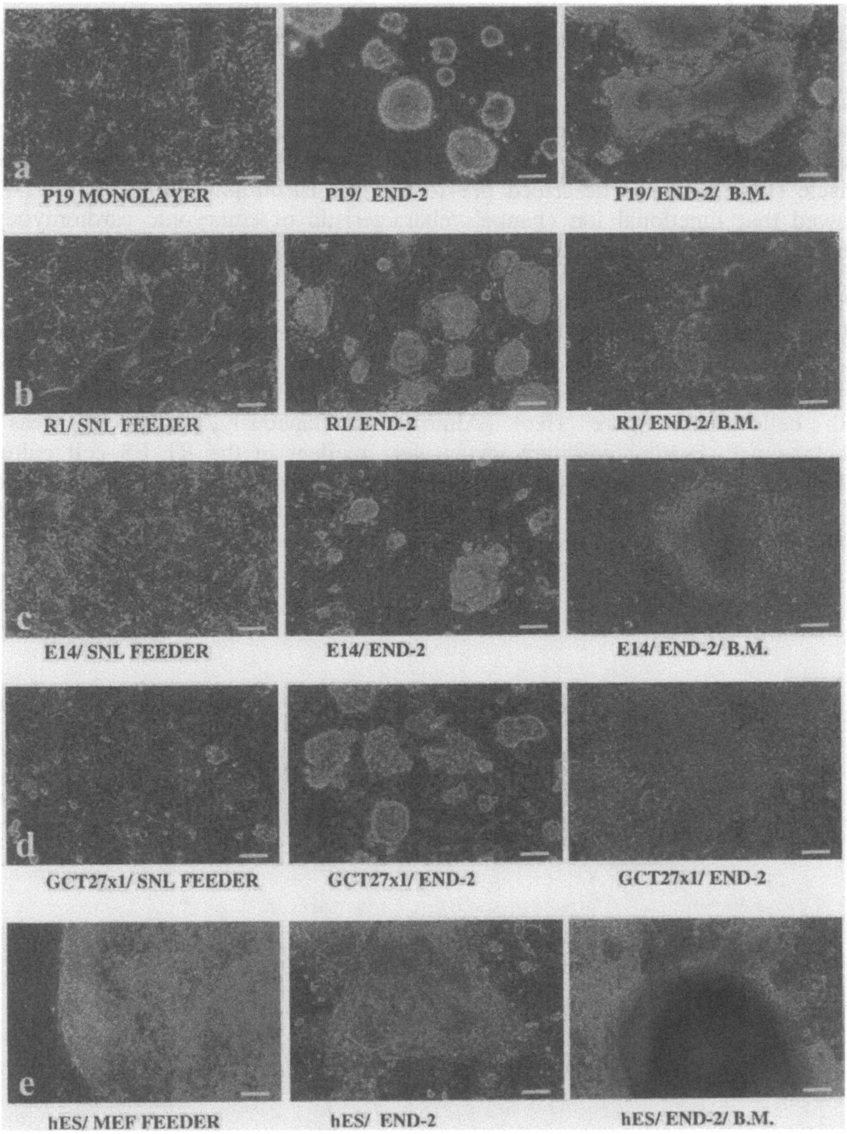

Figure 1 *Co-cultures of stem cells with the mouse visceral endoderm-like cell line END-2.* ***a.*** *P19 EC in normal monolayer culture, 3 days after initiation of co-culture with END-2 cells and after 10 days, when beating muscle (B.M.) is evident.* ***b.*** *mES cell line R1 in monolayer on its normal "feeder" cells (SNL), 3 days after initiation of co-culture and 2 days later, when beating muscle is evident.* ***c.*** *as b, with the exception that B.M. is evident on day 7 after aggregation.* ***d.*** *GCT27X human EC cell line on mouse embryonic fibroblast (MEF) feeder cells, 3 days after initiation of co-culture and after 16 days. No beating muscle is present.* ***e.*** *hES cells on MEF feeders. 3 days after initiation of END-2 co-cultur and beating muscle formed after 11 days.*

Results

mEC – END-2 co-cultures

Two days after initiation of co-cultures with END-2 cells,, P19 EC cells aggregated spontaneously and 7-10 days later many of the aggregates contained areas of beating muscle (figure 1a), as described previously [4]. Electrophysiology and RT-PCR showed that functional ion channels characteristic of embryonic cardiomyocytes were expressed in these cells (figure 2, table 2).

mES –END-2 co-cultures

Two independent mouse ES cell lines (E14 and R1) were tested for their response to co-culture conditions. Although the cultures were not initiated as single cell suspensions, within 3 days larger aggregates than initially seeded were evident for both cell lines (figure 1b,c). Almost simultaneously, extensive areas of spontaneously beating cardiomyocytes were evident in the R1 ES cell cultures, although only 7 days later, were (smaller) areas of beating muscle found in the E14 ES cells. Cells in beating areas exhibited the characteristic sarcomeric banding pattern of myocytes when stained with α-actinin (figure 4d).

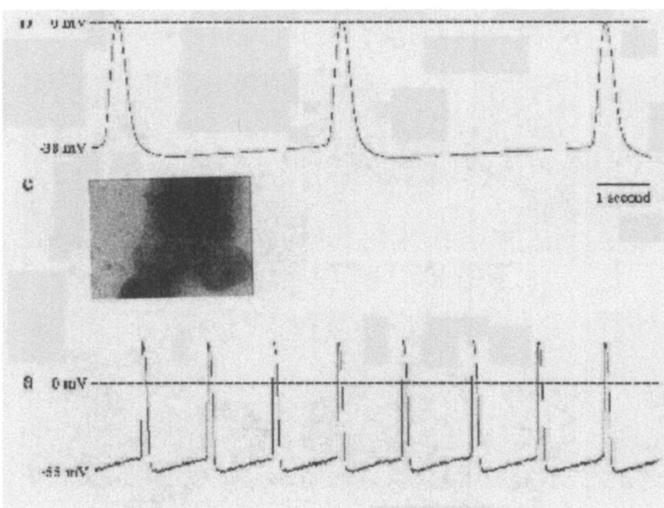

Figure 2 *Electrophysiological characteristics of cardiomyocytes from stem cells. Repetitive action potentials recorded from spontaneously beating areas.* ***a****. In mouse P19 EC cell-derived cardiomyocytes.* ***b****. In an aggregate of hES-derived cardiomyocytes.* ***c****. Phase contrast image of the beating area in the hES culture from which the recording showed in b. was derived. (Note the height of the protruding structure where the beating region is located, 20x objective).*

hEC – and END-2 co-cultures
The human EC cell line, GCT27X is a feeder-dependent, pluripotent EC cell line, with characteristics similar to human ES cells [9]. In co-culture with END-2 cells, formation of large aggregates was observed (figure 1d). However even after 3 weeks, there was no evidence of beating muscle.

hES – END-2 co-cultures.
During the first week of co-culture, the small aggregates of cells gradually spread and differentiated to cells with mixed morphology but with a relatively high proportion of epithelial-like cells. By the second week, these swelled to fluid-filled

Table 2 Relative levels of cardiac marker and ion channel mRNA expression as determined by semi-quantitative RT-PCR. Identical amounts of cDNA of undifferentiated P19 (EC), differentiated P19 cardiomyocytes (CMC) and adult mouse heart (Heart) was PCR amplified for the indicated gene products. Relative levels for each product are indicated.

	EC	CMC	Heart	ion channel and current
cardiac actin	+	++	+++++	
MLC2v	+	+++	+++++	
α1c	+	+++	++++	L-type calcium channel, I_{Ca}
Scn5a	+	+	+++++	Heart specific sodium channel, I_{Na}
ERG	++	++	++++	Delayed rectifier potassium channel, I_{Kr}
Kir2.1	-	+	++++	Voltage-gated potassium channel, I_{K1}
Tubulin	+++	+++	+++	

cysts (not shown). Between these, distinct patches of cells become evident which begin to beat a few days later. Between 12 and 21 days, more of these beating patches appear (e.g. figure 1e). Overall, 15-20% of the wells contain one or more areas of beating muscle. Beating rate is approximately 60 beats per minute and is highly temperature sensitive, compared with mouse ES-derived cardiomyocytes. These cells stain positively with α-actinin, confirming their muscle phenotype (figure 4e). In contrast to mES and P19EC-derived cardiomyocytes, however, the sarcomeric banding patterns were poorly defined but entirely comparable with primary human cardiomycytes grown for only two days in culture (figure 3; 4a-c).
It is clear that while primary human cardiomycytes initially retain the sarcomeric structure, standard culture conditions result in its rapid deterioration (figure 3). It may be assumed that hES culture conditions are not optimal for cardiomyocytes so that the hES-derived cardiomyocytes similarly exhibit a deterioration in their

characteristic phenotype. It will be essential to optimize these conditions to obtain fully functional cardiomyocytes from stem cells in culture

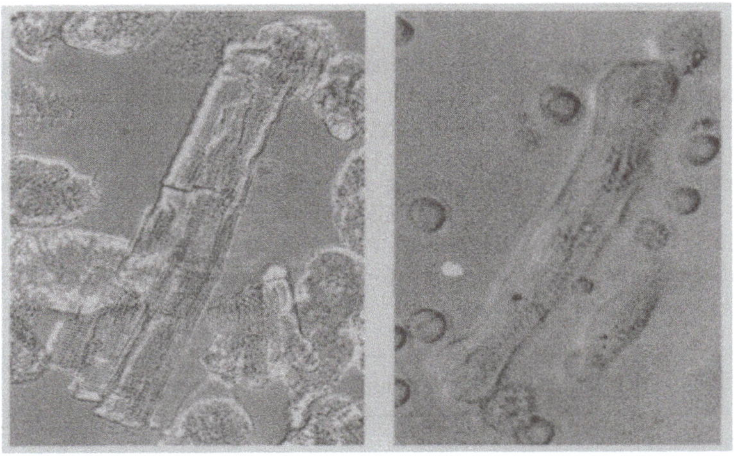

Figure 3 *Isolated cardiomyocytes (a) exhibiting sharp edges and well defined sarcomeres in contrast with cells cultured for two days (b) which had disorganised sarcomeric patterning*

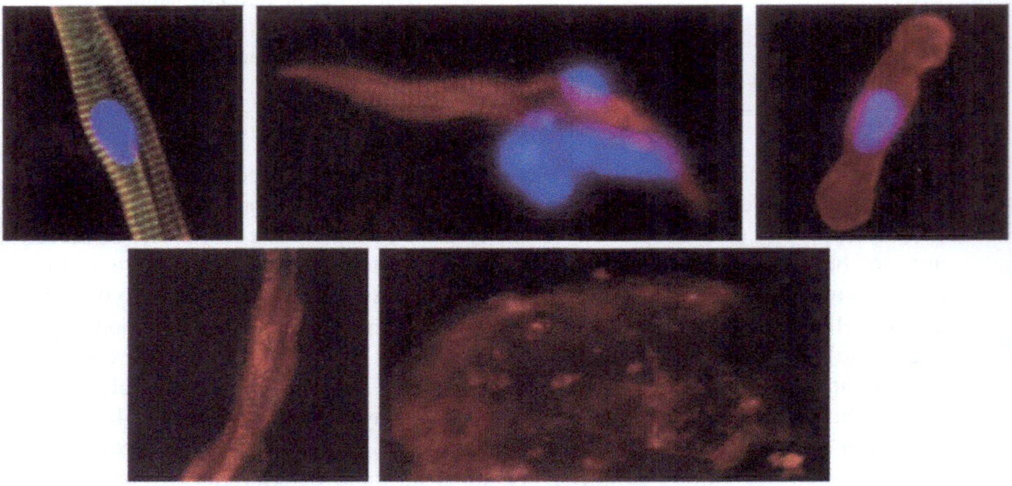

Figure 4 *Immunocytochemistry on adult human primary atrial cardiomyocytes and stem cell-derived cardiomyocytes. Primary atrial cardiomyocytes stained positive for sarcomeric proteins including (green) α-actinin, (red) mlc-2a (a) and tropomyosin (b). Cell DNA was stained with (blue) Hoechst to distinguish normal and apoptotic cells. Cells cultured for two days had a disorganised tropomyosin sarcomeric patterning and diffuse antibody staining (c). mES-derived cardiomyocytes also show sharp banding when stained with α-actinin (d) but in hES-derived cardiomyocytes α-actinin is diffuse and poorly banded (e).*

Despite deterioration in sarcomeric structure, hES-derived cardiomyocytes continued to beat rhythmically over several weeks and action potentials were detectable by current clamp electrophysiology (figure 2b), performed by inserting electrodes into aggregates, shown in figure 2c. However, carrying out electrophysiology in this manner, i.e. in aggregates rather than single cells, yields action potentials that are the accumulated effects of groups of cells. They are therefore difficult to interpret and to attribute to either ventricular, atrial or pacemaker cells. Work is currently in progress to dissociate and replate aggregates to allow single cell determinations.

Cardiac ion channel expression during stem cell differentiation.
The order in which ion currents, responsible for the subsequent phases of the adult action potential, appear during heart development has been established in electrophysiological studies [17,18]. Inward L-type Ca^{2+} currents play a dominant role during early cardiac embryogenesis, whereas inward Na^{2+} currents increase only just before birth [17,19]. Mouse ES and P19 EC cells display similar timing in ion current expression [20]. To unravel the sequence of ion-channel expression at the molecular level during differentiation of P19 EC cells, we performed RT-PCR on RNA isolated from undifferentiated and 16-day old beating clusters of P19 derived cardiomyocytes and compared these with expression in adult mouse heart (table 2). Expression of cardiac markers cardiac actin and MLC2v are detected in EC cell cultures, but increases in P19-derived cardiomyocytes. At the protein level, no MLC2v is detected in EC cells, while prominent expression is found in P19-derived cardiomyocytes (data not shown). Likewise, RNA for calcium, sodium and potassium channels can be detected in EC cells. Calcium channel RNA level (α1c) is upregulated in P19-derived cardiomyocytes, while sodium channel RNA (SCN5a) remains at initial levels. Furthermore, RNA for the delayed rectifier potassium channel (ERG) remains also unchanged. Inward rectifier Kir2.1 RNA can not be detected at EC stage, but becomes expressed after cardiomyocyte differentiation, albeit at low levels compared to adult heart. These data are compatible with the action potential of day16 P19 cardiomyocytes (data not shown), i.e. a low upstroke velocity (I_{Ca} mainly instead of I_{Na}), a relative positive resting membrane potential between -40 to -60 mV (little to no I_{K1}). These results indicate that day 16 P19 cardiomyocytes resemble fetal cardiomyocytes with respect to ion channel expression, as has been described previously for mES-derived cardiomyocytes [21,22].

MI model in mice
In order to test the effectiveness of cardiomyocyte transplantation *in vivo*, it is important to have reproducible animal models with a measurable parameter of cardiac function. The parameters used should clearly distinguish control and experimental animals (see for example [23]) so that the effects of transplantation can be adequately determined. PV relationships are a measure of the pumping capacity of the heart and could be used as a read-out of altered cardiac function following transplantation. Here, we have tested aspects of a procedure towards establishing a MI model in immunodeficient mice as a "universal acceptor" of

cardiomyocytes from various sources. The infarct size obtained through occluding the anterior descending coronary artery encompassed 30-50% of the left ventricular circumference. The septum involvement can be neglected in mice, resulting in more then 94% survival after 4 weeks (1 infarcted animal died). A 30% increase in left ventricular volume was recorded with conserved contractility post MI. There were significant differences in ESPVR (22.6 sham vs 10 post MI, $p < 0.05$), and PRSW (81.2 sham vs 43.5 post MI, $p<0.05$). This is visualised by comparing the shapes of the "PV loops" in sham operated versus infarcted mice in figure 5a and 5 b.
Despite maintained left ventricular contractile function, this mouse MI model provides a reproducible system for studying left ventricular remodeling making it feasible to assess the extent of cardiac repair following transplant interventions.

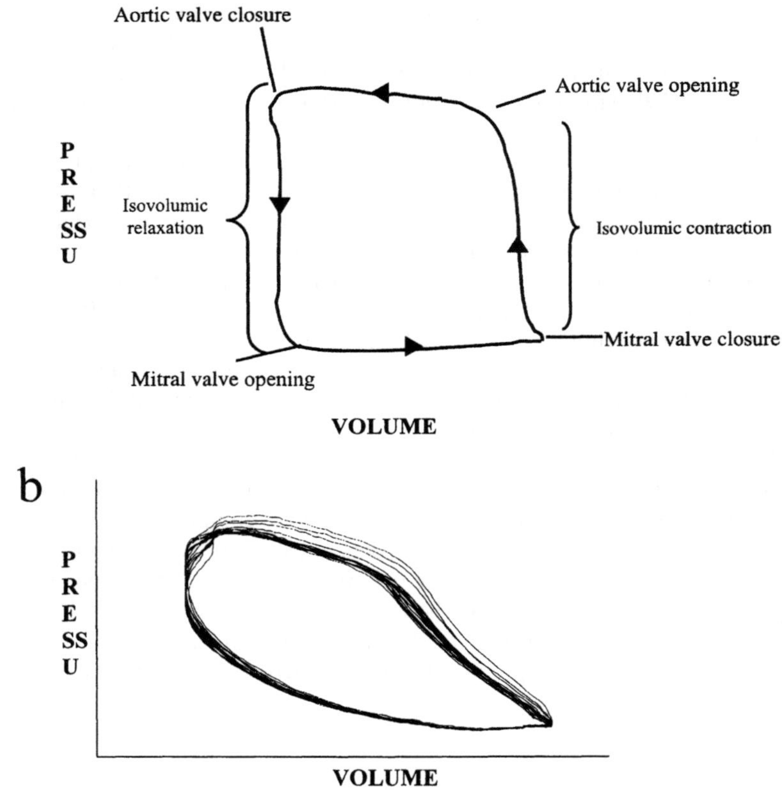

Figure 5 *Hemodynamic assessment of left ventricular function in mice.*
A Normal loop representing the relationship between volume and pressure changes in the mouse heart: indicated are the valvular events and stages during one cycle of contraction and relaxation.
B Pressure volume relationship 4 weeks post myocardial infarction: note the difference in the shape of the loop and the alterations in both contraction and relaxation.

Conclusions

The results of the work described here show that VE–like cells induce/promote differentiation of pluripotent cells to cardiomyocytes. These cells include pluripotent mouse EC cells, mouse ES as well as human ES cells, shown here for the first time. The results also showed that the capacity of different mES cell lines to differentiate to cardiomyocytes is variable. We have so far tested only one hES cell line for endoderm co-culture responses and it is not unlikely that different hES cell lines equally show variable capacities for differentiation. In a recently published study of "spontaneous" human ES cell differentiation, for example, it appears that only one of several subclones tested of one particular hES cell line (known as H9), actually had the capacity to form spontaneously beating cardiomyocytes [24]. The uncloned "mother" cell line did not exhibit this property. It remains to be established whether this subclone actually represents a precursor of the somatic or mesoderm lineage rather than a truly pluripotent cell.

Pluripotent mouse stem cells differentiate to cardiomyocytes with embryonic rather than mature characteristics. It is unclear what the phenotype of hES-derived cardiomyocytes precisely is and, indeed, the phenotypic characteristics of primary human cardiomyocytes (ventricular versus atrial, fetal versus adult) are insufficiently detailed for comparisons to be made directly between various human sources. Mouse and rat cardiomycytes are at present the best reference tissues. We have also shown here that primary adult human cardiomyocytes have a sharply defined morphology and sarcomeric banding pattern in culture but within a few days this deteriorates, presumably because culture conditions are suboptimal for these highly sensitive cells. Likewise, hES-derived cardiomycytes express appropriate markers and display action potentials but under present conditions have a relatively poor morphology. This contrasts with mouse ES-derived cardiomyocytes, which maintain morphology under standard culture conditions. An immediate aim is then to optimize culture conditions for primary cardiomyocytes and apply these to the hES derivatives. The MI model in mice provides a means of functional analysis of cardiomyocytes after transplantation. Transfer to immunodeficient mice will make it suitable for comparing human and mouse cardiomyocytes from different sources for their ability to restore cardiac function after infarct. Once the efficiency of cardiomyocyte differentiation and culture conditions for cardiomyocytes have been improved, we will use this model to evaluate effects on cardiac function *in vivo*.

Acknowledgements

Part of this study (MvdH) is financed by NWO-MW, Embryonic Stem Cell International (S.D.B.), Netherlands Interuniversity Cardiology Institute (D.W.,P.D, D.L.). hES cells (line 1 and 2) were supplied by Embryonic Stem Cell International. We thank Teun de Boer for figure 5b and contributions to the electrophysiology, respectively.

References

1. Yatskievych, TA, Ladd, AN, Antin, PB. Induction of cardiac myogenesis in avian pregastrula epiblast: the role of the hypoblast and activin. Development 1997;124:2561-70.
2. Nascone N, Mercola M. An inductive role for the endoderm in Xenopus cardiogenesis. Development 1995;121:515-23.
3. Mummery CL, Feijen A, van der Saag PT, van den Brink CE, de Laat SW. Clonal variants of differentiated P19 embryonal carcinoma cells exhibit epidermal growth factor receptor kinase activity. Dev Biol.1985;109:402-10.
4. Mummery CL, van Achterberg TA, van den Eijnden-van Raaij AJ, van Haaster L, Willemse A, de Laat SW, Piersma AH. Visceral-endoderm-like cell lines induce differentiation of murine P19 embryonal carcinoma cells. Differentiation 1991;46:51-60.
5. van den Eijnden-van Raaij AJ, van Achterberg TA, van der Kruijssen CM, Piersma AH, Huylebroeck D, de Laat SW, Mummery CL. Differentiation of aggregated murine P19 embryonal carcinoma cells is induced by a novel visceral endoderm-specific FGF-like factor and inhibited by activin A. Mech Dev. 1991;33:157-65.
6. Rohwedel J, Maltsev V, Bober E, Arnold HH, Hescheler J, Wobus AM. Muscle cell differentiation of embryonic stem cells reflects myogenesis in vivo: developmentally regulated expression of myogenic determination genes and functional expression of ionic currents. Dev Biol.1994;164:87-101.
7. Dyer, MA, Farrington, SM, Mohn, D, Munday, JR, Baron, MH. Indian hedgehog activates hematopoiesis and vasculogenesis and can respecify prospective neurectodermal cell fate in the mouse embryo. Development 2001;128:1717-30
8. Reubinoff BE, Pera MF, Fong CY, Trounson A, Bongso A. Embryonic stem cell lines from human blastocysts: somatic differentiation in vitro. Nat Biotechnol. 2000;18:399-404.
9. Pera MF, Cooper S, Mills J, Parrington JM. Isolation and characterization of a multipotent clone of human embryonal carcinoma cells. Differentiation 1989;42:10-23.
10. Slager HG, Van Inzen W, Freund E, Van den Eijnden-Van Raaij AJ, Mummery CL. Transforming growth factor-beta in the early mouse embryo: implications for the regulation of muscle formation and implantation. Dev Genet.1993;14:212-24.
11. Peeters G.A., Sanguinetti, M.C, Eki, Y, Konarzewska H, Renlund D.G, Karwande S.V , Barry W.H. Method for isolation of human ventricular myocytes from single endocardial and epicardial biopsies. Am J. Physiol.1995;268:H1757-64.
12. Rudnicki, M.A, McBurney M.W. Teratocarcinomas and embryonic stem cells, a practical approach (ed. E.J. Robertson). pp19-49. Oxford: IRL Press 1987.
13. Lanson Jr N.A, Glembotski C.C, Steinhelper M.E, Field L.J, Claycomb W.C. Gene expression and atrial natriuretic factor processing and secretion in cultured AT-1 cardiac myocytes. Circulation 1992;85:1835-41.
14. Meyer N, Jaconi M, Landopoulou A, Fort P, Puceat M. A fluorescent reporter gene as a marker for ventricular specification in ES-derived cardiac cells.FEBS Lett.2000; 478:151-8.
15. Lees-Miller J.P, Kondo C, Wang L, Duff H.J. Electrophysiological characterization of an alternatively processed ERG K^+ channel in mouse and human hearts. Circ. Res. 1997;81:719-26.
16. Vandorpe D.H, Shmukler B.E, Jiang L, Lim B, Maylie J, Adelman J.P, De Franceschi L, Domenica Cappelline M, Brugnara C, Alper S.L. cDNA cloning and functional characterization of the mouse Ca^{2+}-gated K^+ channel, mIK1. J. Biol. Chem. 1998;273:21542-53.

17. Davies M.P, An R.H, Doevendans P, Kubalak S, Chien K.R, Kass R.S. Developmental changes in ionic channel activity in the embryonic murine heart. Circ. Res. 1996;78:15-25.
18. Yasui K, Liu W, Opthof T, Kada K, Lee J-K, Kamiya K, Kodama I. I_f current and spontaneous activity in mouse embryonic ventricular myocytes. Circ. Res. 2001;88:536-42.
19. An R.H, Davies M.P, Doevendans P.A, Kubalak S.W, Bangalore R, Chien K.R. Kass R.S. Developmental changes in beta-adrenergic modulation of L-type Ca2+ channels in embryonic mouse heart. Circ.Res.1996;78:371-8.
20. Wobus A.M, Kleppisch T, Maltsev V, Hescheler J. Cardiomyocyte-like cells differentiated in vitro from embryonic carcinoma cells P19 are characterized by functional expression of adrenoceptor and Ca^{2+} channels. In Vitro Cell. Dev. Biol. 1994;30A, 425-34.
21. Doevendans P.A, Daemen M, de Muinck E, Smits J. Cardiovascular phenotyping in mice. Cardiovasc. Res 1998;39:34-49.
22. Doevendans P.A, Kubalak S.W, An R.H, Becker D.K, Chien K.R, Kass R.S. Differentiation of cardiomyocytes in floating embryoid bodies is comparable to fetal cardiomyocytes. J Mol Cell Cardiol 2000;32:839-51.
23. Palmen M, Daemen M, Bronsaer R, Smits J, Doevendans P.A. Cardiac remodeling but not cardiac function is impaired in IGF-1 deficient mice after chronic myocardial infarction. Cardiovasc Res 2001;50:516-24.
24. Kehat I, Kenyagin-Karsenti D, Snir M, Segev H, Amit M, Gepstein A, Livne E, Binah O, Itskovitz-Eldor J, Gepstein L. Human embryonic stem cells can differentiate into myocytes with structural and functional properties of cardiomyocytes. J.Clin.Invest. 2001;108: 407-14.

21. USE OF MESENCHYMAL STEM CELLS FOR REGENERATION OF CARDIOMYOCYTES AND ITS APPLICATION TO THE TREATMENT OF CONGESTIVE HEART FAILURE

K. Fukuda

Introduction

Cardiomyocytes do not regenerate after birth, and they respond to mitotic signals by cell hypertrophy rather than by cell hyperplasia [1]. Loss of cardiomyocytes leads to regional contractile dysfunction, and necrotized cardiomyocytes in infarcted ventricular tissues are progressively replaced by fibroblasts to form scar tissues. Recent studies revealed that transplanted fetal cardiomyocytes could survive in this heart scar tissue, and that these transplanted cells limited scar expansion and prevented post-infarction heart failure [2-4]. The transplantation of cultured cardiomyocytes into the damaged myocardium has been proposed as a future method for the treatment of heart failure. Although this is a revolutionary idea, it remains unfeasible in the clinical setting, since it is difficult to obtain donor fetal heart. A cardiomyogenic cell line could potentially substitute for fetal cardiomyocytes in this therapy.

Recent reports demonstrated the existence of pluripotent stem cells in adult tissues. Figure 1 showed the classification of pluripotent stem cells in adult tissues. Roy et al. reported the existence of neural stem cells in the brain, which can differentiate into neurons, oligodendrocytes and astrocytes in vitro [5], and marrow stromal cells have been shown to have many characteristics of mesenchymal stem cells [6]. Pluripotent progenitor marrow stromal cells may differentiate into various types of cell types including bone [7,8], muscle [9], fat [10], tendon and cartilage [11]. Based on these findings, we hypothesized that marrow stromal cells might also differentiate into cardiomyocytes, and repeatedly screened marrow stromal cells which began spontaneous beating after exposing them to 5-azacytidine, a cytosine analog capable of altering expression of certain genes that may regulate differentiation. We finally established a cell line which differentiates into cardiomyocytes in vitro, named CMG (cardiomyogenic), from adult marrow stromal cells [12]. The use of adult tissues as a source of cardiomyocytes makes this system particularly appropriate for the development of gene therapy strategies for heart disease. In this chapter, I will introduce the characteristics of bone marrow- derived cardiomyocytes, and discuss the possibility of the use of these cells for cardiovascular tissue engineering.

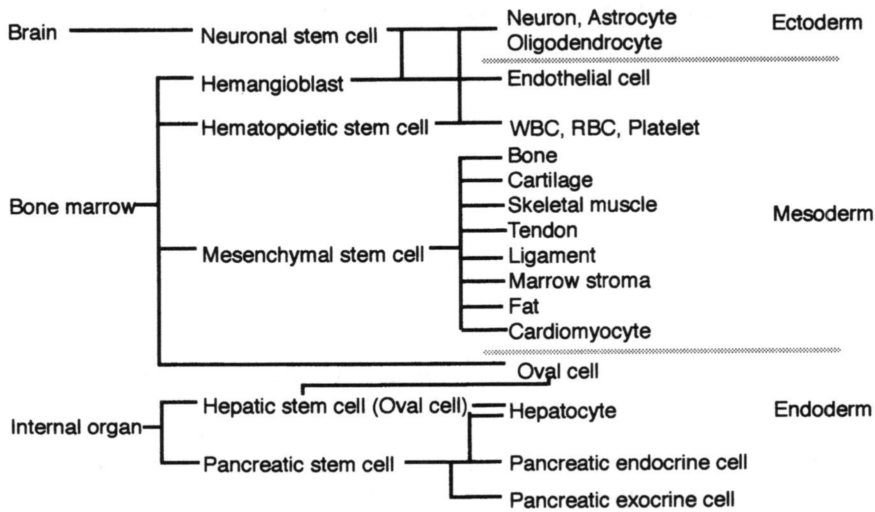

Figure. 1 *Classification of pluripotent stem cells in adult tissues. Bone marrow contains various kinds of stem cells. Mesenchymal stem cells may differentiate into various mesoderm-derived cells such as osteoblasts, chondroblasts, adipocytes, skeletal muscle cells and possibly cardiomyocytes.*

Materials and Methods

Cell culture

Female C3H/He mice (n=10) were anesthetized with ether, thighbones were excised, and bone marrow cells were obtained. The procedures were performed in accordance with the guidelines for animal experimentation of Keio University. Primary culture of the marrow cells was performed according to Dexter's method as described [13]. Cells were cultured in Iscove's modified Dulbecco's medium (IMDM) supplemented with 20% fetal bovine serum and penicillin (100 μg/ml)/streptomycin (250 ng/ml)/amphotericin B (85 μg/ml) at 33 ^{O}C in humid air with 5% CO2. After a series of passages, attached marrow stromal cells became homogeneous, and were devoid of hematopoietic cells. The marrow stromal cells basically did not require co-culture of blood stem cells. Immortalized cells were obtained by frequent subculture for more than 4 months. Cell lines from different dishes were subcloned by limiting dilution. To induce cell differentiation, cells were treated with 3 μmol/L of 5-azacytidine for 24 hours. Subclones that included spontaneous beating cells were screened by microscopic observation (first screening), and cells surrounding spontaneous beating cells were subcloned by cloning syringes. Subcloned cells were maintained, exposed to 5-azacytidine again for 24 hours, and clones that showed spontaneous beating most frequently were screened (second screening). The clonal cell line thus obtained was named CMG cell.

Transmission electron microscopy

Cells were washed three times with PBS (pH 7.4). The initial fixation was done in PBS containing 2.5% glutaraldehyde for 2 hours. The cells were embedded in epoxy resin. Ultrathin sections cut horizontally to the growing surface were double stained in uranyl acetate and lead citrate, and viewed under a JEM-1200EX transmission electron microscope.

Action potential recording

Electrophysiological studies were performed in IMDM containing (mmol/L) $CaCl_2$ 1.49, KCl 4.23, and HEPES 25 (pH 7.4). Cultured cells were placed on the stage of an inverted phase contrast optic (Diaphoto-300, Nikon) at 23°C. Action potentials were recorded by conventional microelectrode. Intracellular recordings were taken from 2- to 5-week cultured cells. Glass microelectrodes filled with KCl (3 mol/L) having a DC resistance of 15-30 MΩ were selected. Membrane potentials were measured by means of current clamp mode (Nihon Kohden, MEZ-8300) with a built-in 4-pole Bessel filter set at 1 kHz. The data were recorded on a thermal recorder (Nihon Kohden, RTA-1100M) and stored on a digital magnetic tape (Sony Magnescale; frequency range 0 - 20 kHz).

RNA extraction, RT-PCR and Northern blot

Total RNA was extracted from fetal, neonatal and adult mouse heart, skeletal muscle, and differentiated CMG cells by TRIzol Reagent (GIBCO BRL). RT-PCR of cardiomyocyte-specific genes including including atrial natriuretic peptide (ANP), brain natriuretic peptide (BNP), α–myosin heavy chain (MHC), β–MHC, α–skeletal actin, α-cardiac actin, myosin light chain (MLC)-2a, MLC-2v was performed as described previously [12]. For detection of α-skeletal and α–cardiac actin expression, Northern blot analysis was used as described previously [12].

Results

CMG cells form myotubes and show spontaneous contraction

By repeated rounds of limiting dilution, we isolated 192 single clones, several of which could differentiate into cardiomyocytes and show spontaneous beating. These experiments were reproducible, but the percentage of cardiomyocyte differentiation was distinct among these clones.

Phase contrast photography and/or immunostaining with anti-sarcomeric myosin antibodies were used to determine the morphological changes in CMG cells (figure 2). CMG cells showed a fibroblast-like morphology before 5-azacytidine treatment (0 week), and this phenotype was retained through repeated subculturing under non-stimulating conditions. After 5-azacytidine treatment, the morphology of the cells gradually changed. Approximately 30% of the CMG cells gradually increased in size, formed a ball-like appearance, or lengthened in one direction, and formed a stick-like morphology at 1 week. They connected with adjoining cells after 2 weeks, and formed myotube-like structures at 3 weeks. The differentiated CMG myotubes maintained the cardiomyocyte phenotype and beat vigorously for at least 8 weeks after final 5-azacytidine treatment, and did not de-differentiate. Most of the other non-myocytes had an adipocyte-like appearance.

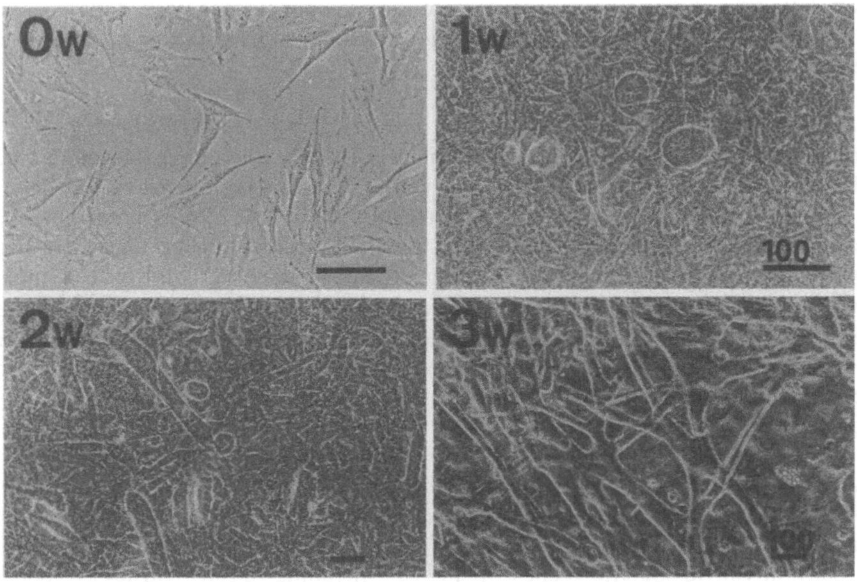

Figure 2. *Phase contrast photographs of CMG cells before and after 5-azacytidine treatment. (upper left) CMG cells have a fibroblast-like morphology before 5-azacytidine treatment (0 week). (upper right) One week after treatment, some cells gradually increased in size, and formed a ball-like or stick-like appearance (Arrows). These cells began spontaneous beating thereafter. (lower left) Two weeks after treatment, ball-like or stick-like cells connected with adjoining cells, and began to form myotube-like structures. (lower right) Three weeks after treatment, most of the beating cells were connected and formed myotube-like structures. Bars indicated 100 μm.*

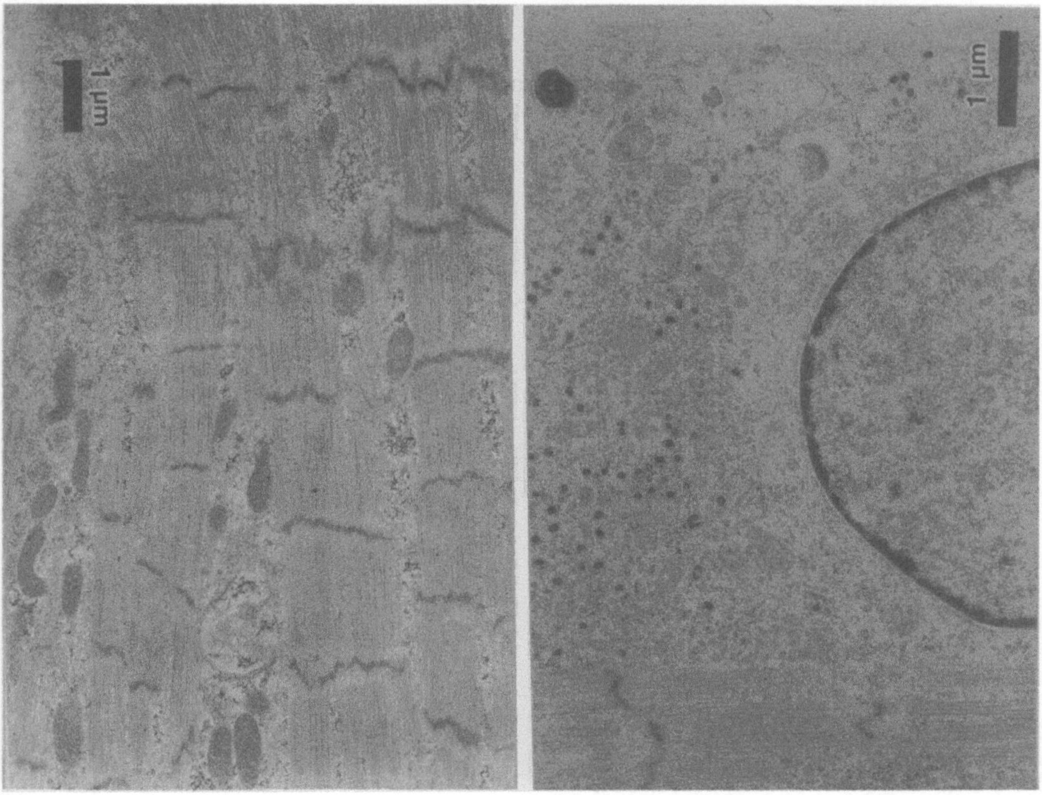

Figure 3. *Transmission electron micrograph of CMG myotubes. (A) Differentiated CMG myotubes had well organized sarcomeres. Rich glycogen granules and a number of mitochondria were observed. (B) Ultrastructural analysis revealed that nuclei (**N**) were oval and positioned in the central part of the cell, not immediately beneath the sarcolemma. Atrial granules (**AG**), measuring 70~130 nm in diameter, are observed in the sarcoplasm, and are concentrated especially in the juxtanuclear cytoplasm. Bar indicates 1 μm.*

CMG cells have a cardiomyocyte-like ultrastructure

Representative transmission electron microscopy photographs are shown in figure 3. A longitudinal section of the differentiated CMG myotubes clearly revealed the typical striation and pale-staining pattern of the sarcomeres. CMG myotube nuclei were positioned in the center of the cell, not beneath the sarcolemma. The most conspicuous feature of the differentiated CMG myotubes was the presence of membrane-bound dense secretary granules measuring 70 - 130 nm in diameter. These granules were thought to be atrial granules, and were especially concentrated in the juxtanuclear cytoplasm, but some were also located near the sarcolemma. These findings indicated that CMG cells had a cardiomyocyte-like rather than skeletal muscle ultrastructure.

CMG myotubes have several types of action potential

An electrophysiological study was performed on differentiated CMG cells at 2 - 5 weeks after 5-azacytidine treatment. There were at least two types of distinguishable morphological action potentials; sinus node-like potentials (figure 4-A), and ventricular myocyte-like potentials (figure 4-B).

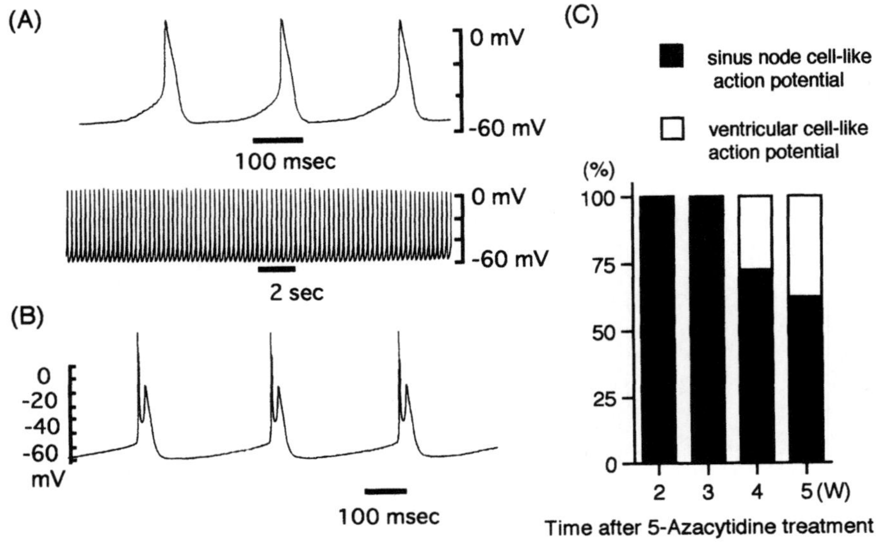

Figure 4 *Representative tracing of the action potential of CMG myotubes. (A, B) Action potential recordings were obtained from the spontaneous-beating cells at day 28 after 5-azacytidine treatment using a conventional microelectrode. We categorized these action potentials into two groups: (A) a sinus node-like action potential, or (B) a ventricular cardiomyocyte-like action potential. (C) The percentage of CMG cells with a sinus node-like or ventricular cardiomyocyte-like action potential after 5-azacytidine treatment. A ventricular cardiomyocyte-like action potential was first recorded 4 weeks after 5-azacytidine treatment, and then rapidly became more prevalent.*

The sinus node-like action potential showed a relative shallow resting membrane potential with late diastolic slow depolarization, like a pacemaker potential. Peak and dome-like morphology were observed in ventricular myocyte-like cells.

The cardiomyocyte-like action potential recorded from these spontaneous beating cells was characterized by (1) a relatively long action potential duration or plateau, (2) a relatively shallow resting membrane potential, and (3) a pacemaker-like late diastolic slow depolarization. Figure 4-C shows a time course of the percentage of the sinus node-like and ventricular myocyte-like action potentials. All the action potentials recorded from the CMG cells until 3 weeks revealed sinus node-like action potential. The ventricular myocyte-like action potentials could be recorded after 4 weeks, and the percentage of these action potentials gradually increased thereafter. It is possible that the percentage of the ventricular myocyte-like action potentials at 5 weeks was underestimated. Most of the action potentials recorded from differentiated CMG myotubes had a ventricular myocyte-like appearance, but the action potential of the differentiated CMG myotubes was difficult to record. The glass microelectrode was frequently damaged, because the spontaneous contraction of the differentiated myotube at 5 weeks was too big.

Cardiomyocyte-specific gene expression
RT-PCR was performed to detect the expression of cardiomyocyte-specific genes in differentiated CMG cells. Total RNA obtained from cardiomyocytes (*in vivo* heart) and skeletal muscles (soleus muscle) were used as positive and negative controls, respectively. Differentiated CMG myotubes expressed both ANP and BNP genes (data not shown).
Table 1 summarizes the expression of cardiac contractile protein isoforms. Fetal, neonatal and adult ventricle and atrium were uses as controls. Both α- and β-MHC expression could be detected by RT-PCR in differentiated CMG cells, but β-MHC expression was overwhelmingly stronger than that of α–MHC. CMG cells expressed both α-cardiac and α-skeletal actin. On Northern blot analysis the α–skeletal actin gene was expressed at markedly higher levels than the α-cardiac actin gene in CMG cells. Interestingly, CMG cells expressed MLC-2v, but not MLC-2a. These patterns of gene expression in the differentiated CMG cells were consistent with a fetal ventricular phenotype.

Table 1. Expression of the isoforms of the cardiac contractile proteins in CMG cell
MHC: myosin heavy chain, MLC: myosin light chain

	atrium		ventricle			CMG
	fetal	adult	fetal	neonatal	adult	
α-actin	skeletal	cardiac	skeletal> cardiac	skeletal	cardiac	skeletal> cardiac
MHC	α>β	α	β>α	α>β	α	β>α
MLC	2a	2a	2v	2v	2v	2v

Discussion

Development of cardiomyocytes in vitro
We have established a cardiomyogenic cell line from mouse bone marrow stromal cells that can be induced to differentiate into cardiomyocytes *in vitro* by 5-azacytidine treatment. A number of lines of evidence confirmed the cardiomyocyte characteristics of these CMG cells. These cells expressed a number of cardiomyocyte-specific genes including ANP, BNP, GATA4 and Nkx2.5. In ventricular muscle of small mammals, there is a developmental switch from expression of β-MHC, which is the predominant fetal form, to that of α-MHC around the time of birth. There is also a developmental switch from expression of α-skeletal actin, which is the predominant fetal and neonatal form, to that of α–cardiac actin, the predominant adult form. Differentiated CMG cells mainly expressed β-

MHC and α-skeletal actin. Expression of α-MHC and α-cardiac actin was detected, but only at low levels. MLC-2 genes are specifically expressed in the chamber. MLC-2v is specifically expressed in ventricular cells, while MLC-2a was specifically expressed in atrial cells. Differentiated CMG cells expressed MLC-2v, but not MLC-2a. Moreover, skeletal muscle cells do not express α-MHC or MLC-2v. These results indicated that differentiated CMG cells had the specific phenotype of fetal ventricular cardiomyocytes.

Differentiated CMG cells expressed Nkx2.5, GATA4, TEF-1 and MEF-2C before final 5-azacytidine treatment. The MEF-2A and MEF-2D genes were expressed after final 5-azacytidine treatment. This pattern of gene expression in CMG cells was similar to that of *in vivo* developing cardiomyocytes [14]. These results indicated that the stage of differentiation of CMG cells is between cardiomyocyte-progenitor and differentiated cardiomyocytes.

Differentiated CMG myotubes have a cardiomyocyte-like ultrastructure including atrial granules. Tagoe *et al.* [15] reported that the most common size of atrial granules observed in the adult mice atrium was 150 to 200 nm in diameter, but they also found that approximately 35% of the atrial granules in adult mice atrium ranged between 50 to 150 nm in diameter. The atrial granules observed in the differentiated CMG myotubes were 70 to 130 nm in diameter. A previous report found that almost all atrial myocytes expressed ANP in fetal heart, whereas in the ventricular wall, cells containing immunoreactive granules were scattered [16]. The high density granules observed in the differentiated CMG cells might correspond to those in fetal ventricular cardiomyocytes.

CMG myotubes have either sinus node-like or ventricular myocyte-like action. Although action potentials can be seen in non-myocyte cells such as skeletal muscle cells or nerve cells, the action potential in CMG cells is characterized by duration [17,18]. The duration of action potentials in skeletal muscle cells or nerve cells is less than 5 milliseconds [19,20]. The most diastolic potential, action potential amplitude, and the overshoot potential of the Sinus-nodal-like CMG cells were close to the equivalent values reported in *in vivo* rabbit sinus nodal cells [21]. In rabbit ventricular cells, the most diastolic potential and action potential amplitude were reported to be -90~-95 mV and 120 mV, respectively. Although the most diastolic potential and action potential amplitude of the ventricular cardiomyocyte-like CMG cells was slightly shorter than these values, the shape of the action potential was very close to that of *in vivo* ventricular cardiomyocytes. The observation of several distinct patterns of action potential in CMG cells may reflect different developmental stages.

Future directions

CMG cells provide a powerful tool for the further investigation of cardiomyocyte differentiation and cardiomyocyte transplantation. We have already transplanted these cells into normal adult mice heart, and observed that the transplanted cell could survive in the recipient heart for at least several weeks. Cell transplantation into scar tissue of the *in vivo* heart caused by experimental myocardial infarction was initially performed using fibroblasts, smooth muscle cells or skeletal muscle cells. Transplantation of these cells into the scar tissue might improve cardiac remodeling or diastolic function, but are unlikely to improve systolic function. Transplantation of cardiomyocytes, however, could potentially rescue systolic function. The only potential sources of regenerated cardiomyocytes to date are embryonic stem (ES) cells and mesenchymal stem cells. ES cells differentiate into cardiomyocytes *in vitro*, and have both advantages and disadvantages for cardiomyocyte regeneration. A major advantage of ES cells is that the method for induction of differentiation into cardiomyocytes is already well established. However, transplanted ES cells can potentially form teratomas if some undifferentiated totipotent cells are still present. In addition, recipients must receive immuno-suppressants because ES cells are allogeneic. In contrast, mesenchymal stem cells do not carry any inherent risks of tumor formation, and are syngeneic. However, there is a need to both improve the current methods of identification and culture of mesenchymal stem cells, and of induction of CMG cell differentiation, which remain inefficient and slow. The identification of specific growth factors, cytokines or extracellular matrix factors that regulate cardiomyocyte differentiation may help to make this process faster and more efficient.

The number of candidates for heart transplantation remains much larger than the number of donor hearts world wide, but the possibilities of transplantation of regenerated cardiomyocytes may provide a premising option to current therapies.

Acknowledgments

This study was supported by Research Grant from the Ministry of Education, Science and Culture, Japan, and Health Science Research Grant for Advanced Medical Technology from the Ministry of Welfare, Japan.

References

1. Pan J, Fukuda K, Saito M, Matsuzaki J, Kodama H, Sano M, Takahashi T, Kato T, Ogawa S. Mechanical stretch activates the JAK/STAT pathway in rat cardiomyocytes. Circ Res 1999; 84:1127-36.
2. Leor J, Patterson M, Quinones MJ, Kedes LH, Kloner RA. Transplantation of fetal myocardial tissue into the infarcted myocardium of rat. A potential method for repair of infarcted myocardium? Circulation 1996;94(Suppl II):332-6.
3. Soonpaa MH, Koh GY, Klug MG, Field LJ. Formation of nascent intercalated disks between grafted fetal cardiomyocytes and host myocardium. Science 1997;264:98-101.
4. Delcarpio JB, Claycomb WC. Cardiomyocyte transfer into the mammalian heart. Cell-to-cell interactions in vivo and in vitro. Ann NY Acad Sci 1997;52:267-85.
5. Roy NS, Wang S, Jiang L, Kang J, Benraiss A, Harrison-Restelli C, Fraser RA, Couldwell WT, Kawaguchi A, Okano H, Nedergaard M, Goldman SA.: In vitro neurogenesis by progenitor cells isolated from the adult human hippocampus. Nat Med 2000;6:271-7.
6. Prockop DJ. Marrow stromal cells as stem cells for nonhematopoietic tissues. Science 1997;276:71-4.
7. Rickard DJ, Sullivan TA, Shenker BJ, Leboy PS, Kazhdan I. Induction of rapid osteoblast differentiation in rat bone marrow stromal cell cultures by dexamethasone and BMP-2. Dev Biol 1994;161:218-28.
8. Friedenstein AJ, Chailakhyan R, Gerasimov UV. Bone marrow osteogenic stem cells: in vitro cultivation and transplantation indiffusion chambers. Cell Tissue Kinet 1987;20:263-72.
9. Ferrari G, Angelis GC, Colleta M, Paolucci E, Stornaiolo A, Cossu G, Mavilio F. Muscle regeneration by bone marrow-derived myogenic progenitors. Science. 1998;279:1528-30.
10. Umezawa A, Maruyama T, Segawa K, Shadduck RK, Waheed A, Hata J. Multipotent marrow stromal cell line is able to induce hematopoiesis in vivo. J Cell Physiol 1992;151:197-205.
11. Ashton BA, Allen T.D, Howlett C.R, Eaglesom CC., Hattori A, Owen M. Formation of bone and cartilage by marrow stromal cells in diffusion chambers in vivo. Clin Orthop 1980;151:294-307.
12. Makino S, Fukuda K, Miyoshi S, Konishi F, Kodama H, Pan J, Sano M, Takahashi T, Hori S, Abe H, Hata J, Umezawa A, Ogawa S. Cardiomyocytes can be generated from marrow stromal cells in vitro. J Clin Invest 1999;103:697-705.
13. Dexter TM, Allen TD, Lajtha LG. Conditions controlling the proliferation of haemopoietic stem cells in vitro. J Cell Physiol 1977;91:335-44.
14. Edmondson DG, Lyons GE, Martin JF, Olson EN. Mef2 gene expression marks the cardiac and skeletal muscle lineages during mouse embryogenesis. Development 1994;120:1251-63.
15. Tagoe CN, Ayettey AS, Yates RD. Comparative ultrastructural morphometric analysis of atrial specific granules in the bat, mouse, and rat. Anat Rec 1993;235:87-94.
16. Venance SL, Pang SC. Ultrastructure of atrial and ventricular myocytes of newborn rats: evidence for the existence of specific atrial granule-like organelles in the ventricle. Histol Histopathol 1989;4:325-33.
17. Irisawa H. Electrophysiology and contractile function. In: Piper HM, Isenberg G, eds. Isolated adult cardiomyocytes. CRC Press, Boca Raton, Florida; pp. 1-11. 1989.
18. Carmeliet E. Cardiac transmembrane potentials and metabolism. Circ Res 1978;42:577-87.
19. Nicholls JG, Martin AR, Wallace BG. In: Nicholls JG, Martin AR, Wallace BG. From Neuron to Brain. a cellular and molecular approach to the function of the nervous system. 3rd eds. Sinauer Associates, Inc. Sunderland, MA; pp. 90-120. 1992.

20. Brinley FJ Jr. Excitation and conduction in nerve fibers. In: Mountcastle VB, et al. Medical Physiology 14th Eds. Mosby, St. Louis; pp. 46-81. 1980.
21. Noma A, Irisawa H. Membrane currents in the rabbit sinoatrial node cell as studied by the double microelectrode method. Pflügers Archiv 1976;364:45-52.

22. CELLULAR CARDIAC REINFORCEMENT

P. Menasché

Introduction

Because they are terminally differentiated cells, adult cardiomyocytes cannot regenerate and there is no myocardial pool of stem cells to replace those which have suffered irreversible ischemic injury. In fact, this dogma has been recently challenged by some experimental (for a review, see [1]) and clinico-pathological studies [2,3] suggesting that cardiomyocytes of infarcted or failing human hearts had actually retained a capacity of reentering a cell cycle. Whereas these observations are interesting from a cognitive standpoint, their clinical relevance is probably limited because the number of "new" cells that can be generated through this mechanism is by far too low to compensate for the loss of cardiomyocytes resulting from an infarct (or at least an infarct large enough to cause heart failure).

Thus, in clinical practice, the usual responses to myocardial infarction involve evolution of the infarct zone toward a fibrous noncontractile scar and hypertrophy of cells harboured in the still viable segments of the heart. At best, these compensatory responses can temporarily maintain an adequate contractile function. At worst, the combination of interstitial fibrosis and inappropriate remodelling promote deterioration of systolic and diastolic functions and lead to heart failure.

This condition has now become a major problem of public health because of its prevalence (approximately 5 million patients in the U.S.), incidence (3-500,000 cases per year), high mortality (up to 30%) and the related costs due to drugs and repeated hospitalizations. When heart failure becomes refractory to medical therapy, patients can be offered a wide variety of surgical therapies which culminate in cardiac transplantation. However, organ shortage still results in a substantial mortality (approximately 30%) while on the waiting list; on the other hand, dynamic cardiomyoplasty and partial resections of the left ventricle have yielded overall disappointing clinical results whereas permanently implanted assist devices are still in a developmental stage. Thus, there is clearly room left for alternate therapies and, over the past decade, experience has accumulated that suggests that "cardiac reinforcement" by transplantation of contractile cells might be one of them.

Fetal cardiomyocytes

Initial experiments with fetal cardiomyocytes, conducted by Soonpa and co-workers [4] in transgenic mice expressing the gene of β-galactosidase have shown that fetal cardiomyocytes could form stable intramyocardial grafts up to 2 months and develop connexions, identified as intercalated disks, with host cardiomyocytes. Additional studies conducted in animals suffering from Duchenne de Boulogne dystrophy (which has allowed to trace the dystrophin-positive injected cells) have supported these data by demonstrating gap junctions between transplanted fetal cardiomyocytes and the surrounding host myocardium.

An important step has then been the demonstration that these morphological patterns translated into improved functional outcomes. Thus, Li and coworkers [5] have transplanted fetal cardiomyocytes into fibrotic scars created by cryoinjury in rats and demonstrated that, two months later, the Langendorff-perfused hearts of these animals developed better functional indices than controls only receiving the culture medium. Likewise, in our group, Scorsin and coworkers [6] have used a rat model of reperfused infarct and shown that injection of fetal cardiomyocytes 30 minutes after the onset of reperfusion resulted in an increase in ejection fraction and cardiac output, as assessed by echocardiography 1 month later. Of note, the comparison by the group of Weisel and Li in Toronto [7] of three types of fetal cells (cardiomyocytes, smooth muscle cells and fibroblasts) has clearly established the functional superiority of the cardiomyocyte cell line, thereby supporting the concept that the intrinsic contractile properties of the grafted cells are critical for the procedure to be functionally optimized.

However, in a clinical perspective, the use of fetal cells raises several issues associated with ethics, rejection and procurement, which has motivated the interest for another type of contractile cells, i.e., satellite cells or skeletal myoblasts.

Skeletal myoblasts

Satellite cells are normally present in a quiescent state under the basal membrane of muscular fibers. After an injury, they are rapidly mobilized, actively proliferate and ultimately fuse to regenerate the damaged myotubes. Several of their characteristics make them attractive for use in the setting of clinical cellular transplantation: (1) an autologous origin, which overcomes problems associated with availability and immunology, (2) the ease with which they can be expanded in vitro so that the problem of scale-up can be satisfactorily addressed, (3) their commitment to an exclusively myogenic differentiation, which should guarantee against tumorigenicity, (4) a high resistance to ischemia, reflected by their survival after engraftment in nonperfused fibrotic scars, and (5) a potential for self-regeneration, which raises the attractive albeit still purely speculative hypothesis that if the new intramyocardial myotubes happened to be damaged by a recurrent ischemic event, some regeneration might still be possible from this pool of precursors.

The functional benefits of autologous skeletal myoblast transplantation have first been demonstrated by Taylor and coworkers [8] in a rabbit model of cryonecrosis.

Left ventricular function, assessed 6 weeks after the procedure by micromanometry and sonomicrometry, was found to have significantly improved in the transplanted group compared with control rabbits and those in which grafted cells could not be detected, thereby strongly suggesting the causal relationship between the presence of myoblasts and the functional benefits. Subsequent studies by the same group [9] have then pointed out that the posttransplant improvement primarily involved diastolic function whereas that of systolic performance was less consistent. In our group [10], we have compared skeletal myoblasts and fetal cells and found that, after one month, the echocardiographically assessed left ventricular ejection fraction was improved to a similar extent in the two groups, with myoblasts being identified in all grafted animals by positive staining for myosin heavy chain. A subsequent study [11] has then established that the functional improvement yielded by skeletal myoblast transplantation was linearly related to the number of injected cells but was independent of the baseline ejection fraction, i.e., was still present in case of severely depressed postinfarct left ventricular function (ejection fraction <0.25), thereby suggesting that myoblast transplantation could be relevant to the most severe forms of ischemic heart failure. The clinical relevance of the procedure is further strenghtened by our more recent observations made in rats that the benefits of autologous myoblast transplantation are additive to those of angiotensin-converting enzyme inhibitors [12] and remain stable over a 1-year period of follow-up (unpublished data). Importantly, all these findings made in rodents and rabbits have now been confirmed in large animal models involving dogs [13] and sheep (unpublished data).

Although there is consistent evidence that skeletal myoblast transplantation improves postinfarct left ventricular function, several basic issues remain unanswered. First, it is unclear whether, and how, grafted myoblasts interact with host cardiomyocytes. Histological studies by Chiu [14] , Taylor [8], and their co-workers have shown, in the core of the transplanted area (in the dog and rabbit hearts, respectively) structures resembling cardiac-specific intercalated disks. Conversely, we have consistently failed to detect a positive staining for connexin-43 which is the major gap junction protein. This observation is consistent with those of Reinecke et al [15] who have reported that only undifferentiated rat skeletal myoblasts express N-cadherin (the major adhesion protein of the intercalated disk) and connexin-43 whereas these proteins are markedly downregulated after differentiation into myotubes, both in culture and after myoblast implantation in normal or cryoinjured hearts. One possibility is that coupling develops through connexins different from connexin 43; alternatively, one can hypothesize that even if the grafted myoblasts remain electrically insulated, they might respond to the mechanical stimulation exerted by the surrounding myocardium and thus beat synchronously with host cardiomyocytes. Clearly, this area requires further investigations to better understand the mechanism by which myoblast transplantation improves function. A second issue pertains to the possible phenotypic changes of the grafted myoblasts over time. Chiu et al. [14] have advocated the theory of a "milieu-induced" differentiation which would cause myoblasts to acquire a cardiac-like phenotype. The observations of Atkins et al [16]

partly support this concept in that these authors have observed, at the periphery of cryoinfarcts, clusters of nonskeletal (myogenin-negative) muscle cells resembling immature cardiomyocytes. In contrast, we and others [17] have consistently observed that engrafted myoblasts remained committed to a skeletal muscle type-phenotype and did not "turn" to cardiac cells. This, however, does not rule out a possible role of the local environment since these myotubes express a slow-type myosin which could account for the ability of these transplanted hearts to sustain a cardiac-type workload in the long term. The potential impact of transplanted cells on the induction of angiogenesis is another issue which remains to be addressed.
Elucidation of these various problems should help in defining the mechanism(s) responsible for the functional benefits of skeletal myoblast transplantation, and which possibly include direct contribution to contractility, limitation of ventricular remodeling due to the elastic properties of the grafted cells, release of growth and/or angiogenic factors.
Notwithstanding these problems related to basic research, it appears that the functional benefits of skeletal myoblast transplantation have been demonstrated convincingly enough to consider riping to phase I clinical trials. This, in turn, raises specific problems related to both regulatory and technical constraints. The former require cell cultures to be performed in GMP facilities, with clinical-grade media and additives. Technically, scale-up techniques have to be developed to yield large number of cells, among which a high percentage of myoblasts, within the shortest time period so that cell transplantation can be performed reasonably close to the preceding muscular biopsy. Our recent efforts to make this transition "from bench to bedside" have allowed us to undertake a phase I human trial using autologous skeletal myoblast transplantation, the first case of which (June 15, 2000) has been previously reported [18]. This study includes patients with a major impairment of left ventricular function, a discrete postinfarction scar into which autologous myoblasts (harvested from the thigh a few weeks earlier) are implanted and an indication for concomitant coronary artery bypass in remote, ischemic but viable myocardial areas. Feasibility and safety are the primary end points whereas efficacy will become the primary outcome measurement of the future placebo-controlled phase 2 trial planned for the year 2002. All injections of myoblasts are made intraoperatively under direct vision but concurrent studies are now assessing the percutaneous endoventricular approach for cell delivery. Regardless of the mode of myoblast administration, it should be emphasized that the expected functional benefits are most likely hampered by the high rate of cell death occurring shortly after the injections. This major issue needs to be addressed by improvements in delivery devices and, possible, more "biological" interventions targeted at enhancing cell survival (for example by inhibition of apoptosis and/or matrix proteases). It is also conceivable that for some indications, cardiac reinforcement could be achieved by seeding cells onto biocompatible scaffolds.

Stem cells

In parallel to the experimental and clinical studies on myoblasts, it is worth pursuing to investigate other cell types, in particular bone marrow and embryonic stem cells.

The choice of bone marrow stem cells arises from their "pluripotentiality" which should allow them to differentiate into various cell types including cardiomyocytes. If the clinical use of these cells is attractive because of their autologous origin and the ease with which they can be retrieved through bone marrow aspiration (or even simple blood collection after pharmacologic stimulation of the bone marrow), several fundamental issues still remain to be addressed. Thus, it is yet unclear whether these cells should first be cultured to differentiate into "cardiac" cells before intramyocardial implantation or whether one should transplant them "fresh" and rely upon local signals to drive them towards the target cardiac lineage. Of note, all experiments based on the former approach have used the compound 5-azacytidine which "turns on" a wide array of genes and one can then question the ultimate consequences of injecting a human heart with these genetically modified cells. A second major question is to sort out whether these bone marrow cells should be implanted "as a whole" or following the selection of a given subpopulation. It may look sound to use the $CD34^+$ progenitors but because these precursors only represent a few percent of the total bone marrow cell population, the risk exists that their small number would not allow for any significant improvement in cardiac function. To address this issue, one could try to expand them in vitro but this, in turn, might compromise to some extent their "pluripotentiality". Another option could be to inject bone marrow stromal (mesenchymal) cells which, once implanted into a myocardial environment, have been reported to receive signals driving them towards a cardiomyogenic differentiation [19]. However, it remains to carefully validate the "cardiac" transformation of all these bone marrow-derived cells as if they only turn to muscular-type cells, the use of native myoblasts might be easier ! In this context, the recent paper by Orlic et al. [20] showing that injection of a selected subpopulation of bone marrow cells (Lin^- $c\text{-}kit^{POS}$) into an infarcted mouse myocardium resulted in their transformation into cardiac, smooth muscle and endothelial cells associated with an improvement in function, is extremely important from a cognitive standpoint but should be cautiously analyzed in a clinical perspective, primarily because the previously mentioned small number of these progenitors led these authors to use several mice to inject a single one, thereby raising all the issues associated with allografting, and also injections were made in a fresh infarct (3 to 5 hours following coronary artery ligation) so that the microenvironmental signals which drove the cells towards the various reported lineages may be well different from those present in the clinically relevant setting of an old postinfarction scar, as encountered in patients with advanced heart failure.

The challenge of using embryonic cells is still greater. Suffice is to say that apart from the ethical and regulatory problems still under debate, therapeutic cloning of mammalian cells is still fraught with major technical problems which casts serious doubts about the potential availability of these "personalized" cell lineages in a near future. Alternative approaches are to "reprogramme" the patient's cells (instead of cloning them) so as to rewind them back to an embryonic stem cell-like phenotype which could then be oriented towards the desired cell lineage or to genetically engineer allogeneic embryonic stem cells to make them match the intended graft recipient (and thus overcome an immune response from the host). These approaches, however, are still in an experimental stage and nobody can predict to what extent they may ultimately be useful for repairing damaged tissue in the clinical setting.

Conclusion

Although it is quite possible that skeletal myoblasts currently used for reinforcing the heart may only represent a first step on the long journey of cellular therapy it should, be well appreciated that stem cells still raise a lot of unsettled issues so that it may take some time before we end up with a reasonable trade-off between attractiveness of this concept and clinical applicability.

References

1. Soonpaa MH, Field LJ. Survey of studies examining mammalian cardiomyocyte DNA synthesis. Circ Res 1998; 83: 15-26.
2. Kajstura J, Leri A, Finato N, Di Loreto C, Beltrami CA, Anversa P. Myocyte proliferation in end-stage cardiac failure in humans. Proc. Natl. Acad. Sci. USA 1998; 95: 8801-5.
3. Beltrami AP, Urbanek K, Kajstura J, Yan SM, Finato N, Bussani R, Nadal-Ginard B, Silvestri F, Leri A, Beltrami A, Anversa P. Evidence that human cardiac myocytes divide after myocardial infarction. N Engl J Med 2001; 344: 1750-7.
4. Soonpaa MH, Koh GY, Klug MG, Field LJ. Formation of nascent intercalated disks between grafted fetal cardiomyocytes and host myocardium. Science 1994; 264: 98-101.
5. Li R-K, Jia Z-Q, Weisel RD, Mickle DAG, Zhang J, Mohabeer MK, Rao V, Ivanov J. Cardiomyocyte transplantation improves heart function. Ann Thorac Surg 1996; 62: 654-61.
6. Scorsin M, Hagège AA, Marotte F, Mirochnik N, Copin H, Barnoux M, Sabri A, Samuel J-L, Rappaport L, Menasché P. Does transplantation of cardiomyocytes improve function of infarcted myocardium. Circulation 1997; 96: II-188-93.
7. Sakai T, Li RK, Weisel RD, Mickle DAG, Jia ZQ, Tomita S, Kim EJ, Yau TM. Fetal cell transplantation : a comparison of three cell types. J Thorac Cardiovasc Surg 1999; 118: 715-25.
8. Taylor DA, Atkins BZ, Hungspreugs P, Jones TR, Reedy MC, Hutcheson KA, Glower DD, Kraus WE. Regenerating functional myocardium: improved performance after skeletal myoblast transplantation. Nature Med 1998; 4: 929-33.
9. Atkins BZ, Hueman MT, Meuchel JM, Cottman MJ, Hutcheson KA, Taylor DA. Myogenic cell transplantation improves in vivo regional performance in infarcted rabbit myocardium. J Heart Lung Transplant 1999; 18: 1173-80.
10. Scorsin M, Hagège AA, Vilquin J-T, Fiszman M, Marotte F, Samuel J-L, Rappaport L, Schwartz K, Menasché P. Comparison of the effects of fetal cardiomyocytes and skeletal myoblast transplantation on postinfarct left ventricular function J Thorac Cardiovasc Surg 2000; 119: 1169-75.
11. Pouzet B, Vilquin J-T, Messas E, Scorsin M, Fiszman M, Hagège AA, Schwartz K, Menasché P. Factors affecting functional outcome following myoblast cell transplantation. Ann Thorac Surg 2000; 71: 844-51.
12. Pouzet B, Ghostine S, Vilquin JT, Garcin I, Scorsin M, Hagège AA, Duboc D, Schwartz K, Menasché Ph. Is skeletal myoblast transplantation clinically relevant in the era of angiotensin-converting enzyme inhibitors ? Circulation 2000; 102: II 682 (Abstract).
13. Rao RL, Chin TK, Ganote CE, Hossler FE, Li C, Browder W. Satellite cell transplantation to repair injured myocardium. Cardiovasc –Res. 2000; 1: 31-42.
14. Chiu RC-J, Zibaitis A, Kao RL. Cellular cardiomyoplasty: myocardial regeneration with satellite cell implantation. Ann Thorac Surg 1995; 60: 12-18.
15. Reinecke H, McDonald GH, Hauschka SD, Murry CE.. Electromechanical coupling between skeletal and cardiuac muscle. Implications for infarct repair. J Cell Biol 2000; 149: 731-40.
16. Atkins BZ, Lewis CW, Kraus WE, Hutcheson KA, Glower DD, Taylor DA. Intracardiac transplantation of skeletal myoblasts yields two populations of striated cells in situ. Ann Thorac Surg 1999; 67: 124-9.
17. Murry CE, Wiseman RW, Schwartz SM, Hauschka SD. Skeletal myoblast transplantation for repair of myocardial necrosis. J Clin Invest 1996; 98: 2512-23.
18. Menasché Ph, Hagège AA, Scorsin M, Pouzet B, Desnos M, Duboc D, Schwartz K, Vilquin JT, Marolleau JP. Myoblast transplantation for heart failure. The Lancet 2001; 357: 279-80.

19. Wang JS, Shum-Tim D, Galipeau J, Chedrawy E, Eliopoulos N, Chiu RCJ. Marrow stromal cells for cellular cardiomyoplasty: feasibility and potential clinical advantages. J Thorac Cardiovasc Surg 2000; 120: 999-1006.
20. Orlic D, Kajstura J, Chimenti S, Jakonluc I, Anderson SM, Li B, Pickel J, McKay R, Nadal-Ginard B, Bodline DM, Leri A, Anversa P. Bone marrow cells regenerate infarcted myocardium. Nature 2001; 410: 701-5.

APPENDIX TO SESSION 3 HYPERTENSION

ADDUCIN PARADIGM: AN APPROACH TO THE COMPLEXITY OF HYPERTENSION GENETICS

G. Bianchi

Many approaches to the genetics of hypertension have been proposed. Here I will illustrate our approach that consists of an integration of different disciplines. In particular:

a] pathophysiology of hypertension caused by an experimental renal injury in dogs and rats or by a genetic mechanism in an inbred strain of rats or kidney transplantation in rats, followed by cellular and biochemical studies aimed at detecting a primary protein alteration responsible of the sequence of events leading to hypertension.

b] detection of polymorphisms in the genes coding for the protein involved in the pathophysiological phenomena mentioned above and performance of the appropriate genetic studies in rats and humans to assess their role in causing arterial hypertension.

Following the demonstration that after some weeks from the initial blood pressure rise produced by renal artery constriction, the changes more directly linked to the renal injury disappear, it was hypothesized that the same pattern of changes may occur also in primary or polygenic forms of human hypertension where a subtle kidney abnormality may cause hypertension without overt renal alterations [1-3]. This hypothesis was proven with kidney cross-transplantation between MHS rats (with "genetic" form of hypertension) and its normotensive control strain (MNS) [4,5]. Also in humans, recipients of kidney from donors with hypertensive parents require higher dosage of antihypertensive therapy than the recipients of kidney removed from donors with normotensive parents [6,7]. These results stimulated a series of studies both in human and rats aimed at clarifying the cellular and molecular mechanisms underlying the "genetic message" travelling with the kidney [8-15].

These studies led to the identification of adducin polymorphism, as a candidate "pressor" gene, through the following steps:

1) The comparisons between MHS and MNS carried out using different approaches (Na balance in metabolic cages, isolated kidneys, tubuli and tubular cells, erythrocytes, isolated cell membrane from both renal and red blood cells, genetic crosses) yielded data consistent with the notion that a genetic alteration in cell membrane ion transport, due to a cytoskeleton protein abnormality, could be responsible for an abnormality in renal Na handling and the development of at least a portion of hypertension in MHS [16].
2) Cross immunization between MHS and MNS with cell membrane proteins stimulated the development of an antibody against an 105 KD protein subsequently identified as adducin [17].

Adducin participates in the assembly of the spectrin-actin cytoskeleton [18,19], it modulates actin polymerization and bundling [20], it binds calmodulin [21], it is phosphorylated by PKC Rho-kinase [22] and tyrosine kinase, it regulates cell signal transduction and cell ion transports [23]. These cellular properties of adducin may be the basis of its regulatory effect on tubular sodium reabsorption and blood pressure demonstrated by our research group.

Adducin functions within the cell as a heterodimer composed by related but not identical subunits (alfa, beta and gamma) coded by specific genes mapping on different chromosomes.

Human alpha adducin (HaAd) maps in 4p16.3 [24]. Rat alpha adducin (RaAd) maps in chr 14 [25]. In HaAd two polymorphisms Gly46Trp [26] and Ser586Cys have been associated with hypertension. Between MNS and MHS rats a Phe316Tyr (F316Y) polymorphism has been found [27], aAd mRNA is differently expressed in all tissues investigated [28].

Human beta adducin (HbAd) maps in 2p14-p13 [29]. Rat beta adducin (RbAd) maps in chr 4 [25]. Three alternative spliced isoforms b2, b3 and b4 have been found. The HbAd4 isoform results in an out-of-frame insertion of 63 aminoacids (exon 15), with an alternative stop codon leading to a truncated protein with a different C-terminus amino acid sequence [30]. RbAd4 isoform results in the expression of additional in-frame 18 aminoacids. HbAd4 is polymorphic in humans at codon 599 (silent polymorphism C599T). Between MNS and MHS rat a Gln529Arg polymorphism [27] has been found in RbAd1. bAd mRNA shows restricted expression mainly in brain and erythropoietic tissues [28] even if, by PCR analysis, it can be detected in additional tissues.

Human Gamma adducin (HgAd) maps in 10q24.2-10q24.3 [31]. Rat gamma adducin (RgAd) maps in chr 1 [32]. In humans an A/G polymorphism in intron 11 has been identified. Between MNS and MHS rats a Gln572Lys polymorphism has been found [32]. gAd mRNA is differently expressed in all tissues investigated [28].

3) In an F2 hybrid population (MHSXMNS), rat alpha-adducin a F316Y mutation cosegregated with blood pressure (ADD1) [33]. Mutations in rat beta-adducin (ADD2) (Q529R) [33] and rat gamma adducin (ADD3) (Q527K) [34] were not "per se" associated with blood pressure variation but epistatically

interacted with the rat ADD1 mutation in determining blood pressure level. A wide genome search on the same F2 population using 245 DNA markers confirmed these data [35].

4) MHS rat ADD1 stimulates actin polymerization and enhances actin bundling in a cell free system [36]. Transfection of MHS ADD1 in kidney cell cultures increases the surface expression of Na-K pump while the transfection of the MNS ADD1 variant is without effect [36].

5) We and others detected polymorphisms in human ADD1 [37,38]. The 460Trp allele showed a positive association to hypertension in 8 out of 15 populations examined [39].

6) Hypertensives with the 460Trp allele, compared to those homozygous for the 460Gly one, are salt-sensitive [37], have less steep pressure natriuresis relationship [40], increased proximal tubular reabsorption of lithium [41], larger blood pressure fall on diuretic treatment [37,42], lower PRA [37,42]. These differences, consistent with an increased tubular reabsorption in 460Trp carriers, are also present in a population where no difference in the 460Trp frequency was found between hypertensives and normotensives [42].

7) Linkage analysis with a highly polymorphic DNA marker mapping 20Kb from the human ADD1 locus yielded positive results in two studies [43,44]. Negative results were reported in two other studies [45,46] but DNA markers mapping 400Kb from the ADD1 locus were used.

8) Compared to the "wild" variants, both human and rat "mutated" adducins (either extracted from tissues or prepared by recombinant DNA technology) bind to the Na-K ATPase with greater affinity, in a cell free system [47]. Therefore, in spite of the difference in the mutation sites between rats and humans, the mutated variants similarly modify the interaction of adducin with the Na-K ATPase [48,49], which is the key enzyme for tubular Na transport.

9) Congenic strains are in preparation by introgressing either in MHS or MNS the ADD1, ADD2, ADD3 of the other strain alone or combined. So far we have obtained two congenic strains where the MHS ADD1 locus increases the blood pressure of MNS rats.

10) Targeted disruption of ADD2 locus (carried out by the Baralle's group) increases systolic blood pressure in mice [49].

In summary, the results on physiological genetics in human and rats and those of genetics and genomics on rats are quite consistent for a role of adducin polymorphisms in body sodium and blood pressure regulation. Conversely, the data on statistical genetics, summarized under points 5 and 7 seem to be less consistent compared with many other candidate genes when this type of approach is applied.
There are possible explanations for this discrepancy: statistical genetics is based on differences in allelic frequency between normotensives and hypertensives or on

cosegregation of a given allele with hypertension in families. There are many confounding factors which may be responsible for contrasting results when the methodology of statistical genetics is applied to the dissection of complex diseases like hypertension. Indeed, according to some experts [50,51] statistical genetics is unable to explain the complexity of the variety of genotype-phenotype relationships in arterial hypertension.

As mentioned above, adducin functions within the cell as a heterodimer composed by related but not identical subunits (alfa, beta and gamma) coded by specific genes mapping on different chromosomes. These unique operative conditions justify the search of epistatic interactions among the adducin loci as an additional and robust genetic approach for the assessment of the role of adducin. In fact, if such interactions are found, it would be more plausible that they are really due to the polymorphisms in the adducin genes rather than to still undetected variants in other nearby genes in linkage disequilibrium with the adducin genes.

Preliminary findings obtained in a Sardinian population (in collaboration with N. Glorioso) and in a Belgian population (in collaboration with J. Staessen) support the hypothesis of an interaction between HaAd and HbAd polymorphisms [unpublished] and between HaAd and ACE I/D polymorphisms [52].

References:

1. Bianchi G, Tenconi LT, Lucca R. Effect in the conscious dog of constriction of the renal artery to a sole remaining kidney on haemodynamics, sodium balance, body fluid volume, plasma renin concentration and pressor responsiveness to angiotensin. Cli Sci 1970; 38:741-766.
2. Bianchi G, Baldoli E, Lucca E et al. Pathogenesis of arterial hypertension after the constriction of the renal artery leaving the opposite kidney intact both in the anesthetized and in the conscious dog. Cli Sci 1972; 42:651-664.
3. Caravaggi AM, Bianchi G, Brown JJ et al. Blood pressure and plasma angiotensin II concentation after renal artery constriction and angiotensin in the dog (5-Isoleucine) angiotensin II and its breakdown fragments in dog blood. Circ Res 1976;38:315-21.
4. Bianchi G, Fox U, Di Francesco GF et al. The hypertensive role of the kidney in spontaneously hypertensive rats. Clin Sci Mol Med 1973;45:135s-9.
5. Bianchi G, Fox U, Di Francesco GF et al. Blood pressure changes produced by kidney cross-transplantation between spontaneously hypertensive rats and normotensive rats. Clin Sci Mol Med 1974;47:435-48.
6. Guidi E , Bianchi G, Rivolta E et al. Hypertension in man with a kidney transplant: role of familial versus other factors. Nephron 1985; 41:14-21.
7. Guidi E, Menghetti D, Milani Pharm S et al. Hypertension may be transplanted with the kidney in humans. J Am Soc Nephrol 1996;7:1131-8.
8. Bianchi G, Cusi D, Gatti M et al. A renal abnormality as a possible cause of "essential" hypertension. Lancet 1979; I:173-7.
9. Bianchi G, Cusi D, Barlassina C et al. Renal dysfunction as a possible cause of essential hypertension in predisposed subjects. Kidney Int 1983; 23:870-5.
10. Bianchi G, Baer PG, Fox U et al. Changes in renin, water balance, and sodium balance during development of high blood pressure in genetically hypertensive rats. Circ Res 1975;36&37 [suppl I]:153-61.
11. Baer PC, Bianchi G. Renal micropuncture study of normotensive and Milan hypertensive rats before and after development of hypertension. Kidney Int 1978; 13:452-66.
12. Persson AE, Bianchi G, Boberg U. Tubuloglomerular feedback in hypertensive rats of the Milan strain. Acta Physiol Scand 1985;123:139-46.
13. Parenti P, Hanozet G, Bianchi G. Sodium and glucose transport across renal brush-border membranes of Milan hypertensive rats. Hypertension 1986; 8:932-9.
14. Ferrari P, Barber BR, Torielli L et al. The Milan hypertensive rat as a model for studying cation transport abnormality in genetic hypertension. Hypertension 1987;10 [suppl I]:32-6.
15. Bianchi G, Ferrari P, Trizio D et al. Red blood cell abnormalities and spontaneus hypertension in the rat: a genetically determined link. Hypertension 1985;7:319-25.
16. Bianchi G et al. The Milan Hypertensive strain of rats. In "Textbook of hypertension", ed. Swales JD, Blackwell Scientific Publications [Oxford, UK], 1994; 457-60.
17. Salardi S, Saccardo B, Borsani G et al. Erythrocyte adducin differential properties in normotensive and hypertensive rats of the Milan strain (characterization of spleen adducin m-RNA) Am J Hypertens 1989;2:229-37.
18. Bennett V. Spectrin-based membrane cytoskeleton: a multipotential adaptor between plasma nembrane and cytoplasm. Physiol Rev 1990; 70:1029-65.
19. Hughes CA, Bennet V. Adducin: a physical model with implications for function in assembly of spectrin-actin complexes. J Biol Chem 1995;270:18990-6.
20. Kuhlman PA, Hughes CA, Bennet V et al. A new function for adducin. J Biol Chem 1996;271:7986-91.
21. Gardner K., and Bennet V. A new erythrocyte membrane-associated protein with calmodulin binding activity: identification and purification. J Biol Chem 1986;261:1339-48.

22. Fukata Y, Oshiro N, Kinoshita N et al. Phosphorylation of adducin by Rho-kinase plays a crucial role in cell motility. J Cell Biol 1999; 19;145:347-61.
23. Kurana S. Role of actin cytoskeleton in regulation of ion transport: Examples from epithelial cells. J Membrane Biol 2000;178:73-87.
24. Lin B, Nasir J, McRonald H et al. Genomic organization of the human a-adducin gene and its alternately spliced isoforms. Genomics 1995;25:93-9.
25. Tripodi G, Casari G, Tisminetzky S et al. Characterisation and chromosomal localisation of the rat a- and b- adducin- encoding genes. Gene 1995;166:307-11.
26. Cusi D, Barlassina C, Azzani T et al. Polymorphisms of alpha-adducin and salt sensitivity in patients with essential hypertension. Lancet 1997; 349:1353-7.
27. Bianchi G, Tripodi G, Casari G et al. Two point mutations in the adducin genes are involved in blood pressure variation. Proc Natl Acad Sci USA. 1994;91:3999-4003.
28. Gilligan D, Lozovatsky L, Gwynn B et al. Targeted disruption of the b-adducin gene (Add2) causes red blood cell spherocytosis in mice. Proc Natl Acad Sci USA 1999;96:10717-22.
29. Gilligan D, Lieman J, Bennett V. Assignement of the human b-adducin gene (Add2) to 2p13-p14. Genomics 1995;28:610-2.
30. Sinard J, Stewart GW, Stabach PR et al. Utilization of an 86bp exon generates a novel adducin isoform (b4) lacking the MARCKS homology domain. Biochem Biophys Acta 1998;1396:57-66.
31. Citterio L, Azzani T, Duga S et al. Genomic organization of the human g-adducin gene. Biochem Biophys Res Com 1999;266:110-4.
32. Tripodi G, Szpirer C, Reina C et al. Polymorphism of g-adducin gene in genetic hypertension and mapping of the gene to rat chromosome 1p55. Biochem Biophys Res Com 1997;237:685-9.
33. Bianchi G, Tripodi G, Casari G et al. Two point mutations within the adducin genesa are involved inblood pressure variation. Proc Natl Acad Sci USA.1994; 91:3999-4003.
34. Tripodi G, Szpirer C, Reina C et al. Polymorphism of g-adducin gene in genetic hypertension and mapping of the gene to rat chromosome 1q55. Biochem Biophys Res Com 1997;237:685-9.
35. Zagato L, Modica R, Florio M et al. Genetic mapping of blood pressure quantitative trait loci in Milan hypertensive rats. Hypertension 2000;36:734-9.
36. Tripodi G, Valtorta F, Torielli L et al. Hypertension-associated point mutations in the adducin alpha and beta subunits affect actin cytoskeleton and ion transport. J Clin Invest 1996;97:2815-22.
37. Cusi D, Barlassina C, Azzani T et al. Polymorphism of alpha-adducin and salt sensitivity in patients with essential hypertension. Lancet 1997;349:1353-7.
38. Halushka MK, Fan JB, Bentley K et al. Patterns of single-nucleotide polymorphisms in candidate genes for blood-pressure homeostasis. Nat Genet 1999;22:239-47.
39. Bianchi G, Cusi D. Association and linkage analysis of alpha-adducin polymorphism. Is the glass half full or half empty? Am J Hypertens 2000;13:739-43.
40. Barlassina C, Schork NJ, Manunta P et al. Synergetic effect of alpha-adducin and ACE genes in causing blood pressure changes with body sodium and volume expansion. Kidney Int 2000;57/3:1083-90.
41. Manunta P, Burnier M, D'Amico M et al. Adducin polymorphism affects renal proximal tubule reabsorption in hypertension. Hypertension1999;33:694-7.
42. Glorioso N, Manunta P, Filigheddu F et al. The role of alpha-adducin polymorphism in blood pressure and sodium handling regulation may not be excluded by a negative association study. Hypertension 1999;34:649-54.
43. Casari G, Barlassina C, Cusi D et al. Association of the alpha-adducin locus with essential hypertension. Hypertension 1995, 25:320-6.
44. Busjajn A., Aydin A., von Treuenfels Net al. Linkage but lack of association for blood pressure and the alpha-adducin locus in normotensive twins. J Hypertens 1999;17:1437-41.

45. Bray M, Li L, Turner S, Kardia SL et al. Association and linkage analysis of the alpha-adducin gene and blood pressure. Am J Hypertens 2000;13:699-703.
46. Niu T, Xu X, Cordell HJ et al. Linkage analysis of candidate genes and gene-gene interactions in Chinese hypertensive sib pairs. Hypertension 1999;33:1332-7.
47. Ferandi M, Salardi S, Tripodi G et al. Evidence for an interaction between adducin and Na-K-ATPase: relation to genetic hypertension. Am J Physiol 1999;277:H1338-49.
48. Tripodi G, Ferrandi M, Modica R et al. Effect of adducin gene transfer on blood pressure and Na-K pump activity in Milan Hypertensive rats (MHS). J Mol Med 2000;79(2-3):B30.
49. Marro ML, Scremin OU, Jordan MC et al. Hypertension in beta-adducin-deficient mice. Hypertension 2000;36:449-53.
50. Tyerwilliger JD, Goering HHH. Gene mapping in the 20th and 21st centuries: statistical methods, data anlysis and experimenta design. Hum Biol 2000; 72:63-132.
51. Shork NJ et al. Molecular genetics of hypertension. Eds. Dominiczak, Connell, Soubrier; Bios S.P. Ltd, Oxford 1999; 9.
52. Staessen J, Wang GJ, Brand E et al. Effects of three candidate genes on prevalence and incidence of hypertension in a Caucasian population. J Hypertens 2001;19:1349-58.

Index

2D-gel, 21

Action potential, 159, 169, 201, 215, 234, 247
Adeno-associated virus, 70, 164, 228
Adenoviral vector, 70, 165
Aldosterone, 14, 55
Allele, 4, 55, 120, 213
Angiotensinogen, 9, 55, 68
Angiotensin-converting enzyme, 55, 87, 259
Antisense, 55, 234
Arachidonic acid, 86
Arrhythmias, 159, 168, 199
Atrial fibrillation, 2, 168, 199, 213
Atrial natriuretic factor, 97, 132, 242
AV-node, 163
Azacytidine, 245, 260

Beta-adrenergic receptor kinase, 109
Brugada syndrome, 165

Catechlamines, 77, 109, 135
cDNA microarray, 43, 97
Clinical scientist, 3
CMG cells, 246
Contractile function, 107, 115, 147, 175, 204, 240, 257
CyaY, 30
Cytokine, 44, 72, 87, 122, 134, 146, 253

Diabetes mellitus, 44, 77
Dilated cardiomyopathy, 47, 107, 115, 137

EDHF, 84
Education, 3
Electrical remodeling, 175, 199
Electrophysiology, 159, 232
Embryonic stem cell, 225, 231, 260
Endothelin-1, 87, 135
Endothelium, 44, 65, 71, 78, 83
Expression profile, 43, 99, 120, 133

Focal adhesion, 118, 138
Fraxatin, 29
Friedreichs' ataxia, 29

Gene therapy, 2, 65, 162, 225, 245
Genotype, 56, 215
Growth factor, 23, 46, 74, 87, 123, 131, 163, 225, 253

HscB, 30
HUVEC, 44
Hybridization, 44, 174
Hypertension, 7, 23, 55, 65, 77, 83, 119, 147, 176, 199, 213
salt-sensitive, 7, 56, 66
Hypertrophy, 90, 97, 107, 115, 131, 175, 214, 245, 257
adaptive, 207
maladaptive, 150
Hypertrophic cardiomyopathy, 1, 97
Hypoxia, 24, 29, 72, 87, 164
Hypoxia-inducing factor, 23

IGF-1, 88, 149
Inhibitory G-protein, 134, 163
Insulin, 10, 77, 88, 131
Ion channel, 8, 86, 159, 201, 232
Iron sulfer cluster, 32

JAC1, 30

Kallikrein, 65
Kinetics, 163, 167

Labeling, 22, 101
L-arginine, 67, 90
Leptin, 77
Liddle syndrome, 7
Long QT syndrome, 1, 160, 168

MAPK, 83, 108, 123, 132
Microarray, 2, 21, 43, 97
Mouse, 10, 37, 97, 115, 144, 169, 232, 247, 261
Myosin light chain, 45, 97, 143, 233, 247
Myoblast, 258
Myotube, 247, 258

Nitric oxide, 46, 65, 78, 84, 132

Oligonucleotides, 68, 99

Peptide map, 22
Phosphoinositide-3, 107

Phylogenetic profile, 29
Polymorphism, 9, 56, 213
Pressure-volume relationship, 234
Prostacyclin, 65, 84
Proteomics, 2, 15, 21

RAC1, 83, 108, 123, 132
Renin, 9, 55, 77, 88, 109, 214
RFLP, 56

SAGE, 7, 11
Shear stress, 44, 72, 83
Signal transduction, 23, 79, 115, 131, 213
Sudden cardiac death, 175
Superoxide anion, 88
Superoxide dismutase, 66, 88
Stem cells, 225, 231, 245, 264
Stromal cells, 245

Tandem mass spectometry, 22
Training, 3, 146
Transduction level, 227
Transgenic, 97, 108, 115, 135, 258
Transplantation, 49, 115, 225, 231, 245, 257
Transverse aortic banding, 98
Tropomodulin, 117

VEGF, 23, 46, 88, 163

MIX
Papier aus verantwortungsvollen Quellen
Paper from responsible sources
FSC® C105338

If you have any concerns about our products,
you can contact us on
ProductSafety@springernature.com

In case Publisher is established outside the EU,
the EU authorized representative is:
Springer Nature Customer Service Center GmbH
Europaplatz 3, 69115 Heidelberg, Germany

Printed by Libri Plureos GmbH
in Hamburg, Germany